Current Topics in Microbiology 213/II and Immunology

Editors

Springer
Berlin
Heidelberg
New York
Barcelona
Budapest
Hong Kong
London
Milan
Paris
Santa Clara
Singapore
Tokyo

Attempts to Understand Metastasis Formation II

Regulatory Factors

Edited by U. Günthert and W. Birchmeier

With 33 Figures

Springer

URSULA GÜNTHERT, Ph.D., PD
Basel Institute for Immunology
Grenzacherstraße 487
CH-4005 Basel
Switzerland

Professor WALTER BIRCHMEIER, Ph.D.
Max-Delbrück-Centrum für Molekulare Medizin
Robert-Rössle-Straße 10
D-13122 Berlin
Germany

Cover illustration: The graph shows intercellular junctions between epithelial cells mediated by the cell adhesion molecule E-cadherin (yellow). These junctions are disturbed in many types of carcinomas which leads to invasive and metastatic cells (see articles by Birchmeier et al. and Bracke et al.). β-Catenin or plakoglobin mediate the interaction between E-cadherin and the cytoskeleton associated αcatenin. Also, the tumor suppressor gene product APC (adeno-matous polyposis colt) forms a similar complex with catenins.

Cover design: Künkel+Lopka, Ilvesheim

ISBN-13: 978-3-642-64698-0 e-ISBN-13: 978-3-642-61109-4
DOI: 10.1007/978-3-642-61109-4

Softcover reprint of the hardcover 1st edition 1996
Library of Congress Catalog Card Number 15-12910

Typesetting: Thomson Press (India) Ltd., Madras

SPIN: 10495477 27/3020/SPS – 5 4 3 2 1 0 – Printed on acid-free paper

Preface

In metastasis, tumor cells disseminate from the primary lesion and home to secondary organs where they may remain dormant for a long time. Metastasis formation is still the most feared manifestation for tumor patients and clinicians. Although improvements have been made concerning earlier detection and specific therapy, most of the cancer patients still die of distant metastases. The purpose of these three volumes is to review the recent progress in molecular metastasis research and to attempt to further understand the biology of this multifocal process.

With respect to present day molecular biology, the pioneers of metastasis research established the basic concepts of metastasis formation in the 1970s and 1980s, namely, clonal selection of metastatic cells, heterogeneity of metastatic subpopulations, organ specificity of metastasis and the importance of angiogenesis (Fidler, Kripke, Nicolson, Folkman and others). In the 1980s and 1990s, several of the molecules involved were identified and their network interactions elucidated. These three volumes of *Current Topics in Microbiology and Immunology* compile the most recent developments on these metastasis-related molecules; their interactions, regulation, and ways to interfere with their action. It became evident that metastasis-related molecules are confined to distinct cellular compartments, such as the extracellular space, the cell membrane, the cytoplasmic signalling network, and the nuclear regulatory system.

For the complex metastatic cascade, proteolysis and alterations in adhesive functions are the most obvious and thus one of the most thoroughly investigated processes. Various proteases and precursors (metalloproteinases and serine proteases) and their inhibitors (tissue inhibitors of metalloproteases, plasminogen activator inhibitors and serpins) exhibit a sensitive complex of interplay – we are particularly fascinated by their highly regulated nature. Not only the proteases and their inhibitors are important in all the different stages of metastasis formation, but also to the same extent adhesive and "de-adhesive" interactions: metastatic cells must constantly detach themselves from their old partners and reattach to new ones, as mainly outlined in the first volume and partly in the second volume. Among the widespread members of the adhesion molecule families, certain immunoglobulins, integrins, cadherins, selectins, and hyaluronic acid receptors as well as their ligands are implicated in the spread of metastatic cells. The control of the metastatic cascade by these extracellularly acting molecules is delicately balanced, and slight changes could affect the establishment of the normal cellular organization and consequently promote metastasis formation. Strikingly, some genes of adhesion molecules have recently been identified as tumor suppressor genes in model organisms (e.g. Drosophila) and are in fact mutated in metastasizing human tumors.

Growth of primary tumors and metastases is strictly dependent on angiogenesis, the formation of new blood vessels. How this process is regulated by cytokines is another topic of the second volume. Cytokines are not only important in angiogenesis but are essential for the direct migration of metastatic cells. Cytokines act through specific receptors which mediate signals by different means, e.g., tyrosine phosphorylation. A recent discovery is that cytoplasmic signal transduction components, transcription factors, and cell cycle regulators are also metastasis-related. Many of the presently described genes in metastasis were known as activated oncogenes for several years, but apparently the encoded gene products have a broader spectrum of action than was originally assumed.

We have recently learned that the spread of metastatic cells, especially of micrometastases, is far more extensive than previously expected. A successful antimetastatic therapy therefore requires new strategies: for this reason the third volume comprises novel approaches such as immunotherapy, transfer of tumor-inhibiting genes and anti-sense constructs, as well as interference with signal transduction pathways. Promising new therapeutic approaches also involve the use of anti-angiogenic factors or of recombinant soluble metastasis-related molecules which interfere with ligand interactions.

As the process of metastatic spread is presently regarded as a multifactor event which is yet to be sufficiently understood in the multitude of its aspects, approaches to clinical treatment have to be polypragmatic. Methods of treatment are based on chemotherapy and radiotherapy, refined and adapted to the type of tumor pertaining and the pattern of metastatic spread. Increasingly, therapies which incorporate new insights from immunology and molecular biology are adopted for clinical use. To present a rounded scope of the topic, these current strategies are covered by the third volume in particular. Surgical treatment options are indicated in cases where a curative intervention is feasible e.g. in solitary metastases of colorectal carcinoma, soft tissue, and kidney tumors.

We hope that the reader of these volumes is impressed by the quality of the contents. Metastasis has obviously emerged as a serious discipline of natural sciences due to the fact that the molecular biology of various metastasis-related molecules and their complex interplay became transparent. We are, nevertheless, still in the beginning phase and await further progress from which patients will finally benefit.

Most, if not all of the metastasis-specific processes described are also known to be involved in embryonic development and pattern formation, as well as in leukocyte biology. The disciplines of metastasis research, developmental biology, and immunology can, therefore, profit from and stimulate each other. The genetic analysis of candidate molecules and their interplays in transgenic mice will certainly further broaden our understanding of the molecular basis of metastasis formation.

We would like to thank the authors who have spent their valuable time in writing a chapter for this series. Without their expertise and cooperation, this compilation of newest developments in metastasis research would not have been attainable. Leslie Nicklin (Basel) assisted the edition of this series with her competent skills; we are most grateful for her contribution.

Basel URSULA GÜNTHERT
Berlin WALTER BIRCHMEIER
Berlin PETER M. SCHLAG

List of Contents

R.K. Singh and I.J. Fidler
Regulation of Tumor Angiogenesis
by Organ-Specific Cytokines 1

R. Giavazzi
Cytokine-Mediated Tumor-Endothelial
Cell Interaction in Metastasis 13

M.S. Pepper, S.J. Mandriota, J.-D. Vassalli, L. Orci,
and R. Montesano
Angiogenesis-Regulating Cytokines: Activities
and Interactions 31

D.G. Tang and K.V. Honn
Adhesion Molecules and Tumor Cell-Vasculature
Interactions: Modulation by Bioactive Lipid
Molecules 69

G.L. Nicolson, D.G. Menter, J.L. Herrmann, Z. Yun,
P. Cavanaugh, and D. Marchetti
Brain Metastasis: Role of Trophic, Autocrine,
and Paracrine Factors in Tumor Invasion
and Colonization of the Central Nervous System 89

W. Birchmeier, J. Behrens, K.M. Weidner, J. Hülsken,
and C. Birchmeier
Epithelial Differentiation and the Control
of Metastasis in Carcinomas 117

S. Silletti and A. Raz
Regulation of Autocrine Motility Factor
Receptor Expression in Tumor Cell Locomotion
and Metastasis 137

E.M. Lukanidin and G.P. Georgiev
Metastasis-Related *mts1* Gene 171

R.J. Muschel and W.G. McKenna
Alterations in Cell Cycle Control
During Tumor Progression: Effects on Apoptosis
and the Response to Therapeutic Agents 197

J.M.P. Freije, N.J. MacDonald, and P.S. Steeg
Differential Gene Expression
in Tumor Metastasis: Nm23 215

E.H. Postel
NM23/Nucleoside Diphosphate Kinase
as a Transcriptional Activator of *c-myc* 233

J.G. Collard, G.G.M. Habets, F. Michiels, J. Stam,
R.A. van der Kammen, and F. van Leeuwen
Role of Tiam 1 in Rac-Mediated Signal
Transduction Pathways 253

P.H. Watson, R. Singh, and A.K. Hole
Influence of *c-myc* on the Progession
of Human Breast Cancer 267

Subject Index 285

List of Contents of Companion Volume 213/I

W.C. POWELL and L.M. MATRISIAN
Complex Roles of Matrix Metalloproteinases in Tumor Progression

M. SEIKI
Membrane Type-Matrix Metalloproteinase and Tumor Invasion

P. MIGNATTI and D.B. RIFKIN
Plasminogen Activators and Angiogenesis

R. SAGER, S. SHENG, P. PEMBERTON, and M.J.C. HENDRIX
Maspin: A Tumor Suppressing Serpin

B.M. MUELLER
Different Roles for Plasminogen Activators and Metalloproteinases in Melanoma Metastasis

M.J. REED and E.H. SAGE
SPARC and the Extracellular Matrix: Implications for Cancer and Wound Repair

J.P. JOHNSON, M.M. RUMMEL, U. ROTHBÄCHER, and C. SERS
MUC18: A Cell Adhesion Molecule with a Potential Role in Tumor Growth and Tumor Cell Dissemination

R.C. ELBLE and B.U. PAULI
Lu-ECAM-1 and DPP IV in Lung Metastasis

M.E. BRACKE, F.M. VAN ROY, and M.M. MAREEL
The E-cadherin/Catenin Complex in Invasion and Metastasis

S. BRAUN and K. PANTEL
Biological Characteristics of Micrometastatic Carcinoma Cells in Bone Marrow

B. BOYER, A.M. VALLÉS, and J.P. THIERY
Model Systems of Carcinoma Cell Dispersion

B.A. IMHOF, L. PIALI, R.H. GISLER, and D. DUNON
Involvement of α6 and αv Integrins in Metastasis

N. MONTUORI and M.E. SOBEL
The 67-kDa Laminin Receptor
and Tumor Progression

M. ZÖLLER
Joint Features of Metastasis Formation
and Lymphocyte Maturation and Activation

L. SHERMAN, J. SLEEMAN, P. DALL, A. HEKELE, J. MOLL,
H. PONTA, and P. HERRLICH
The CD44 Proteins in Embryonic Development
and in Cancer

U. GÜNTHERT
CD44 in Malignant Disorders

Subject Index

List of Contents of Companion Volume 213/III

K. PANTEL and G. RIETHMÜLLER
Micrometastasis Detection and Treatment with Monoclonal Antibodies

I.R. HART
Transcriptionally Targeted Gene Therapy

R.A. REISFELD and S.D. GILLIES
Recombinant Antibody Fusion Proteins for Cancer Immunotherapy

Z. QIN and T. BLANKENSTEIN
Influence of Local Cytokines on Tumor Metastasis: Using Cytokine Gene-Transfected Tumor Cells as Experimental Models

J.A. HENDRZAK and M.J. BRUNDA
Antitumor and Antimetastatic Activity of Interleukin-12

L. EISENBACH
Curing Metastases? Gene and Peptide Therapy

L. OSSOWSKI
Effect of Antisense Inhibition of Urokinase Receptor on Malignancy

W. WELS, B. GRONER, and N.E. HYNES
Intervention in Receptor Tyrosin Kinase-Mediated Pathways: Recombinant Antibody Fusion Proteins Targeted to ErbB2

M.-S. SY, D. LIU, R. SCHIAVONE, J. MA, H. MORI, and Y. GUO
Interactions Between CD44 and Hyaluronic Acid: Their Role in Tumor Growth and Metastasis

R. RISTAMÄKI, H. JOENSUU, and S. JALKANEN
Does Soluble CD44 Reflect the Clinical Behavior of Human Cancer?

R. Alessandro, J. Spoonster, R.P. Wersto,
and E.C. Kohn
Signal Transduction as a Therapeutic Target

V. Schirrmacher, V. Umansky, and M. Rocha
Immunotherapy of Metastases

R.U. Hilgenfeld and E.-D. Kreuser
Immunological and Biochemical Modulation
in the Treatment of Advanced Colorectal Cancer:
Update and Future Directions

P.M. Schlag, M. Hünerbein, and P. Hohenberger
Surgical and Multimodality Approaches
to Liver Metastases

Subject Index

List of Contributors

(Their addresses can be found at the beginning of their respective chapters.)

Behrens, J. 117
Birchmeier, C. 117
Birchmeier, W. 117
Cavanaugh, P. 89
Collard, J.G. 251
Fidler, I.J. 1
Freije, J.M.P. 213
Georgiev, G.P. 171
Giavazzi, R. 13
Habets, G.G.M. 251
Herrmann, J.L. 89
Hole, A.K. 265
Honn, K.V. 69
Hülsken, J. 117
Lukanidin, E.M. 171
MacDonald, N.J. 213
Mandriota, S.J. 31
Marchetti, D. 89
McKenna, W.G. 197
Menter, D.G. 89
Michiels, F. 251
Montesano, R. 31
Muschel, R.J. 197
Nicolson, G.L. 89
Orci, L. 31
Pepper, M.S. 31
Postel, E.H. 231
Raz, A. 137
Silletti, S. 137
Singh, R. 267
Singh, R.K. 1
Stam, J. 251
Steeg, P.S. 213
Tang, D.C. 69
van der Kammen, R.A. 251
van Leeuwen, F. 251
Vassalli, J.-D. 31
Watson, P.H. 265
Weidner, K.M. 117
Yun, Z. 89

Regulation of Tumor Angiogenesis by Organ-Specific Cytokines

R.K. SINGH and I.J. FIDLER

1 Introduction 1
2 Pathogenesis of Cancer Metastasis 1
3 Tumor Angiogenesis 2
4 Patterns of Metastasis 3
5 Host Microenvironment and Angiogenic Phenotype 4
6 Organ Site-Dependent Expression of Angiogenic Molecules 4
7 Regulation of Expression of Angiogenic Molecules by Organ-Derived Cytokines 6
8 Conclusion 8
References 9

1 Introduction

The outcome of cancer metastasis is dependent on a series of interactions between tumor cells and host factors (FIDLER 1990). A crucial step in tumor progression and metastasis is vascularization of the tumor and its immediate surroundings (FOLKMAN 1992). Without the ability to recruit new blood vessels, it is likely that most tumors would not grow beyond 1–2 mm and thus remain localized at the primary site. The induction of neovascularization is mediated by both tumor cells and host cells and the interaction of the two. In this article, we will discuss these interactions in the regulation of the angiogenic phenotype and how organ-derived cytokines modulate the expression of angiogenic factors.

2 Pathogenesis of Cancer Metastasis

Cancer metastasis consists of a series of sequential interrelated steps (FIDLER 1990). The major steps are as follows: after the initial transforming event, either

Department of Cell Biology, The University of Texas M.D. Anderson Cancer Center, 1515 Holcombe Boulevard, Box 173, Houston, TX 77030, USA unicellular or multicellular neoplastic cells grow progressively. If a tumor mass is to exceed 2 mm in diameter, extensive vascularization must occur. Several angiogenesis factors play key roles in establishing a neocapillary network from the surrounding host tissue. Once vascularization occurs, local invasion of the host stroma by some tumor cells could occur by several mechanisms that are not mutually exclusive: thin-walled venules, like lymphatic channels, offer very little resistance to penetration by tumor cells and provide the most common pathways for tumor cell entry into the circulation. Detachment and embolization of small tumor cell aggregates occur next; those that survive the circulation must then arrest in the capillary beds of organs. Extravasation then occurs, probably by the same mechanisms that influence initial invasion. Finally, proliferation within the organ parenchyma completes the metastatic process. To produce detectable lesions, these metastases must develop their own vascular network, evade the host immune system, and respond to organ-specific factors that influence their growth. Once they do so, the cells can invade host stroma, penetrate blood vessels, and enter the circulation to produce secondary metastases, the so-called metastasis of metastases (FIDLER 1995).

3 Tumor Angiogenesis

Tumor angiogenesis refers to the directional sprouting of new blood vessels toward a solid tumor. Tumors that are less than 1–2 mm in diameter can receive all nutrients by diffusion, but further growth depends on the development of an extensive blood supply (FOLKMAN and COTRAN 1976; FOLKMAN 1992). The prevascular stage of a tumor is associated with local, noninvasive, benign tumors, whereas the vascular stage is associated with aggressive, invasive, and metastatic tumors (FOLKMAN and KLAGSBRUN 1987). Indeed, the extent of neovascularization in different malignancies, such as melanoma, breast carcinoma, and prostate cancer, correlates with their potential for invasion and metastasis (FOLKMAN 1984; WEIDNER et al. 1992, 1993).

Angiogenesis entails a cascade of processes emanating from microvascular endothelial cells. To generate capillary sprouts, endothelial cells must proliferate, migrate, and invade host stroma, the direction of migration generally pointing toward the source of angiogenic molecules (FOLKMAN and KLAGSBRUN 1987). The capillary sprout subsequently expands and undergoes morphogenesis to yield a capillary. Although most solid tumors are highly vascular, their vessels are not identical to normal vessels of normal tissue. The distinction between tumor and normal tissue blood vessels includes differences in the cellular composition, differences in permeability, differences in blood vessel stability, and differences in the regulation of growth (FOLKMAN and KLAGSBRUN 1987).

The stimuli for tumor angiogenesis may be produced directly by tumor cells themselves or by host inflammatory cells at the tumor site. The induction of angiogenesis is mediated by several angiogenic factors (FOLKMAN 1992). At present, more than a dozen purified molecules have been reported, including fibroblast growth factor (FGF) family members, vascular endothelial cell growth factor (VEGF) or vascular permeability factor (VPF), interleukin-8 (IL-8), angiogenin, angiotropin, epidermal growth factor (EGF), fibrin, nicotinamide, platelet-derived endothelial cell growth factor (PD-ECGF), transforming growth factor (TGF)-α, TGF-β, and tumor necrosis factor-α (TNF-α) (reviewed in FOLKMAN 1992). Although these molecules have been shown to induce angiogenesis, most of them were first purified on the basis of some other unrelated activity. One question that arises regarding expression of angiogenic molecules is their redundancy, suggesting that the process under in vivo conditions utilizes multiple molecules.

4 Patterns of Metastasis

During the last two decades, studies of the pathogenesis of cancer metastasis have clearly established that the outcome of metastasis depends on the interactions of tumor cells with various host factors and that the pattern of metastasis is predictable, not random (FIDLER and KRIPKE 1977; POSTE and FIDLER 1980). These metastatic patterns are determined by factors that are independent of vascular anatomy, rate of blood flow, and the number of tumor cells delivered to each organ (FIDLER 1994).

Although this concept and the search for the mechanisms that regulate organ-specific metastasis have only recently reached prominence, they are not new. A century ago, PAGET questioned whether the distribution of metastases was due to chance. He analyzed the autopsies of a large number of patients with metastatic breast cancers (PAGET 1889). The nonrandom pattern of visceral metastases suggested to him that metastasis to a particular site was not due to chance, but rather that certain tumor cells (the seed) had a specific affinity for certain organ (the soil). Metastasis resulted only when both seed and soil matched. In the last few years, PAGET's hypothesis has received considerable experimental and clinical support (HART 1982; FIDLER 1990, 1995; PRICE 1994). Site-specific metastasis occurs with many transplantable experimental tumors and has been reported in autochthonous human tumors in patients with peritoneovenous shunts (TARIN et al. 1984a,b; FIDLER 1994).

The modern version of the seed and soil hypothesis consists of three principles:

1. Metastasis represents a highly selective nonrandom process favouring the survival of minor subpopulations of metastatic cells that preexist in the parent neoplasms (FIDLER and KRIPKE 1977; POSTE AND FIDLER 1980).
2. Neoplasms exhibit heterogeneous biologic and metastatic properties. Indeed, at the time of diagnosis, neoplasms contain different subpopulations of cells with diverse biologic characteristics (FIDLER 1990). Tumor cells isolated from individual tumors have been shown to differ with respect to many phenotypic characteristics, including antigenicity, immunogenicity, growth rate, karyotype, cell surface receptors,susceptibility to cytotoxic drugs, production of cytokine

growth factors, and propensity for invasion and metastasis. Biologic heterogeneity is not confined to cells in primary tumors; it is equally true of cells polluting distant metastases. Multiple metastases proliferating in different organs or even in the same organ of a cancer patient exhibit diversity in many biologic characteristics.

3. The outcome of metastasis depends on the balance of both tumor (seed) and host factors (soil). The contribution of host factors may vary in tumors arising from different tissues and in tumors of similar histologic origin in different patients (Fidler 1995).

5 Host Microenvironment and Angiogenic Phenotype

The tissue-specific metastatic pattern seen clinically and in experimental models highlights the importance of the local microenvironment for the development of metastases. Recent studies suggest that metastases can utilize host haemostatic mechanisms to survive and grow preferentially in a particular organ microenvironment. The organ environment can influence the behaviour of tumor cells in several different ways. For this reason, the implantation site of tumor cells in nude mice influences not only the growth of local tumor, but also the production of metastases. Further, organ-specific host factors have been shown to enhance or suppress the growth, angiogenesis, invasion, and metastasis of human tumors implanted into nude mice (Fidler 1994).

6 Organ Site-Dependent Expression of Angiogenic Molecules

Previous studies from our laboratory suggested that human renal cell carcinoma (HRCC) cells implanted into the different organ sites of nude mice have different metastatic phenotypes (Naito et al. 1986). HRCC implanted into the kidney produced a high incidence of lung metastasis, whereas those implanted subcutaneously did not. Histopathologic examination of the tissues stained with hematoxylin and eosin and immunohistochemistry revealed that the HRCC growing in the subcutis of nude mice had few blood vessels, whereas the tumors in the kidney had many (Singh et al. 1994a). Northern blot analysis of the tumors suggested a differential pattern of expression of an angiogenic molecule, basic FGF (bFGF). HRCC tumors in the subcutis or tunica muscularis of nude mice showed fewer mRNA transcripts for bFGF than did continuously cultured cells, whereas HRCC in the kidney of nude mice had ten- to 20-fold higher levels of bFGF mRNA than HRCC subcutaneous tumors. Immunohistochemical analysis of

HRCC tumors growing subcutaneously and in the kidney revealed significant differences in staining intensity: the tumors in the kidney were highly positive for bFGF protein, whereas subcutis tumors were not. The differential expression of bFGF mRNA and protein correlated with the cellular bFGF protein levels. The bFGF and mRNA levels revert to normal with prolonged culture of HRCC tumors, suggesting that the change in bFGF levels in HRCC tumors growing in the kidney and subcutis of nude mice is transient. Collectively, these data indicated that the organ microenvironment influences the expression level of bFGF in HRCC (SINGH et al. 1994a).

Similar observations were made in human melanoma, another excellent example of hematogenous malignancies. Melanoma cells secrete a variety of angiogenic molecules, e.g. VEGF, bFGF, and IL-8 (HERLYN 1990), and recent reports from our laboratory show that IL-8 is an important molecule in melanoma growth and progression. Constitutive expression of IL-8 directly correlated with the metastatic potential of the human melanoma cell lines examined. Further, IL-8 induced proliferation, migration, and invasion of endothelial cells and, hence, neovascularization (SINGH et al. 1994b). Several organ-derived cytokines (produced by inflammatory cells) are known to induce expression of IL-8 in normal and transformed cells (SINGH et al. 1994b). Since IL-8 expression in melanocytes and melanoma cells can be induced by inflammatory signals, the question of whether specific organ microenvironments could influence the expression of IL-8 was analyzed. Melanoma cells were implanted into the subcutis, the spleen (to produce liver metastasis), and the lateral tail vein (to produce lung metastasis) of athymic nude mice. Northern blot and immunohistochemical analyses determined that subcutaneous tumors, lung lesions, and liver lesions expressed high, intermediate, and no IL-8 mRNA and protein, respectively (GUTMAN et al. 1995). This differential expression was not due to the selection of a subpopulation of cells (GUTMAN et al. 1995). Melanoma cell lines established from tumors growing in vivo exhibited similar levels of IL-8 mRNA transcripts as continuously cultured cells. The crossover experiment suggested that the IL-8 mRNA level was always higher in skin and low in liver tumors regardless of whether the melanoma cells had first been harvested from subcutaneous or liver tumors (GUTMAN et al. 1995). These observations suggest that the organ microenvironment modulates the expression of IL-8 in human melanoma cells.

IL-8 expression was upregulated in coculture of melanoma cells with keratinocytes (skin), whereas it was inhibited in cells cocultured with hepatocytes (liver). Similar results were obtained with conditioned media from keratinocyte and hepatocyte cultures (GUTMAN et al. 1995). These data suggest that organ-derived factors modulate the expression of IL-8 in human melanoma cells.

Recent reports from our laboratory suggest an important role for host-infiltrating leukocytes and their products in the induction of tumor neovascularization (GUTMAN et al. 1994). This idea was first proposed by COMAN and SHELDON (1946) and later supported by SIDKEY and AUERBACH (1976). We investigated angiogenesis around melanomas growing subcutaneously and their correlation with tumor size in normal and myelosuppressed mice. The subcutaneous growth of the weakly immunogenic B16 melanoma cells was slow in myelosuppressed mice, and neovascularization was likewise low. Interestingly, the reconstitution of myelosuppressed mice with normal splenocytes resulted in rapid vascularization around the tumor implants and then rapid growth of the tumors (GUTMAN et al. 1994). Whether this lymphoid-mediated angiogenesis was due to the production of angiogenic molecules is still unclear.

Taken together, these observations suggest that angiogenesis is controlled by a local balance between factors that stimulate or inhibit new blood vessel growth (ZHANG et al. 1994; O'REILLY et al. 1994). In most normal tissues, the inhibitory influence predominates. In contrast, cells of the malignant phenotype switch from an angiogenesis-inhibiting to an angiogenesis-stimulating phenotype, as was observed in cultured fibroblasts from Li-Fraumeni patients (DAMERON et al. 1994). The switch to the angiogenic phenotype coincides with the loss of wild-type allele of the p53 tumor suppressor gene and is the result of reduced production of antiangiogenic factor TSP-1 (DAMEERON et al. 1994). As one possible explanation, UEBA et al. (1994) showed that p53 regulate the promoter activity of the bFGF gene at the transcriptional level, and expression of bFGF was also activated by mutant p53. Wild-type p53 repressed expression of bFGF and its mutant activated it, suggesting that p53 may play a role as an angiogenic switch.

Another excellent example of the regulation of angiogenic molecules by host factors was demonstrated in psoriases (NICKOLOFF et al. 1994), a common, inherited skin disease characterized by hyperproliferation of keratinocytes and excessive dermal angiogenesis. Media conditioned by keratinocytes from both symptomatic and psoriatic plaques induced a vigorous angiogenic response, whereas media conditioned from normal keratinocytes did not (NICKOLOFF et al. 1994). Furthermore, keratinocytes from psoriatic skin expressed a ten- to 20-fold increased level of IL-8 and a seven fold reduction of TSP-1. These data suggest that aberrant angiogenesis in psoriatic skin might to due to the defect in production of the angiogenic stimulator IL-8 and a deficiency in production of the angiogenic inhibitor TSP-1 (NICKOLOFF et al. 1994).

7 Regulation of Expression of Angiogenic Molecules by Organ-Derived Cytokines

Many different cytokines and growth factors that stimulate or inhibit angiogenesis are present in the host tissue. Presumably, their relative availability and activity determines the angiogenic phenotype. The endogenous stimulators include acidic FGF (aFGF), bFGF, VEGF, TGF-α and -β, EGF, PD-ECGF, angiogenin, pleiotropin, hepatocyte growth factor (HGF), IL-1, IL-6, IL-8, and heparinase. The antagonists/inhibitors include IFN-α and -β, angiostatic cartilage-derived inhibitors, platelet factor IV, protamine, thrombospondin, tissue inhibitor of metalloproteinase (TIMP), and plasminogen activator inhibitor (PAI) (FOLKMAN 1992, 1995;

FIDLER and ELLIS 1994). These factors may be released by a number of different cell types within solid tumors. The remainder of the chapter will concentrate on the role of organ-derived cytokines in the modulation of angiogenesis.

In the HRCC model, we examined the balance between the expression of bFGF and IFN-β. The two cytokines are differentially expressed in tumor cells and host cells. As stated earlier, HRCC tumors implanted in the subcutis of nude mice expressed low levels of bFGF, whereas HRCC tumors in the kidney expressed high levels. In sharp contrast, the expression of IFN-β was high in and around the subcutaneous tumors, whereas no immunohistochemical reactivity was found in the HRCC tumors growing in the kidney (R.K. SINGH, C.D. BUCANA, and I.J. FIDLER, unpublished).

There is now increasing evidence that fibroblasts derived from different anatomical sites in the adult tissue display heterogeneity in interaction with cytokines, hormones, and growth factors. A unique interaction of fibroblasts from specific organs with tumor cells has been demonstrated in in vitro systems utilizing both rodent and human samples (CAMPS et al. 1990; CULLEN et al. 1991). Diffusible factors produced by fibroblasts subsequent to their interaction with tumor cells have been shown to stimulate the growth of mouse mammary tumor cells, human breast cancer cells, and HRCC (CAMPS et al. 1990) and to modulate invasion and migration of tumor cells (CHEW et al. 1988).

IFN-β is one cytokine produced by fibroblasts (TORRENCE and CLERCQ 1977). IFNs have been shown to influence the invasive and metastatic potential of malignant cells by a variety of mechanisms (RAMANI and BALKWILL 1989; GOHJI et al. 1994). In animal models, they inhibit tumor-induced angiogenesis. Unlike other angiogenic inhibitors which bind to angiogenic molecules, it was speculated that IFNs modulate the signal for angiogenesis produced by the tumor cells. Support for this idea is provided by a report that inhibition of angiogenesis is independent of IFN antiproliferative activity. To substantiate this, SIDKEY and BORDEN (1987) treated L1210R leukemia cells resistant to the antiproliferative effect to IFNs with IFN-β. IFN-β had no effect on their proliferation, but L1210R cells were impaired in their ability to induce angiogenesis.

This study, recent clinical observations suggesting an antiangiogenice role of IFNs (EZEKOWITZ et al. 1992), and our own in vivo observations led us to investigate whether IFNs could modulate the expression of bFGF, an angiogenic signal in HRCC cells. IFN-α and IFN-β, but not IFN-γ, downregulated the expression of bFGF mRNA and protein (SINGH et al. 1995). This effect was independent of the antiproliferative effects of IFNs. The downregulation of bFGF required a long exposure of cells to a low concentration of IFN. Moreover, once IFN was withdrawn, cells resumed production of bFGF. These observations are consistent with clinical experience indicating that IFN-α must be given for many months to induce a response (EZEKOWITZ et al. 1992). Whether IFNs affect the expression of bFGF in other tumor types was analyzed. The incubation of human bladder, prostate, colon, and breast carcinoma cells with noncytostatic concentrations of IFN-α or IFN-β also inhibited bFGF production (SINGH et al. 1995). The underlying mechanism for this modulation remains unclear, however.

Chronic immune/inflammatory disorders elicit a myriad of organ microenvironmental perturbations that contribute to the regulation of angiogenesis (Queluz et al. 1990; Tsuboi et al. 1990). A complex cytokine network is involved, and IFN play a major role (Granstein et al. 1990). The interaction of IFN with other inflammatory cytokines determines whether neovascularization will be suppressed or induced (Stout et al. 1993). For example, IFN-α and IL-2 together induce markedly hemorrhagic and exudative lesions in the experimental model of Goodpasture pneumonitis (Queluz et al. 1990). Administration of IFN-α/IL-2 induced release of bFGF, which in turn elicited endothelial cell (EC) production of proteases (Cozzolino et al. 1993), potentially facilitating blood vessel invasion as occurs in rheumatoid pannus. Preincubation of cultured human EC with IFN-α followed by exposure to IL-2 resulted in production and release of bFGF (Tsuboi et al. 1990; Cozzolino et al. 1993). These observations raised questions as to why IFN-α would both cause regression of hemangiomas with an abnormal vascular structure and induce angiogenesis in normal vasculature in vitro and in the rabbit cornea model (Maheshwari et al. 1991; Bhartiya et al. 1992; Ezekowitz et al. 1992; Cozzolino et al. 1993). The neoplastic nature of hemangioma cells may account for the regression in that case. HRCC can be cited as an example: IFN-α and IL-2 gene transfer inhibits the invasive, angiogenic, and metastatic potential in in vitro and in vivo HRCC (Halthorn et al. 1994). Future studies will be required to fully elucidate the tumor phenotype induced by IFN and its interaction with other cytokines in physiology and pathology.

To explore the complex situation of multiple cytokines, we investigated the effects of two cytokines produced by keratinocytes (IL-1, IFN-β) and two produced by hepatocytes (TGF-α, TGF-β) on regulation of expression of IL-8 in human melanoma cells. IL-1 upregulated the expression of IL-8 in human melanoma cells at both the mRNA and protein levels if de novo protein synthesis occured. The effect of IL-1 on IL-8 induction was dose and duration dependent. Preliminary date suggest that upregulation of Il-8 by IL-1 in melanoma cells occurs at the transcriptional level. IFN-β did not affect constitutive IL-8 mRNA and protein production in human melanoma cells, but it blocked the induction of IL-8 by IL-1. Oliveira et al. (1992) also reported that inflammatory cytokines such as TNF and IL-1 are potent inducers of IL-8; they found that this induction was inhibited by IFN-β. We found that TGF-β inhibited IL-8 expression, but TGF-α did not affect it. What other cytokines may be involved in the downregulation of IL-8 in liver remains unclear (Gross et al. 1995).

8 Conclusion

Clearly, the complement of positive and negative regulators of angiogenesis may vary among different physiologic and pathologic settings. Recognition of this dual mechanism of control is necessary if we are to gain a better understanding of this complex process and its significance in disease. The recent elucidation of the cooperative interaction among positive and negative regulatory molecules during normal physiology and the apparent disruption of this program in pathology suggest that future studies of pathologic angiogenesis must focus on the interaction of both positive and negative regulators of this process.

The cross-talk between the tumor cells and neighboring stromal cells occurs mainly via a complex network of extracellular signals, including a large number of cytokines and growth factors, their antagonists/inhibitors, and soluble receptors, all of which are released by and act on cells within the tumor microenvironment. The activities of these pleiotropic cytokines have yet to be fully elucidated. From what we do know, we propose that their net effect is to regulate the proliferation of metastatic cells, suppress the activity of infiltrating host cells, and enhance the establishment of the stromal environment by inducing development of a new blood supply for the tumors.

Tumor cells produce more than one angiogenic molecule. Whether these molecules share the same regulatory mechanism remains unclear. Conceivably, each factor may be regulated by different sets of host-derived cytokines. Understanding these events will allow us to design potent antiangiogenic approaches to address pathologic situations such as cancer.

Acknowledgment. This work was supported in part by Cancer Center Support Core grant CA 16672 and grant R35-CA 42107 from the National Cancer Institute, National Institutes of Health, and by the Josef Steiner Foundation.

References

Bhartiya D, Sklarsh JW, Maheshwari RK (1992) Enhanced wound healing in animal models by interferon and an interferon inducer. J Cell Physiol 150: 312–319

Camps JL, Chang SM, Hsu TC, Freeman MR, Hong SJ, Zhau HE, von Eschenbach AC, Chung LWK (1990) Fibroblast-mediated acceleration of human epithelial cancer growth in vivo. Proc Natl Acad Sci USA 87: 75–79

Chew EC, Mok CH, Tsao SW, Liu WK, Riches DJ (1988) Scanning electron microscopic study of invasiveness between normal and neoplastic cells and fibroblasts in monolayer culture. Anticancer Res 8: 275–280

Coman DR, Sheldon WF (1946) The significance of hypermia around tumor implants. Am J Pathol 22: 821–831

Cozzolino F, Torcia M, Lucibello M, Morbidolli L, Ziche M, Platt J, Fabini S, Brett J, Stern D (1993) Interferon-α and interleukin-2 synergistically enhance basic fibroblast growth factor synthesis and induce release, promoting endothelial cell growth. J Clin Invest 91: 2504–2515

Cullen KJ, Smith HS, Hill S, Rosen N, Lippman ME (1991) Growth factor mRNA expression by human breast fibroblasts from benign and malignant lesions. Cancer Res 51: 4978–4985

Dameron KM, Volpert OV, Tainsky MA, Bouk N (1994) Control of angiogenesis in fibroblasts by p53, regulation of thrombospondin-1. Science 265: 1502–1504

Ezekowitz RAB, Mulliken JB, Folkman J (1992) Interferon-α-2a therapy for life threatening hemangiomas of infancy. N Engl J Med 326: 1456–1463

Fidler IJ (1990) Critical factors in the biology of human cancer metastasis: twenty-eighth GHA Clowes Memorial Award lecture. Cancer Res 50: 6130–6138

Fidler IJ (1994) Experimental orthotopic models of organ-specific metastasis by human neoplasms. Adv Mol Cell Biol 9: 191–215

Fidler IJ (1995) Cancer biology: invasion and metastasis. In: Abeloff MD, Armitage JO, Lichter AS, Niederhuber JE (eds) Clinical oncology. Churchill Livingstone, New York, pp 55–76
Fidler IJ, Ellis LM (1994) The implications of angiogenesis for the biology and therapy of cancer metastasis. Cell 79: 185–188
Fidler IJ, Kripke ML (1977) Metastasis results from preexisting variant cells within a malignant tumor. Science 197: 893–895
Folkman J (1984) Angiogenesis: initiation and modulation. In: Nicolson GL, Milas L (eds) Cancer invasion and metastasis: biologic and therapeutic aspects. Raven, New York, pp 201–209
Folkman J (1992) The role of angiogenesis in tumor growths. Semin Cancer Biol 3: 65–71
Folkman J (1995) Angiogenesis in cancer, vascular, rheumatoid and other disease. Nature Med 1: 27–31
Folkman J, Cotran R (1976) Relation of vascular proliferation to tumor growth. Int Rev Exp Pathol 16: 207–248
Folkman J, Klagsbrun M (1987) Angiogenic factors. Science 235: 444–447
Gohji K, Fidler IJ, Fabra A, Bucana CD, von Eschenbach AC, Nakajima M (1994) Regulation of gelatinase production in metastatic renal cell carcinoma by organ-specific fibroblasts. Jpn J Cancer Res 85: 152–160
Granstein RD, Flotle TJ, Amento EP (1990) Interferons and collagen production. J Invest Dermatol 95: 75S–80S
Gross V, Andus T, Daig R, Aschenbrenner E, Scholmerich J, Falk W (1995) Regulation of interleukin-8 production in a human colon epithelial cell lije (HT-29). Gastroenterology 108: 653–661
Gutman M, Singh RK, Yoon S, Xie K, Bucana CD, Fidler IJ (1994) Leukocyte-induced angiogenesis and subcutaneous growth of B16 melanoma. Cancer Biother 9: 163–170
Gutman M, Singh RK, Xie K, Bucana CD, Fidler IJ (1995) Regulation of IL-8 expression in human melanoma cells by the organ environment. Cancer Res 55: 2470–2475
Halthorn RW, Tso CL, Kaboo R, Pang S, Figlin R, Sayers C, de Kernion JB, Belldegrun A (1994) In vitro modulation of metastatic potentials of human renal cell carcinoma by interleukin-2 and/or interferon alpha gene transfer. Cancer 74: 1904–1911
Hart IR (1982) "Seed and soil" revisited: mechanisms of site-specific metastasis. Cancer Metastasis Rev 1: 5–7
Herlyn M (1990) Human melanoma: development and progression. Cancer Metastasis Rev 9: 101–109
Maheshwari RK, Srikantan V, Bhartiya D, Kleinman HK, Grant DS (1991) Differential effect of interferon gamma and alpha on in vitro model angiogenesis. J Cell Physiol 146: 164–169
Naito S, von Eschenbach AC, Giavazzi R, Fidler IJ (1986) Growth and metastasis of tumor cells isolated from a human renal cell carcinoma implanted into different organs of nude mice. Cancer Res 46: 4109–4115
Nickoloff BJ, Mitre RS, Varoni J, Dixit VM, Polvenini PJ (1994) Aberrant production of interleukin-8 and thrombospondin-1 by psoriatic keratinocytes mediates angiogenesis. Am J Pathol 144: 820–828
Oliveira IC, Sciavolino PJ, Lee TH, Vilcek J (1992) Downregulation of interleukin-8 gene expression in human fibroblast: unique mechanisms of transcriptional inhibition by interferon. Proc Natl Acad Sci USA 89: 9049–9053
O'Reily MS, Homgren L, Shing Y, Chen C, Rosenthal RA, Moses M, Lane WS, Cao Y, Sago EH, Folkman J (1994) Angiostatin: a novel angiogenesis inhibitor that mediates the suppression of metastases by a Lewis lung carcinoma. Cell 79: 315–328
Paget S (1889) The distribution of secondary growths in cancer of the breast. Lancet 1: 571–573
Poste G, Fidler IJ (1980) The pathogenesis of cancer metastasis. Nature 283: 139–146
Price JE (1994) Host tumor interaction in progression of breast cancer metastasis. In Vivo 8: 145–154
Queluz TH, Pawlowski I, Brunda MJ, Brentjens JR, Vladutiu AO, Andres G (1990) Pathogenesis of an experimental model of Goodpasture's hemorrhagic pneumoniates. J Clin Invest 85: 1507–1515
Ramani P, Balkwill FR (1989) Action of recombinant alpha interferon against experimental and spontaneous metastasis in a murine model. Int J Cancer 43: 140–146
Sidkey YA, Auerbach R (1976) Lymphocyte-induced angiogenesis in tumor bearing mice. Science 192: 1237–1238
Sidkey YA, Borden EC (1987) Inhibition of angiogenesis of interferons: effect on tumor and lymphocyte-induced vascular responses. Cancer Res 47: 5155–5161
Singh RK, Bucana CD, Gutman M, Fan D, Wilson MR, Fidler IJ (1994a) Organ site-dependent expression of basic fibroblast growth factor in human renal cell carcinoma cells. Am J Pathol 145: 365–374
Singh RK, Gutman M, Radinsky R, Bucana CD, Fidler IJ (1994b) Expression of interleukin-8 correlates with the metastatic potential of human melanoma cells in nude mice. Cancer Res 54: 3242–3247

Singh RK, Gutman M, Bucana CD, Sanchez R, Llansa N, Fidler IJ (1995) Interferons alpha and beta downregulate expression of basic fibroblast growth factor in human carcinomas. Proc Natl Acad Sci USA 92: 4562–4566

Stout AJ, Gresser I, Thompson WD (1993) Inhibition of wound healing in mice by local interferon α/β injection. Int J Exp Pathol 74: 79–85

Tarin D, Price JE, Kettlewell MGW, Souter RG, Vass ACR, Crossley B (1984a) Clinicopathological observations on metastasis in man studied in patients treated with peritoneovenous shunts. Br Med J 288: 749–751

Tarin D, Price JE, Kettlewell MGW, Souter RG, Vass ACR (1984b) Mechanism of human tumor metastasis studied in patients with peritoneovenous shunts. Cancer Res 44: 3584–3592

Torrence PF, Clercq ED (1977) Inducers and induction of interferon. Pharmacol Ther 2: 1–88

Tsuboi R, Sato R, Rifkin DB (1990) Correlation of cell migration, cell invasion, receptor number, proteinase production, and basic fibroblast growth factor level in endothelial cells. J Cell Biol 110: 511–517

Ueba T, Nosaka T, Takahashi JA, Shibata F, Florkiewicz RZ, Vogelstein B, Oda Y, Kikuchi H, Hatanak U (1994) Transcriptional regulation of basic fibroblast growth factor gene by p53 in human glioblastoma and hepatocellular carcinoma cells. Proc Natl Acad Sci USA 91: 9009–9013

Weidner N, Folkman J, Pozza F, Bevilacqua P, Allred EN, Moore DH, Meli S, Gaspanni G (1992) Tumor angiogenesis: a new significant and independent prognostiic indicator in early stage breast carcinoma. J Natl Cancer Inst 84: 1875–1887

Weidner N, Carroll PR, Flax J, Blumenfeld W, Folkman J (1993) Tumor angiogenesis correlates with metastasis in invasive prostate carcinoma. Am J Pathol 143: 401–409

Zhang Y, Deng Y, Luther T, Muller M, Ziegler R, Waldherr R, Stern DM, Nawroth PP (1994) Tissue factor controls the balance of angiogenic and antiangiogenic properties of tumor cells in mice. J Clin Invest 94: 1320–1327

Cytokine-Mediated Tumor-Endothelial Cell Interaction in Metastasis

R. GIAVAZZI

1 Introduction 13
2 Cytokines Affect Adhesive Properties of Endothelial Cells and Metastasis 15
3 Role of Cytokine-Induced Adhesion Molecules 17
3.1 Selectin-Mediated Adhesion and Metastasis 18
3.2 Integrin/Immunoglobulin Superfamily-Mediated Adhesion and Metastasis 21
3.3 Other Adhesion Molecules 23
4 Conclusions 23
References 24

1 Introduction

A variety of adhesive interactions take place between tumor cells and host vasculature in tumor progression and metastasis (MCCORMICK and ZETTER 1992; HONN and TANG 1992; BELLONI and TRESSLER 1990). Perhaps one of the most important adhesion events in the process of metastasis is the interaction of circulating tumor cells with the endothelium and subendothelial matrix (LIOTTA 1986; KRAMER and NICOLSON 1979). The arrest of tumor cells in the capillary bed of secondary organs, through the interaction with vascular or lymphatic endothelium and subendothelial basement membrane, precedes their migration into the tissue parenchyma; once extravasation has occurred, micrometastasis proliferation will start (CRISSMAN et al. 1988). The adhesion to vascular endothelium is a necessary step preceding the extravasation of metastatic tumor cells. However, it is becoming increasingly clear that the preferential interaction between tumor cells and the microvascular endothelium and extracellular matrix of different areas contributes to the homing of metastases in specific organs (PAULI et al. 1990; AUERBACH et al. 1987; PAULI and LEE 1988; ROOS et al. 1984; NICOLSON 1988). Attempts to identify endothelial cell surface components responsible for endothelial cell organ-specific interactions have been reported (ZHU et al. 1991; BELLONI and NICOLSON 1988).

Mario Negri Institute for Pharmacological Research, Via Gavazzeni 11, 24125 Bergamo, Italy

Methodological development of microvascular endothelial cell cultures, together with progress in biochemical and molecular biology techniques, has allowed the identification of surface molecules responsible for mediating tumor–host interactions. Thus the adhesion response of tumor cells can be mediated by several receptors belonging to four major families (ALBELDA 1994): cadherins (TAKEICHI 1990), immunoglobulins (WILLIAMS and BARCLAY 1988), integrins (DIAMOND and SPRINGER 1994), and selectins (MCEVER 1991). Many other unclassified molecules, such as carbohydrates, lectins, and proteoglycans, mediate tumor–host cell or tumor–host substrate interactions. Among these, the glycoprotein CD44 has been described to play a major role in metastasis (HERRLICH et al. 1993; GUNTHERT et al. 1991). The augmented expression (or down regulation) of adhesion molecules has been associated with the malignant phenotype and implicated in tumor–host cell recognition. All these aspects are covered by more comprehensive reviews elsewhere in these book.

The mechanism and speed of extravasation varies in different compartments in relation to the endothelial structure (LAPIS et al. 1988).These can be either a rapid migration through preexisting pores and fenestrations that interrupt the continuity of endothelial cells or a slower migration between and underneath the endothelial cells through intercellular junctions (CRISSMAN et al. 1988; DINGEMANS et al. 1978). Therefore, the same tumor can use different mechanisms of extravasation in different areas. However, tumor types with different malignant proprieties can differ even at the same target vessel (CRISSMAN et al. 1988; KINJO 1978).

Vascular changes can easily influence the interaction of tumor cells with endothelium and thus the whole metastatic process (BELLONI and TRESSLER 1990; PAULI et al. 1990; WEISS et al. 1989). The formation of platelet–fibrin thrombi, for example, influences the arrest of tumor cells (GASIC 1984; WEISS et al. 1989); the release of chemotactic factors from the vascular wall can induce tumor cells motility (ORR et al. 1988; WANG et al. 1990). Furthermore, extracellular matrix components may modulate endothelial cell-recognition structures responsible for tumor–endothelial cell interactions (PAULI and LEE 1988). Endothelial cell damage and retraction induced by different modalities, followed by exposure to the subendothelial-basement membrane, results in an increased adhesion of tumor cells and platelets to the exposed matrix. X-rays and some conventional antineoplastic drugs may possibly increase tumor cell localization and metastasis formation through endothelial cell damage (DAO and YOGO 1967; NICOLSON and CUSTEAD 1985; VANDENBRENK et al. 1973; WITHERS and MILAS 1973; ORR et al. 1986). Inflammatory stimuli can also influence the adhesion mechanisms that regulate tumor–endothelial cell interaction and metastasis formation. There are connections between metastasis formation and activation of inflammation. For example, clinical and experimental observations indicate preferential tumor cell arrest and metastasis growth at sites of injury, healing, and inflammation (SUGARBAKER 1981; MURPHY et al. 1988; LEVINE and SALTZMAN 1990). Intravenous injection of a inflammatory stimuli results in an increase of experimental metastases, associated with an intravascular release of inflammatory mediators, leukocytosis, and pulmonary edema (VAN DEN BRENK et al. 1974).

Endothelial cells represent one of the major targets of inflammatory stimuli, to which they respond with specific morphological, molecular, and functional changes (COTRAN 1987;MANTOVANI and DEJANA 1989). The role of inflammatory cytokines, specifically interleukin-1 (IL-1) and tumor necrosis factor α (TNF), in tumor cell adhesion to endothelium and in metastasis formation is presented in this chapter, with emphasis placed on the adhesion molecules expressed on cytokine-activated endothelial cells and their relevance in metastasis formation. The significance of these findings and implications for the treatment of metastases will be discussed. Additional functions of these and other cytokines regulating the tumorigenic and malignant response of tumors will not be addressed.

2 Cytokines Affect Adhesive Properties of Endothelial Cells and Metastasis

The activation of endothelial cells is required for the binding and transendothelium migration of leukocytes to sites of injury and infection (SHIMIZU et al. 1992; ZIMMERMAN et al. 1992; SPRINGER 1994). It has been shown that endothelial cells treated with inflammatory stimuli, such as IL-1, TNF, and lipopolysaccharide (LPS), exhibit increased adhesiveness for leukocytes; other cytokines can also cooperate in this process (OSBORN 1990; ALBELDA et al. 1994).The adhesive behavior of tumor cells to cytokine-activated endothelial cells has been investigated with the belief that the adhesion of tumor cells to microvascular endothelium is necessary for the migration of metastatic cells. Early reports showed that the adhesion of different types of human tumor cells is augmented to IL-1- or TNF-activated human umbilical vein endothelial cells (HUVEC) (DEJANA et al. 1988; RICE et al. 1988; LAURI et al. 1990). Subsequent in vitro studies have shown that IL-1 treatment can increase the adhesiveness of cultured hepatic sinusoidal endothelium for the B16 melanoma (VIDAL-VANACLOCHA et al. 1994), and the treatment of microvascular endothelium of the brain with TNF, IL-1, or LPS also increased adhesiveness in a murine model of mastocytoma (BERETA et al. 1991). Neutralizing antibodies of these cytokines abolished the augmented adhesion of tumor cells to activated endothelial cells (KAJI et al. 1995; BURROWS et al. 1991). Moreover, the augmented adhesion of tumor cells to IL-1-activated endothelial cells was blocked in the presence of the naturally occurring IL-1 receptor antagonist (IL-1ra) (CHIRIVI et al. 1993; VIDAL-VANACLOCHA et al. 1994). At variance with other inducers of cell adhesion, these cytokines enhanced cell binding by altering the endothelial adhesive proprieties, without a direct effect on the adhesiveness of tumor cells (HERBERT 1993).

The mechanism by which tumor cells contribute to endothelial cell activation remains poorly understood. It was believed that tumor cells are able to activate endothelial cells directly by the secretion of factors such as cytokines that are known to contribute to endothelial cell activation. However, in a preliminary report

Hakamuri has shown that tumor cells are incapable of activating endothelial cells directly, but rather that tumor cells readily activate platelets, which in turn activate endothelial cells to express adhesion molecules (HAKOMORI 1994). It is possible that the quantity of cytokines secreted by tumor cells is not always sufficient to stimulate endothelial cells. Tumor-derived IL-1 has been shown to influence tumor–endothelial cell adhesion and cause the induction of adhesion molecules on endothelial cells (BURROWS et al.1991; KAJI et al. 1995). We have observed that melanoma cells transduced with IL-1 release sufficient cytokines to activate endothelial cells and to enhance tumor cell adhesion (R.GIAVAZZI and R.CHIRIVI, submitted), even though they were no more adhesive perse.

The effect of IL-1 and TNF in metastasis formation in vivo has been investigated in a variety of animal models. Work in our laboratory has shown the augmentation of lung colonies in nude mice receiving IL-1α, IL-1β, or TNF (BANI et al. 1991; GIAVAZZI et al. 1990) and injected with human melanoma, colon carcinoma, and renal carcinoma cell lines. The augmentation of spontaneous metastases in C57BL/6 syngeneic mice bearing subcutaneous tumors and treated daily with IL-1 has been described for the murine B16-BL6 melanoma metastatic to the lung and the murine M5076 reticulum cell sarcoma metastatic to the liver (BANI et al. 1991). In these studies, the augmentation of metastases appeared to be independent of the effect of the cytokine on the primary tumor growth. In line with the theory that cytokine treatment affects the interaction of tumor cells with the vasculature of the metastatic organ, the retention of radiolabeled tumor cells was found to be significantly augmented in the lung of IL-1-treated mice at 4 and 24h after intravenous injection (GIAVAZZI et al. 1990; LAURI et al. 1990). Augmentation of tumor spread and metastases by cytokine treatment has been described in different murine tumor models. Injections of recombinant human IL-1 caused the augmentation of experimental metastases to the bone or bone marrow (ARGUELLO et al. 1992) and to the lung (BERTOMEU et al. 1993) from B16 melanoma. The injection of recombinant human or mouse TNF induced the enhancement of metastases to the lung from murine CFS1 fibrosarcoma (OROSZ et al. 1993), B16 melanoma, and Lewis lung carcinoma (MIYATA et al. 1992). The administration of recombinant murine TNF to mice led to significant enhancement of artificial as well spontaneous metastases to the liver from the highly metastatic ESb lymphoma (OROSZ et al. 1995). Interestingly, MALICK and coworkers showed that the intraperitoneal administration of IL-1 or TNF in nude mice bearing human ovarian carcinoma led to the replacement of peritoneal ascitic tumor with solid tumors attached to the peritoneum and intra-abdominal viscera (MALIK et al. 1989,1992).

The augmentation of lung colonies by human melanoma in IL-1-treated mice can be specifically inhibited by treatment with IL-1ra (CHIRIVI et al. 1993). IL-1ra also inhibited the augmentation of experimental metastases caused by LPS treatment (CHIRIVI et al. 1993). LPS is a nonspecific inflammatory stimulus that triggers a cascade of host-derived mediators responsible for its proinflammatory effects. The specific effect of IL-1ra on LPS-induced lung colony augmentation indicates that host-produced IL-1 might play a role in the increase of metastases.

From their work showing that a single dose of IL-1ra reduced the number and volume of hepatic metastases when given prior to intrasplenic injection of melanoma cells, VIDAL-VANACLOCHA et al. (1994) suggested a role for endogenous IL-1, perhaps systematically induced by tumor cells, in metastasis formation and growth. A role of endeogenous TNF to the formation of pulmonary metastases has been demonstrated by using anti murine TNF monoclonal antibodies that caused a significant reduction of endotoxin-induced lung colonies produced by murine fibrosarcoma cells (OROSZ et al. 1993). In contrast, an antimetastatic effect of endogenous TNF induced by interferon-γ (IFN-γ) has been reported (SAIKI et al. 1989).

Altered production and response to cytokines are often involved in neoplasia. Tumor cells can express and release cytokines themselves, which may regulate tumor immunity and growth. Whether this production can influence their malignant behavior is not yet understood. A constitutively TNF-producing fibrosarcoma showed reduced tumorigenicity with less invasiveness (VANHAESEBROECK et al. 1991). However, Chinese hamster ovary (CHO) cells transfected with the gene for human TNF showed an enhanced ability to invade peritoneal surfaces of nude mice, this ability was abrogated by neutralizing antibodies to human TNF (MALIK et al. 1990). Similarly, the production of TNF, after gene transfer, accelerated metastasis formation to the liver by the malignant ESb lymphoma. This effect was blocked by anti-TNF monoclonal antibody (ZHIHAI et al. 1993). In contrast, TNF expression in the nonmetastatic EB parent line caused their rejection when injected in immunocompetent mice. It appears that cytokines neither induced metastasis formation of tumor lines that are not metastatic, nor changed their metastatic homing (BANI et al. 1991). In general, the pretreatment of tumor cells with IL-1 or TNF before injection does not influence their metastatic behavior (OROSZ et al. 1993,1995; BANI et al. 1991; GIAVAZZI et al. 1990). This is in contrast with the results of a study by LOLLINI et al. 1990, who showed that TNF can enhance metastasizing ability by direct action on tumor cells. The augmented metastatic ability of TNF-treated cells has been explained as the result of reduced NK cytotoxicity to tumor cells due to their increased Major histocompatibility Complex (MHC) class I antigen expression (LOLLINI et al. 1990).

These results have implications for immunotherapeutic approaches to cancer treatment. The interested reader may wish to consider recent reviews on the use of cytokine gene-transfected cells as vaccines and how this procedure interferes with the malignant behavior of tumor cells (COLOMBO et al. 1994). The antimetastatic effect of these cytokines used in the therapeutic regimes has also been described elsewhere (BELARDELLI et al. 1989; NAKAMURA et al. 1986; NISHIYAMA et al. 1989).

3 Role of Cytokine-Induced Adhesion Molecules

The fields of cancer metastasis and leukocyte migration are converging rapidly. Several adhesion molecules are shared by cancer cells and leukocytes, and mechanisms proposed for lymphocyte migration appear to play a role in the extravasation of metastatic cells (CHIRIVI et al. 1994; SMITH and ANDERSON et al. 1991; BELLONI and TRESSLER 1990; HONN and TANG 1992; BEVILACQUA 1993). Several endothelial and leukocyte surface proteins responsible for the interaction of leukocytes with vascular endothelium have been molecularly cloned and their functional and immunological activity identified. The molecular and functional characteristics of these determinants are discussed in recent detailed reviews (CARLOS and HARLAN 1994; BEVILACQUA 1993). This group of adhesion molecules includes members of the immunoglobulin superfamily, the integrin family, and the selectin family.

Inflammatory cytokines stimulate leukocyte adhesion on endothelial cells, a process mediated by the induction or augmented expression of multiple adhesion molecules (OSBORN 1990). Three major adhesion molecules on endothelial cells are responsible for this response: the endothelial leukocyte adhesion molecule-1 (ELAM-1), now called E-selectin, the vascular cell adhesion molecule-1 (VCAM-1), and the intercellular adhesion molecule-1 (ICAM-1) BEVILACQUA et al. 1987; OSBORN et al. 1989; DUSTIN et al. 1986). The use of antibodies against these adhesion molecules allows the study of their role in tumor cell adhesion to cytokine-activated endothelial cells.

3.1 Selectin-Mediated Adhesion and Metastasis

The role of carbohydrates and lectins in metastases has been recognized for several years. Changes in the cell surface carbohydrate profile have been associated with the malignant phenotype, and endogenous lectins can select for metastatic properties. These aspects have been covered in detailed reviews (BELLONI and TRESSLER 1990; RAZ and LOTAN 1987; DENNIS and LAFERTE 1987).

Selectins represent a family of three transmembrane glycoproteins, L-selectin, P-selectin, E-selectin, that play a role in leukocyte–endothelial cell interaction (MCEVER 1991; BEVILACQUA and NELSON 1993b). These molecules have a common structure characterized by an N-terminal lectin-like domain. Inflammatory and tumor cells bind the lectin-like region of these molecules via specific carbohydrate structures. At least two of these adhesion molecules, E-selectin and P-selectin (previously called GM-140/PADGEM), expressed on activated endothelial cells, have been shown to mediate tumor cell adhesion (LAURI et al. 1991b; RICE and BEVILACQUA 1989; MARTIN-PADURA et al. 1993). The enhanced binding of colon-related carcinomas to cytokine-activated endothelial cells is blocked by anti-E-selectin antibodies (RICE and BEVILACQUA 1989; LAURI et al. 1991b; DEJANA et al. 1992; ZAIFERT and COHEN 1993; TAKADA et al. 1993; CHIRIVI et al. 1994). E-selectin binding specificity can be demonstrated by showing colon carcinoma cells binding to HUVEC or COS cells transiently transfected with E-selectin or to immobilized recombinant soluble E-selectin (MERWIN et al. 1992; GIAVAZZI et al. 1993). In one study, the efficiency of E-selectin-mediated binding of colon carcinoma to mouse and human endothelial cells correlated with the metastatic potential of the tumor cells (SAWADA et al. 1994a). The ligands recognised by selectins have been identified as fucosylated and sialyl carbohydrate derivatives, including sialyl Lewis^x (sLe^x) and the isomer sialyl Lewis^a (sLe^a) antigens (BERG et al. 1991; PHILLIPS et al. 1990). The expression of the sLe^x and sLe^a in carcinoma cells and their functional role in metastasis formation have been shown (INUFUSA et al. 1991; HOFF et al. 1989). Results from clinical specimens have shown that sLe^x expression in the colon carcinoma may influence the prognosis of the patient (NAKAMORI et al. 1993). Moreover, high levels of the sLe^a and sLe^x ligands have been found in the blood of cancer patients (SINGHAL et al. 1990), and their inhibitory activity to E-selectin binding has been reported (SAWADA et al. 1994b).

Several reports have indicated the contribution of sLe^x and sLe^a to the adhesion of human tumor cells, mostly, colon carcinoma, to E-selectin expressed by endothelial cells (DEJANA et al. 1992; TAKADA et al. 1991; MAJURI et al. 1992). In an extensive study with cultured epithelial cancer cells expressing sLe^x and/or sLe^a, TAKADA et al. showed that sLe^a contributes mainly to the adhesion of cancer cells of colon and pancreas origin, whereas sLe^x contributes to the adhesion of cancer cells of other origins such as lung and liver (TAKADA et al. 1993). The importance of the sLe^a in the adhesion of pancreatic carcinoma cells to activated endothelium has been confirmed (IWAI et al. 1993). The treatment of gastric adenocarcinoma with dimethyl sulfoxide (DMSO) resulted in an enhanced expression of sLe^x on their cell surface; this was associated with increased adhesion to E-selectin mediated by IL-1 activated endothelial cells (MAEHARA et al. 1995). Melanoma cells rarely bind to E-selectin (GAROFALO et al. 1995), but it has been suggested that the surface expression of carbohydrate ligands other than sLe^x, sdiLe^x, or sLe^a enables melanoma cells to adhere to E-selectin (KUNZENDORF et al. 1994).

Efforts are focused on the characterization of the tumor cell surface carbohydrate ligands for E-selectin and the evaluation of potential inhibitors of metastasis. Different strategies for blocking selectin-dependent adhesion can be used. Monoclonal antibodies directed at sLe^x or sLe^a epitopes (DEJANA et al. 1992; TAKADA et al. 1991,1993; MAJURI et al. 1992), oligosaccharide sequences of these epitopes (BEVILACQUA and NELSON 1993a), as well as a variety of soluble molecules containing sLe^x or related structures (SAWADA et al. 1994a; ZHANG et al. 1994; TAKADA et al. 1991) have been shown to block E-selectin-dependent adhesion. Alternatively, E-selectin mediated adhesion can be reduced by blocking the expression of selectin at the surface of endothelial cells or by blocking the expression of sLe^x or sLe^a by inhibitors of the 0-glycosylation of tumor cells (KOJIMA et al. 1992a). Modulation of binding may ultimately affect metastatic potential.

Colon carcinoma cells which recognized cytokine-inducible E-selectin in vitro also showed augmented tumor colonization to the lung or to the liver of cytokine-treated mice (BANI et al. 1991; BRODT et al. 1994). However, the significance of these selectin mediated interactions in metastasis formation is not well understood. For example, the adhesion of HT-29, a human colon carcinoma metastatic in nude mice, to IL-1-activated endothelial cells or to immobilized E-selectin was at least partially inhibited by one monoclonal antibody that binds Lewis fucosylated type I on colon carcinoma cells; the same antibody in vivo inhibited the augmented retention of colon carcinoma cells in the lung of nude mice treated with IL-1 (DEJANA et al. 1992; CHIRIVI et al. 1994). More recently, it has been indicated that injections of monoclonal antibody to E-selectin blocked the augmentation of cytokine-induced liver colonization by colon carcinoma cells (BRODT et al. 1994).

One current concept in leukocyte extravasation is that selectin–carbohydrate interaction occurs first and causes a weak "rolling" attachment of leukocytes to endothelium. Rolling leukocytes then interact strongly with the endothelium via integrin binding by "sticking" onto the surface and then transmigrating (BUTCHER 1991; Hogg 1992; LAWRENCE and SPRINGER 1991). Cell rolling and adherence to endothelial cells can be reproduced in vitro in a dynamic flow environment, and the role of the adhesion molecules that mediate these interactions can be identified (LAWRENCE et al. 1990). We have investigated the interaction of human tumor cells on cultured endothelial cells under dynamic flow conditions using a parallel plate laminar flow chamber (GIAVAZZI et al. 1993). Tumor cell lines studied under flow conditions adhered to IL-1-activated endothelial cells with two different adhesion patterns, depending on the tumor origin. For example, several types of carcinoma cells, including colon, breast, and ovarian carcinoma, adhered to endothelial cells with a rolling mechanism, whereas melanomas and osteosarcomas adhered firmly to activated endothelial cells without rolling (GIAVAZZI et al. 1993). Similarly, some breast and colon carcinoma cells have been shown to accumulate and roll on TNF-stimulated, but not unstimulated endothelial cell monolayers in the presence of flow (TOZEREN et al. 1995). By incubating cytokine-stimulated endothelial cells with monoclonal antibodies against adhesion proteins known to mediate the transient interaction with endothelial cells, it has been demonstrated that E-selectin partially or completely mediates the dynamic interaction of tumor cells (TOZEREN et al. 1995; GIAVAZZI et al. 1993). These findings suggest that under flow conditions, E-selectin has a selective role in tumor cell rolling/adhesion. The finding that colon cells have the capacity to undergo E-selectin-mediated dynamic interaction is consistent with the expression of the carbohydrate motifs on these cells (TOZEREN et al. 1995; GIAVAZZI et al. 1993). KOJIMA et al. have suggested that E-selectin changes its conformation and accommodates different binding specifities in a dynamic flow environment (KOJIMA et al. 1992b). They have shown that the binding of CHO cells to E-selectin under static conditions or low shear stress is largely based on the expression of the classic sLe^x ligand. This structure became less important under high shear stress dynamic conditions, where the major epitope appears to be Le^x with an adjacent unidentified sialosyl substitution (KOJIMA et al. 1992b). The adhesion of melanoma cells to cytokine-activated endothelial cells under shear stress conditions was partially blocked by anti-VCAM-1 monoclonal antibodies (GIAVAZZI et al. 1993). (As we describe below, melanoma adhesion to activated endothelial cells is mediated mainly by VCAM-1/VLA-4). It is worth noting that it has been reported recently that integrin-$\alpha 4$ supports not only adhesion, but lymphocyte tethering and rolling under physiological flow and in the absence of a selectin contribution as well (BERLIN et al. 1995; ALON et al. 1995). Activation of endothelial cells is largely responsible for tumor cell attachment in the presence of flow. However, the fact that some tumor cells adhere to activated endothelial cells under flow conditions without the transient interaction (GIAVAZZI et al. 1993) suggests that the rolling phenomenon is not a necessary prerequisite for the interaction of tumor cells with endothelial cells. Tumor cells can attach themselves to resting endothelial cells under flow conditions. KOJIMA et al. has shown that carbohydrate–carboydrate (GM3/LacCer) interactions are the main mediators of the initial adhesion of murine melanoma cells to nonactivated endothelial cells under flow conditions (KOJIMA et al. 1992c). It is therefore possible that, at least for melanoma, the adhesion mediated by lectin-like molecules initiates metastatic deposition. This may trigger a series of "cascade" reactions leading to activation of endothelial cells and expression of selectin or immunoglobulin family receptors, thus promoting adhesion and migration. Understanding of the dynamic interactions between tumor cells and endothelial cells may have important implications for the specific pattern of tumor metastasis. Intravital videomicroscopy studies should help clarify the interactions of cancer cells with vessel walls during metastasis (MORRIS et al. 1993).

3.2 Integrin/Immunoglobulin Superfamily-Mediated Adhesion and Metastasis

VCAM-1, ICAM-1, and PECAM-1 are immunoglobulin superfamily (Ig-SF) members involved in leukocyte traffic expressed on both leukocyte and endothelial cells (BEVILACQUA 1993; CARLOS and HARLAN 1994). They are also expressed on some types of tumor cells (JOHNSON 1991). RICE and BEVILACQUA originally described an inducible 110-kD[a] endothelial cell surface protein, designated INCAM-110, that is responsible for the adhesion of melanoma cells to activated endothelial cells (RICE and BEVILACQUA 1989). Subsequently, this has been found to be identical to VCAM-1 expressed on activated endothelial cells (CYBULSKY et al. 1991). VCAM-1 has been shown to mediate the adhesion of several melanoma (MARTIN-PADURA et al. 1991; GIAVAZZI et al. 1993; GAROFALO et al. 1995; TAICHMAN et al. 1991) and sarcoma (osteosarcoma and rabdosarcoma, MATTILA et al. 1992; LAURI et al. 1991a) cell lines to cytokine-activated endothelial cells. Other tumor types have been occasionally reported (TAICHMAN et al. 1991; TOMITA et al. 1995). VCAM-1 is recognized by members of the integrin family such as $\alpha_4\beta_1$-integrin (VLA-4; CD49d/CD29; TAICHMAN et al. 1991) and the recently characterized $\alpha_4\beta_7$-integrin (RUEGG et al. 1992). Immunohistochemical analysis has revealed VLA-4 expressed on specimens of human melanoma (ALBELDA et al. 1990), renal (TOMITA et al. 1995), sarcoma (PAAVONEN et al. 1994), and related cell lines. In contrast, $\alpha_4\beta_7$-integrin appeared not to be expressed on nonlymphoid solid tumor (PAAVONEN et al. 1994). The role of $\alpha_4\beta_1$ and $\alpha_4\beta_7$ in leukocyte pathophysiology has been recently reviewed (LOBB and Hemcer 1994).

For those tumor cells that expressed β_1-integrins, antibodies to $\alpha_4\beta_1$-integrin subunits inhibited their adhesion to endothelium substrates (TAICHMAN et al. 1991;

Kawaguchi et al. 1992; Mattila et al. 1992; Martin-Padura et al. 1991). However, other β_1-integrins, such as $\alpha_5\beta_1$-integrin, have been shown to be involved in the adhesion of different tumor types, including melanoma and osteosarcoma, to cytokine-treated endothelial cells (Lauri et al. 1991a). Studies on a panel of melanoma clones have shown differential expression of VLA-4 integrin associated with the ability of the clones to adhere to activated endothelial cells (Martin-Padura et al. 1991). The adhesion of VLA-4-positive cells was abolished by treating tumor cells with monoclonal antibodies to VLA-4-or by treating endothelial cells with anti-VCAM-1 monoclonal antibodies (Martin-Padura et al. 1991). The affinity of VLA-4-expressing tumor cells for VCAM-1 has been confirmed by their preferential adherence to recombinant soluble VCAM-1 protein (Mattila et al. 1992; Garofalo et al. 1995; Mould et al. 1994). These observations suggest that VLA-4/VCAM-1 is the major adhesion pathway involved in melanoma/sarcoma adhesion to activated endothelial cells characterized, so far and is likely responsible for the augmentation of metastases induced by cytokines. In rodents as well as in human tissue, VCAM-1 is constitutively expressed on microvascular endothelial cells, including the lung, and its expression can be increased by inflammatory stimuli (Rice et al. 1991; Fries et al. 1993). Recent observations by Piali et al. are of interest; they showed that VCAM-1 is repressed on tumor-infiltrating vascular endothelial cells in the lung in mice and humans, but not on lung blood vessels distant from the tumor (Piali et al. 1995). They suggest that the downregulation of VCAM-1 is a mechanism by which vascularized tumors avoid invasion by cytotoxic cells of the immune system.

More intriguing is the significance of VLA-4 expression in malignant melanoma in vivo. Expression of VLA-4 on primary cutaneous melanomas has been negatively associated with the disease-free interval and overall survival time (Schadendorf et al. 1995). This is in contrast with the report by Quian et al. showing that the expression of VLA-4 was inversely correlated with the invasive potential of B16 murine melanoma and spontaneous pulmonary metastasis formation; however, the expression of VLA-4 did not affect lung colonization (Qian et al. 1994). Kawaquchi et al. showed that a higher number of fibrosarcoma cells overexpressing VLA-4 were detected in the lung of mice treated with TNF, but this did not correspond to an augmentation of metastases (Kawaguchi et al. 1992). In our laboratory, we have recently investigated the involvement of VLA-4 expressed by human melanoma cells in the augmentation of experimental metastases induced by IL-1. We have shown that only melanoma cells that express VLA-4 produced more lung colonies in nude mice after the administration of IL-1 (Garofalo et al. 1995). Treatment of tumor cells with monoclonal antibody to VLA-4, but not to other VLA integrins, inhibited this augmentation. This treatment did not interfere with the baseline metastatic proprieties of melanoma cells injected in non-cytokine-treated control mice. In a similar study, but with the murine B16 melanoma system in syngeneic mice, TNF-induced augmentation of lung colonies was abolished by treatment of tumor cells with monoclonal antibody to VLA-4 or by the administration of monoclonal antibodies against murine VCAM-1 (Okahara et al. 1994). Taken together, these data on the in vivo efficacy of α_4- and

VCAM-1-directed antibodies are consistent with α_4/VCAM-1-dependent adhesive function causing augmentation of lung colonies induced by cytokines, at least in melanoma.

3.3 Other Adhesion Molecules

Several observations indicate that, in general, solid tumors of nonlymphoid origin lack the other well-characterized integrin/immunoglobulin superfamily interaction (LFA-1/ICAM-1) with endothelial cells, likely because they do not express the β_2-integrins (CD11/CD18) which are receptors for ICAM-1 (DIAMOND et al. 1991). The adhesion of melanoma and sarcoma cells to resting or activated endothelial cells is infact not blocked by ICAM-1 monoclonal antibodies (LAURI et al. 1991a). However, ICAM-1 is expressed on different tumor types. The expression of ICAM-1 in melanoma has been associated with tumor progression and metastasis (JOHNSON et al. 1989; NATALI et al. 1990; KAGHESHITA et al. 1993), and a soluble form of ICAM-1 has been found to be elevated in the serum of patients with malignant disease (GIAVAZZI et al. 1994; HARNING et al. 1991; TSUJISAKI et al. 1991; BANKS et al. 1993). The expression and shedding of ICAM-1 can be induced by inflammatory cytokines and is therefore of critical importance for ICAM-1-mediated immunosurveillance (WEBB et al. 1991; NAGANUMA et al. 1991; ANICHINI et al. 1990). By analogy to these findings, the immunoglobulin family adhesion molecule PECAM-1, which is normally expressed on endothelial cells, has been found to be constitutively expressed on solid tumors and involved in nonstimulated tumor–endothelial cell interaction (TANG et al. 1993).

There are additional adhesion mechanisms that might be upregulated by inflammatory cytokines, some of them not yet characterized.

BUCHANAN and coworkers have shown that the augmented adhesion of human lung carcinoma cells to IL-1-activated endothelial cells is mediated by the expression of endothelial cell vitronectin receptor (LAFRENIE et al. 1992). Furthermore, the enhanced adhesion of tumor cells to endothelium was blocked in presence of the peptide Arg-Gly-Asp (GRGDS; LAURI et al. 1990; LAFRENIE et al. 1992), suggesting that the adhesion of cancer cells to activated endothelial cells may also be related to Arg-Gly-Asp (RGD)-dependent endothelial surface adhesion molecule. In these studies a relationship between vessel wall lypoxygenase metabolites, adhesion molecule expression, and adhesion of tumor cells to endothelium is also suggested (BERTOMEU et al. 1993; for a review, see HONN and TANG 1992).

4 Conclusions

Tumor–endothelial cells and tumor–subendothelial matrix interactions play a role in metastasis formation to secondary sites. Adhesion of tumor cells to endothelium is mediated in part by cytokine-inducible adhesion molecules. Other adhesion molecules may mediate tumor cell adhesion to unstimulated endothelial cells when activation-dependent adhesion molecules are not available.

There appear to be sequential steps in which a particular adhesion mechanism plays a role in tumor cell adhesion to vascular endothelium. Initial tumor cell adhesion to endothelial cells is probably mediated by carbohydrate interactions, which might also be responsible for organ specificity, in metastasis. Activation of endothelial cells leads to the increased expression of adhesion molecules on endothelial cells, which mediates tumor–endothelial cell adhesion. However, specific adhesion pathways seem to be preferentially used by different tumor types. The availability of blocking monoclonal antibodies against these adhesion molecules has greatly facilitated their definition.

Understanding of the involvement and function of adhesion molecules in tumor progression and metastasis generates expectations for novel strategies designed to inhibit tumor cell adhesion.

Acknowledgments. I would like to thank Simona Alborghetti and Laura Arioli for their secretarial assistance. Part of the work reviewed here was supported by a grant from the Italian Association for Cancer Research.

References

Albelda SM (1993) Biology of disease. Role of integrins and other cell adhesion molecules in tumor progression and metastasis. Lab Invest 68: 4–17

Albelda SM, Mette SA, Elder DE, Stewart R, Damjanovich L, Herlyn M, Buck CA (1990) Integrin distribution in malignant melanoma: association of the b_3, subunit with tumor progression. Cancer Res 50: 6757–6764

Albelda SM, Smith CW, Ward PA (1994) Adhesion molecules and inflammatory injury. FASEB J 8: 504–512

Alon R, Kassner PD, Carr MW, Finger EB, Hemler ME, Springer TA (1995) The integrin VLA-4 supports tethering and rolling in flow on VCAM-1. J Cell Biol 128: 1243–1253

Anichini A, Mortarini R, Supino R, Parmiani G (1990) Human melanoma cells with high susceptibility to cell-mediated lysis can be identified on the basis of ICAM-1 phenotype, VLA profile and invasive ability. Int J Cancer 46: 508–515

Arguello F, Baggs RB, Graves BT, Harwell SE, Cohen HJ, Frantz CN (1992) Effect of IL-1 on experimental bone/bone-marrow metastases. Int J Cancer 52: 802–807

Auerbach R, Lu WC, Pardon E, Gumkowski F, Kaminska G, Kaminski M (1987) Specificity of adhesion between murine tumor cells and capillary endothelium: an in vitro correlate of preferential metastasis in vivo. Cancer Res 47: 1492–1496

Bani MR, Garofalo A, Scanziani E, Giavazzi R (1991) Effect of interleukin-1-beta on metastasis formation in different tumor systems. J Natl Cancer Inst 83: 119–123

Banks RE, Gearing AJH, Hemingway IK, Norfolk DR, Perren TJ, Selby PJ (1993) Circulating intercellular adhesion molecule-1 (ICAM-1) E-selectin and vascular cell adhesion molecule-1 (VCAM-1) in human malignancies. Br J Cancer 68: 122–124

Belardelli F, Ciolli V, Testa U, Montesoro E, Bulgarini D, Proietti E, Borghi P, Sestili P, Locardi C, Peschle C, Gresser I (1989) Anti-tumor effects of interleukin-2 and interleukin-1 in mice transplanted with different syngeneic tumors. Int J Cancer 44: 1108–1116

Belloni PN, Nicolson GL (1988) Differential expression of cell surface glycoproteins on various organ-derived microvascular endothelia and endothelial cell cultures. J Cell Physiol 136: 398–410

Belloni PN, Tressler RJ (1990) Microvascular endothelial cell heterogeneity: interactions with leukocytes and tumor cells. Cancer Metastasis Rev 8: 353–389
Bereta M, Bereta J, Cohen S, Zaifert K, Cohen MC (1991) Effect of inflammatory cytokines on the adherence of tumor cells to endothelium in a murine model. Cell Immunol 136: 263–277
Berg EL, Robinson MK, Mansson O, Butcher EC, Magnani JL (1991) A carbohydrate domain common to both sialyl Le^a and sialyl Le^x is recognized by the endothelial cell leukocyte adhesion molecule ELAM-1. J Biol Chem 266: 14869–14872
Berlin C, Bargatze RF, Campbell JJ, von Andrian UH, Szabo MC, Hasslen SR, Nelson RD, Berg EL, Erlandsen SL, Butcher EC (1995) α_4 integrins mediate lymphocyte attachment and rolling under physiologic flow. Cell 80: 413–422
Bertomeu MC, Gallo S, Lauri D, Haas TA, Orr FW, Bastida E, Buchanan MR (1993) Interleukin 1-induced cancer cell/endothelial cell adhesion in vitro and its relationship to metastasis in vivo: role of vessel wall 13-HODE synthesis and integrin expression. Clin Expl Metastasis 11: 243–250
Bevilacqua MP (1993) Endothelial-leukocyte adhesion molecules. Ann Rev Immunol 11: 767–804
Bevilacqua MP, Nelson RM (1993a) Endothelial-leukocyte adhesion molecules in inflammation and metastasis. Thromb Haemost 70: 152–154
Bevilacqua MP, Nelson RM (1993b) Selectins. J Clin Invest 91: 379–387
Bevilacqua MP, Pober JS, Mendrick DL, Cotran RS, Gimbrone MA (1987) Identification of an inducible endothelial leukocyte adhesion molecule. Proc Natl Acad Sci USA 84: 9238–9242
Brodt P, Fallavolitta L, Norton CR, Wolitzky BA (1994) E selectin mediates adhesion of carcinoma cells to the liver endothelium and liver metastasis. Proc Assoc Cancer Res 35: 52 (abstr)
Burrows FJ, Haskard DO, Hart IR, Marshall JF, Selkirk S, Poole S, Thorpe PE (1991) Influence of tumor-derived interleukin 1 on melanoma-endothelial cell interaction in vitro. Cancer Res 51: 4768–4775
Butcher EC (1991) Leukocyte-endothelial cell recognition: three (or more) steps to specificity and diversity. Cell 67: 1033–1036
Carlos TM, Harlan JM (1994) Leukocyte-endothelial adhesion molecules. Blood 84: 2068–2101
Chirivi RGS, Garofalo A, Martin-Padura I, Mantovani A, Giavazzi R (1993) Interleukin 1 receptor antagonist inhibits the augmentation of metastasis induced by interleukin 1 or lipopolysaccharide in a human melanoma/nude mouse system. Cancer Res 53: 5051–5054
Chirivi RGS, Nicoletti MI, Remuzzi A, Giavazzi R (1994) Cytokines and cell adhesion molecules in tumor-endothelial cell interaction and metastasis. Cell Adhes Commun 2: 219–224
Colombo MP, Forni G (1994) Cytokine gene transfer in tumor inhibition and tumor therapy: where are we now? Immunology 15: 48–51
Cotran RS (1987) New roles for the endothelium in inflammation and immunity. Am J Pathol 129: 407–413
Crissman JD, Hatfield JS, Menter DG, Sloane B, Honn KV (1988) Morphological study of the interaction of intravascular tumor cells with endothelial cells and subendothelial matrix. Cancer Res 48: 4065–4072
Cybulsky MI, Fries JWU, Williams AJ, Sultan P, Davis VM, Gimbrone MA Jr, Collins T (1991) Alternative splicing of human VCAM–1 in activated vascular endothelium. Am J Pathol 138: 815–820
Dao TL, Yogo H (1967) Enhancement of pulmonary metastases by X-irradiation in rats bearing mammary cancer. Cancer 20: 2020–2025
Dejana E, Bertocchi F, Bortolami MC, Regonesi A, Tonta A, Breviario F, Giavazzi R (1988) Interleukin 1 promotes tumor cell adhesion to cultured human endothelial cells. J Clin Invest 82: 1466–1470
Dejana E, Martin-Padura I, Lauri D, Bernasconi S, Bani MR, Garofalo A, Giavazzi R, Magnani J, Mantovani A, Menard S (1992) Endothelial leukocyte adhesion molecule-1 dependent adhesion of colon carcinoma cells to vascular endothelium is inhibited by an antibody to Lewis fucosylated type I carbohydrate chain. Lab Invest 66: 324–330
Dennis JW, Laferte S (1987) Tumor cell surface carbohydrate and metastatic phenotype. Cancer Metastasis Rev 5: 185–204
Diamond MS, Springer TA (1994) The dynamic regulation of integrin adhesiveness. Curr Biol 4: 506–971
Diamond MS, Staunton DE, Marlin SD, Springer TA (1991) Binding of the integrin Mac-1 (CD11b-CD18) to the third immunoglobulin-like domain of ICAM-1 (CD54) and its regulation by glycosylation. Cell 65: 961–971
Dingemans KP, Roos E, Van Den Bergh Weerman MA, Van de Pavert IV (1978) Invasion of liver tissue by tumor cells and leukocytes: comparative ultrastructure. J Natl Cancer Inst 60: 583–598
Dustin ML, Rothlein R, Bhan AK, Dinarello CA, Springer TA (1986) Induction by IL 1 and interferon-γ: tissue distribution, biochemistry, and function of a natural adherence molecule (ICAM-1). J Immunol 137: 245–254

Fries JWU, Williams AJ, Atkins RC, Newman W, Lipscomb MF, Collins T (1993) Expression of VCAM-1 and E-selectin in an in vivo model of endothelial activation. Am J Pathol 143: 725–737

Garofalo A, Chirivi RGS, Foglieni C, Pigott R, Mortarini R, Martin-Padura I, Anichini A, Gearing AJ, Sanchez-Madrid F, Dejana E, Giavazzi R (1995) Involvement of the very late antigen 4 integrin on melanoma in interleukin-1-augmented experimental metastases. Cancer Res 55: 414–419

Gasic GJ (1984) Role of plasma, platelets and endothelial cells in tumor metastasis. Cancer Metastasis Rev 3: 99–114

Giavazzi R, Garofalo A, Bani MR, Abbate M, Ghezzi P, Boraschi D, Mantovani A, Dejana E (1990) Interleukin 1-induced augmentation of experimental metastases from a human melanoma in nude mice. Cancer Res 50: 4771–4775

Giavazzi R, Foppolo M, Dossi R, Remuzzi A (1993) Rolling and adhesion of human tumor cells on vascular endothelium under physiological flow conditions. J Clin Invest 92: 3038–3044

Giavazzi R, Nicoletti MI, Chirivi RGS, Hemingway I, Bernasconi S, Allavena P, Gearing AJH (1994) Soluble intercellular adhesion molecule-1 (ICAM-1) is released into the serum and ascites of human ovarian carcinoma patients and in nude mice bearing tumor xenografts. Eur J Cancer 30A: 1865–1870

Gunthert U, Hofmann M, Rudy W, Reber S, Zoller M, Haubmann I, Matzku S, Wenzel A, Ponta H, Herrlich P (1991) A new variant of glycoprotein CD44 confers metastatic potential to rat carcinoma cells. Cell 65: 13–24

Hakomori S-I (1994) Novel endothelial cell activation factor (s) released from activated platelets which induce E-selectin expression and tumor cell adhesion to endothelial cells: a preliminary note. Biochem Biophys Res Commun 203: 1605–1613

Harning R, Mainolfi E, Bystryn JC, Henn M, Merluzzi VJ, Rothlein R (1991) Serum levels of circulating intercellular adhesion molecule 1 in human malignant melanoma. Cancer Res 51: 5003–5005

Herbert J-M (1993) Protein kinase C: a key factor in the regulation of tumor cell adhesion to the endothelium. Biochem Pharmacol 45: 527–537

Herrlich P, Zoller M, Pals ST, Ponta H (1993) CD44 splice variants: metastases meet lymphocytes. Immunol Today 14: 395–399

Hoff SD, Matsushita Y, Ota DM, Clearly KR, Yamori T, Hakomori S-I, Irimura T (1989) Increased expression of sialyl-dimeric Le^x antigen in liver metastases of human colorectal carcinoma. Cancer Res 49: 6883–688

Hogg N (1992) Roll, roll, roll your leucocyte gently down the vein. Immunol Today 13: 113–115

Honn KV, Tang DG (1992) Adhesion molecules and tumor cell interaction with endothelium and subendothelial matrix. Cancer Metastasis Rev 11: 353–375

Inufusa H, Kojima N, Yasutomi M, Hakomori S (1991) Human lung adenocarcinoma cell lines with different lung colonization potential (LCP), and a correlation between expression of sialosyl dimeric Le^x (defined by MAb FH6) and LCP. Clin Exp Metastasis 9: 245–257

Iwai K, Ishikura H, Kaji M, Sugiura H, Ishizu A, Takahashi C, Kato H, Tanabe T, Yoshiki T (1993) Importance of E-selectin (ELAM-1) and sialyl lewis a in the adhesion of pancreatic carcinoma cells to activated endothelium. Int J Cancer 54: 972–977

Johnson JP (1991) Cell adhesion molecules of the immunoglobulin supergene family and their role in malignant transformation and progression to metastatic disease. Cancer Metastasis Rev 10: 11–22

Johnson JP, Stade BG, Holzmann B, Schwable W, Riethmuller G (1989) De novo expression of intercellular-adhesion molecule 1 in melanoma correlates with increased risk of metastasis. Proc Natl Acad Sci USA 86: 641–644

Kagheshita T, Yoshii A, Kimura T, Kuriya N, Ono T, Tsujisaki M, Imai K, Ferrone S (1993) Clinical relevance of ICAM-1 expression in primary lesions and serum of patients with malignant melanoma. Cancer Res 53: 4927–4932

Kaji M, Ishikura H, Kishimoto T, Omi M, Ishizu A, Kimura C, Takahashi T, Kato H, Yoshiki T (1995) E-selectin expression induced by pancreas-carcinoma-derived interleukin-lα results in enhanced adhesion of pancreas-carcinoma cells to endothelial cells. Int J Cancer 60: 712–717

Kawaguchi S, Kikuchi K, Ishii S, Takada Y, Kobayashi S, Uede T (1992) VLA-4 molecules on tumor cells initiate an adhesive interaction with VCAM-1 molecules on endothelial cell surface. Jpn J Cancer Res 83: 1304–1316

Kinjo M (1978) Lodgement and extravasation of tumor cells in blood-borne metastasis: an electron microscope study. Br J Cancer 38: 293–301

Kojima N, Handa K, Newman W, Hakomori S-I (1992a) Inhibition of selectin-dependent tumor cell adhesion to endothelial cells and platelets by blocking O-glycosylation of these cells. Biochem Biophys Res Commun 182: 1288–1295

Kojima N, Handa K, Newman W, Hakomori S-I (1992b) Multi-recognition capability of E-selectin in a dynamic flow system, as evidenced by differential effects of sialidases and anti-carbohydrate antibodies on selectin-mediated cell adhesion at low vs. high wall shear stress: a preliminary note. Biochem Biophys Res Commun 189: 1686–1694

Kojima N, Shiota M, Sadahira Y, Handa K, Hakomori S-I (1992c) Cell adhesion in a dynamic flow system as compared to static system. Glycosphingolipid-glycosphingolipid interaction in the dynamic system predominates over lectin- or integrin-based mechanisms in adhesion of B16 melanoma cells to non-activated endothelial cells. J Biol Chem 267: 17264–17270

Kramer RH, Nicolson GL (1979) Interactions of tumor cells with vascular endothelial cell monolayers: a model for metastatic invasion. Proc Natl Acad Sci USA 76: 5704–5708

Kunzendorf U, Kruger-Krasagakes S, Notter M, Hock H, Walz G, Diamantstein T (1994) A sialyl-le^x-negative melanoma cell line binds to E-selectin but not to P-selectin. Cancer Res 54: 1109–1112

Lafrenie RM, Podor TJ, Buchanan MR, Orr FW (1992) Up-regulated biosynthesis and expression of endothelial cell vitronectin receptor enhances cancer cell adhesion. Cancer Res 52: 2202–2208

Lapis K, Paku S, Liotta LA (1988) Endothelialization of embolized tumor cells during metastasis formation. Clin Exp Metastasis 6: 73–89

Lauri D, Bertomeu MC, Orr FW, Bastida E, Sauder D, Buchanan MR (1990) Interleukin-1 increases tumor cell adhesion to endothelial cells through an RGD dependent mechanism: in vitro and in vivo studies. Clin Exp Metastasis 8: 27–32

Lauri D, Martin-Padura I, Biondelli T, Rossi G, Bernasconi S, Giavazzi R, Passerini F, Van Hinsbergh V, Dejana E (1991a) Role of β_1 integrins in tumor cell adhesion to cultured human endothelial cells. Lab Invest 65: 525–531

Lauri D, Needham L, Martin-Padura I, Dejana E (1991b) Tumor cell adhesion to endothelial cells: ELAM-1 as an inducible adhesive receptor specific for colon carcinoma cells. J Natl Cancer Inst 83: 1321–1324

Lawrence MB, Springer TA (1991) Leukocytes roll on a selectin at physiologic flow rates: distinction from and prerequisite for adhesion through integrins. Cell 65: 859–873

Lawrence MB, Smith CW, Eskin SG, McIntire LV (1990) Effect on venous shear stress on CD18-mediated neutrophil adhesion to cultured endothelium. Blood 75: 227–237

Levine SO, Saltzman A (1990) Lymphatic metastases from the peritoneal cavity are increased in the postinflammatory state. Invasion Metastasis 10: 281–288

Liotta LA (1986) Tumor invasion and metastases-role of the extracellular matrix: Rhoads Memorial Award lecture. Cancer Res 46: 1–7

Lobb RR, Hemler ME (1994) The pathophysiologic role of α_4 integrins in vivo. J Clin Invest 94: 1722–1728

Lollini PL, De Giovanni C, Nicoletti G, Bontadini A, Tazzari PL, Landuzzi L, Scotlandi K, Nanni P (1990) Enhancement of experimental metastatic ability by tumor necrosis factor-alpha alone or in combination with interferon-gamma. Clin Exp Metastasis 8: 215–224

Maehara M, Yagita M, Isobe Y, Hoshino T, Nakagawara G (1993) Dimethyl sulfoxide (DMSO) increases expression of sialyl lewis x antigen and enhances adhesion of human gastric carcinoma (NUGC4) cells to activated endothelial cells. Int J Cancer 54: 296–301

Majuri M-L, Mattila P, Renkonen R (1992) Recombinant E-selectin-protein mediates tumor cell adhesion via sialyl-Lea and sialyl-Lex. Biochem Biophys Res Commun 182: 1376–1382

Malik ST, Griffin DB, Fiers W, Balkwill FR (1989) Paradoxical effects of tumour necrosis factor in experimental ovarian cancer. Int J Cancer 44: 918–925

Malik ST, Naylor MS, East N, Oliff A, Balkwill FR (1990) Cells secreting tumour necrosis factor show enhanced metastasis in nude mice. Eur J Cancer 26: 1031–1034

Malik STA, East N, Boraschi D, Balkwill FR (1992) Effects of intraperitoneal recombinant interleukin-1β in intraperitoneal human ovarian cancer xenograft models: comparison with the effects of tumour necrosis factor. Br J Cancer 65: 661–666

Mantovani A, Dejana E (1989) Cytokines as communication signals between leukocytes and endothelial cells. Immunol Today 10: 370–375

Martin-Padura I, Mortarini R, Lauri D, Bernasconi S, Sanchez-Madrid F, Parmiani G, Mantovani A, Anichini A, Dejana E (1991) Heterogeneity in human melanoma cell adhesion to cytokine activated endothelial cells correlates with VLA-4 expression. Cancer Res 51: 2239–2241

Martin-Padura I, Metzelaar MJ, Menard S, Mantovani A, Dejana E (1993) Enhanced adhesion of colon carcinoma cells to thrombin stimulated endothelial cells. Endothelium 1: 41–45

Mattila P, Majuri ML, Renkonen R (1992) VLA-4 integrin on sarcoma cell lines recognizes endothelial VCAM-1. Differential regulation of the VLA-4 avidity on various sarcoma cell lines. Int J Cancer 52: 918–923

McCormick BA, Zetter BR (1992) Adhesive interactions in angiogenesis and metastasis. Pharmacol Ther 53: 239–260

McEver RP (1991) Selectins: novel receptors that mediate leukocyte adhesion during inflammation. Thromb Haemost 65: 223–228

Merwin JR, Madri JA, Lynch M (1992) Cancer cell binding to E-selectin transfected human endothelia. Biochem Biophys Res Commun 189: 315–323

Miyata K, Kato M, Shikama H, Nishimura K, Sakae N, Kawagoe K, Nishikawa T, Kuroda K, Yamaguchi K, Aoyama Y, Mitsuishi Y, Yamada N (1992) A YIGSR-containing novel mutein without the detrimental effect of human TNF-α of enhancing experimental pulmonary metastasis. Clin Exp Metastasis 10: 267–272

Morris VL, MacDonald IC, Koop S, Schmidt EE, Chambers AF, Groom AC (1993) Early interactions of cancer cells with the microvasculature in mouse liver and muscle during hematogenous metastasis: videomicroscopic analysis. Clin Exp Metastasis 11: 377–390

Mould AP, Askari JA, Craig SE, Garratt AN, Clements J, Humphries MJ (1994) Integrin $\alpha_4\beta_1$-mediated melanoma cell adhesion and migration on vascular cell adhesion molecule-1 (VCAM) and the alternatively spliced IIICS region fibronectin. J Biol Chem 269: 27224–27230

Murphy P, Alexander P, Senior PV, Fleming J, Kirkham N, Taylor I (1988) Mechanisms of organ selective tumour growth by bloodborne cancer cells. Br J Cancer 57: 19–31

Naganuma H, Kiessling R, Patarroyo M, Hansson M, Handgretinger R, Gronberg A (1991) Increased susceptibility of IFN-γ-treated neuroblastoma cells to lysis by lymphokine-activated killer cells: participation of ICAM-1 induction on target cells. Int J Cancer 47: 527–532

Nakamori S, Kameyama M, Imaoka S, Furukawa H, Ishikawa O, Sasaki Y, Kabuto T, Iwanaga T, Matsushita Y, Irimura T (1993) Increased expression of sialyl lewis^x antigen correlates with poor survival in patients with colorectal carcinoma: clinicopathological and immunohistochemical study. Cancer Res 53: 3632–3637

Nakamura S, Nakata K, Kashimoto S, Yoshida H, Yamada M (1986) Antitumor effect of recombinant human interleukin 1 alpha against murine syngeneic tumors. Jpn J Cancer Res 77: 767–773

Natali P, Nicotra MR, Cavaliere R, Bigotti A, Romano G, Temponi M, Ferrone S (1990) Differential expression of intercellular adhesion molecule 1 in primary and metastatic melanoma lesions. Cancer Res 50: 1271–1278

Nicolson GL (1988) Organ specifity of tumor metastasis: role of preferential adhesion, invasion and growth of malignant cells at specific secondary sites. Cancer Metastasis Rev 7: 143–188

Nicolson GL, Custead SE (1985) Effects of chemotherapeutic drugs on platelet and metastatic tumor cell-endothelial cell interactions as a model for assessing vascular endothelial integrity. Cancer Res 45: 331–336

Nishiyama Y, Fuchimoto S, Orita K (1989) Preventive and antiproliferative effects of tumor necrosis factor against experimental hepatic metastases of mouse colon-26 tumor. J Cancer Res 80: 366–372

Okahara H, Yagita H, Miyake K, Okumura K (1994) Involvement of very late activation antigen 4 (VLA-4) and vascular cell adhesion molecule 1 (VCAM-1) in tumor necrosis factor α enhancement of experimental metastasis. Cancer Res 54: 3233–3236

Orosz P, Echtenacher B, Falk W, Ruschoff J, Weber D, Mannel DN (1993) Enhancement of experimental metastasis by tumor necrosis factor. J Exp Med 177: 1391–1398

Orosz P, Kruger A, Hubbe M, Ruschoff J, Von Hoegen P, Mannel DN (1995) Promotion of experimental liver metastasis by tumor necrosis factor. Int J Cancer 60: 867–871

Orr FW, Adamson IYR, Young L (1986) Promotion of pulmonary metastasis in mice by bleomycin-induced endothelial injury. Cancer Res 46: 891–897

Orr FW, Buchanan MR, Tron VA, Guy D, Lauri D, Sauder DN (1988) Chemotactic activity of endothelial cell-derived interleukin 1 for human tumor cells. Cancer Res 48: 6758–6763

Osborn L (1990) Leukocyte adhesion to endothelium in inflammation. Cell 62: 3–6

Osborn L, Hession C, Tizard R, Vassallo C, Luhowskyj S, Chi-Rosso G, Lobb R (1989) Direct expression cloning of vascular cell adhesion molecule 1, a cytokine-induced endothelial protein that binds to lymphocytes. Cell 59: 1203–1211

Paavonen T, Tiisala S, Majuri M-L, Bohling T, Renkonen R (1994) In vivo evidence of the role of $\alpha_4\beta_1$-VCAM-1 interaction in sarcoma, but not in carcinoma extravasation. Int J Cancer 58: 298–302

Pauli BU, Lee CL (1988) Organ preference of metastasis. The role of organ-specifically modulated endothelial cells. Lab Invest 58: 379–387

Pauli BU, Augustin-Voss HG, El-Sabban ME, Johnson RC, Hammer DA (1990) Organ-preference of metastasis: the role of endothelial cell adhesion molecules. Cancer Metastasis Rev 9: 175–189

Phillips ML, Nudelman E, Gaeta FCA, Perez M, Singhal AK, Hakomori S-I, Paulson JC (1990) ELAM-1 mediates cell adhesion by recognition of a carbohydrate ligand, sialyl-Le^x. Science 250: 1130–1132

Piali L, Fichtel A, Terpe H-J, Imhof BA, Gisler RH (1995) Endothelial vascular cell adhesion molecule 1 expression is suppressed by melanoma and carcinoma. J Exp Med 181: 811–816
Qian F, Vaux DL, Weissman IL (1994) Expression of the integrin $\alpha_4\beta_1$ on melanoma cells can inhibit the invasive stage of metastasis formation. Cell 77: 335–347
Raz A, Lotan R (1987) Endogenous galactoside-binding lectins: a new class of functional tumor cell surface molecules related to metastasis. Cancer Metastasis Rev 6: 433–452
Rice GE, Gimbrone MA Jr, Bevilacqua MP (1988) Tumor cell-endothelial interactions. Increased adhesion of human melanoma cells to activated vascular endothelium. Am J Pathol 133: 204–210
Rice GE, Bevilacqua MP (1989) An inducible endothelial cell surface glycoprotein mediates melanoma adhesion. Science 246: 1303–1306
Rice GE, Munro JM, Corless C, Bevilacqua MP (1991) Vascular and nonvascular expression of INCAM-110. A target for mononuclear leukocyte adhesion in normal and inflamed human tissues. Am J Pathol 138: 385–393
Roos E, Tulp A, Middelkoop Op, Van de Pavert IV (1984) Interactions between lymphoid tumor cells and isolated liver endothelial cells. J Natl Cancer Inst 72: 1173–1180
Ruegg C, Postigo AA, Sikorski EE, Butcher EC, Pytela R, Erle DJ (1992) Role of integrin $\alpha_4\beta_7/\alpha_4\beta p$ in lymphocyte adherence to fibronectin and VCAM-1 and in homotypic cell clustering. J Cell Biol 117: 179–189
Saiki I, Maeda H, Murata J, Yamamoto N, Kiso M, Hasegawa A, Azuma I (1989) Antimetastatic effect of endogenous tumor necrosis factor induced by the treatment of recombinant interferon gamma followed by an analogue (GLA-60) to synthetic lipid A subunit. Cancer Immunol Immunother 30: 151–157
Sawada R, Tsuboi S, Fukuda M (1994a) Differential E-selectin-dependent adhesion efficiency in sublines of a human colon cancer exhibiting distinct metastatic potentials. J Biol Chem 269: 1425–1431
Sawada T, Ho JJL, Chung Y-S, Sowa M, Kim YS (1994b) E-selectin binding by pancreatic tumor cells is inhibited by cancer sera. Int J Cancer 57: 901–907
Schadendorf D, Heidel J, Gawlik C, Suter L, Czarnetzki BM (1995) Association with clinical outcome of expression of VLA-4 in primary cutaneous malignant melanoma as well as P–selectin and E-selectin on intratumoral vessels. J Natl Cancer Inst 87: 366–371
Shimizu Y, Newman W, Tanaka Y, Shaw S (1992) Lymphocyte interactions with endothelial cells. Immunol Today 13: 106–112
Singhal AK, Orntoft TF, Nudelman E, Nance S, Schibig L, Stroud MR, Clausen H, Hakomori S (1990) Profiles of Lewis-containing glycoproteins and glycolipids in sera of patients with adenocarcinoma. Cancer Res 50: 1375–1380
Smith CW, Anderson DC (1991) PMN adhesion and extravasation as a paradigm for tumor cell dissemination. Cancer Metastasis Rev 10: 61–78
Springer TA (1994) Traffic signals for lymphocyte recirculation and leukocyte emigration: the multistep paradigm. Cell 76: 301–314
Sugarbaker EV (1981) Patterns of metastasis in human malignancies. Cancer Biol Rev 2: 235–278
Taichman DD, Cybulsky MI, Diaffar I, Longenecker MR, Teixido J, Rice EG, Aruffo A, Bevilacqua MP (1991) Tumor cell surface $\alpha_4\beta_1$ integrin mediates adhesion to vascular endothelium: demonstration of an interaction with the N-terminal domains of INCAM-110/ VCAM-1. Cell Reg 2: 347–355
Takada A, Ohmori K, Takahashi N, Tsuyuoka K, Yago A, Zenita K, Hasegawa A, Kannagi R (1991) Adhesion of human cancer cells to vascular endothelium mediated by a carbohydrate antigen, sialyl Lewis A. Biochem Biophys Res Commun 179: 713–719
Takada A, Ohmori K, Yoneda T, Tsuyuoka K, Hasegawa A, Makoto K, Kannagi R (1993) Contribution of carbohydrate antigens sialyl Lewis A and sialyl Lewis X to adhesion of human cancer cells to vascular endothelium. Cancer Res 53: 354–361
Takeichi M (1990) Cadherins: a molecular family important in selective cell-cell adhesion. Ann Rev Biochem 59: 237–252
Tang DG, Chen YQ, Newman PJ, Shi L, Gao X, Diglio CA, Honn KV (1993) Identification of PECAM-1 in solid tumor cells and its potential involvement in tumor cell adhesion to endothelium. J Biol Chem 268: 22883–22894
Tomita Y, Saito T, Saito K, Oite T, Shimizu F, Sato S (1995) Possible significance of VLA-4 (α_4 β_1) for hematogenous metastasis of renal-cell cancer. Int J Cancer 60: 753–758
Tozeren A, Kleinman HK, Grant DS, Morales D, Mercurio AM, Byers SW (1995) E-selectin-mediated dynamic interactions of breast- and colon-cancer cells with endothelial-cell monolayers. Int J Cancer 60: 426–431

Tsujisaki M, Imai K, Hirata H, Hanzawa Y, Masuya J, Nakano T, Sugiyama T, Matsui M, Hinoda Y, Yachi A (1991) Detection of circulating intercellular adhesion molecule-1 antigen in malignant diseases. Clin Exp Immunol 85: 3–8

Van Den Brenk HAS, Burch WM, Orton C, Sharpington C (1973) Stimulation of clonogenic growth of tumour cells and metastases in the lungs by local x-radiation. Br J Cancer 27: 291–306

Van Den Brenk HAS, Stone M, Kelly H, Orton C, Sharpington C (1974) Promotion of growth of tumour cells in acutely inflamed tissues. Br J Cancer 30: 246–260

Vanhaesebroeck B, Mareel M, Van Roy F, Grooten J, Fiers W (1991) Expression of the tumor necrosis factor gene in tumor cells correlates with reduced tumorigenicity and reduced invasiveness in vivo. Cancer Res 51: 2229–2238

Vidal-Vanaclocha F, Amezaga C, Asumendi A, Kaplanski G, Dinarello CA (1994) Interleukin-1 receptor blockade reduces the number and size of murine B16 melanoma hepatic metastases. Cancer Res 54: 2667–2672

Wang JM, Taraboletti G, Matsushima K, Van Damme J, Mantovani A (1990) Induction of haptotactic migration of melanoma cells by neutrophil activating protein/interleukin-8. Biochem Biophys Res Commun 169: 165–170

Webb DS, Mostowski HS, Gerrard TL (1991) Cytokine-induced enhancement of ICAM-1 expression results in increased vulnerability of tumour cells to monocyte-mediated lysis. J Immunol 146: 3682–3686

Weiss L, Orr FW, Honn KV (1989) Interactions between cancer cells and the microvasculature: a rate-regulator for metastasis. Clin Exp Metastasis 7: 127–167

Williams AF, Barclay AN (1988) The immunoglobulin super-family domains for cell surface recognition. Ann Rev Immunol 6: 381–405

Withers HR, Milas L (1973) Influence of preirradiation of lung on development of artificial pulmonary metastases of fibrosarcoma in mice. Cancer Res 33: 1931–1936

Zaifert K, Cohen MC (1993) Colo 205 utilizes E-selectin to adhere to human endothelium. Clin Immunol Immunopathol 68: 51–56

Zhang K, Baeckstrom D, Hansson GC (1994) A secreted mucin carrying sialyl-lewis a from colon carcinoma cells binds to E-selectin and inhibits HL-60 cell adhesion. Int J Cancer 59: 823–829

Zhihai Q, Kruger-Krasagakes S, Kunzendorf U, Hock H, Diamantstein T, Blankenstein T (1993) Expression of tumor necrosis factor by different tumor cell lines results either in tumor suppression or augmented metastasis. J Exp Med 178: 355–360

Zhu OZ, Cheng CF, Pauli BU (1991) Mediation of lung metastasis of murine melanomas by a lung-specific endothelial-cell adhesion molecule. Proc Nat Acad Sci USA 88: 9568–9572

Zimmerman GA, Prescott SM, McIntyre TM (1992) Endothelial cell interactions with granulocytes: tethering and signaling molecules. Immunol Today 13: 93–100

Angiogenesis-Regulating Cytokines: Activities and Interactions

M.S. PEPPER, S.J. MANDRIOTA, J.-D. VASSALLI, L. ORCI, and R. MONTESANO

1 Introduction ... 31
2 Angiogenesis: A Balance Between Positive and Negative Regulators ... 33
3 Angiogenesis-Regulating Cytokines ... 35
4 In Vitro Models for the Study of Angiogenesis ... 37
5 Vascular Endothelial Growth Factor and Basic Fibroblast Growth Factor ... 39
6 Transforming Growth Factor-β1 ... 46
7 Other Cytokines ... 53
7.1 Interferon-α_{2a} ... 53
7.2 Leukemia-Inhibitory Factor, Oncostatin M, and Interleukin-6 ... 55
8 Conclusion and Perspectives ... 58
References ... 59

1 Introduction

The establishment of a vascular supply is an absolute requirement for the growth of normal and neoplastic tissues and, as might be expected, the cardiovascular system is the first organ system to develop and to become functional during embryogenesis. It has been known since the beginning of this century (EVANS 1909) that, both during development and in postnatal life, all blood vessels begin as simple endothelial-lined capillaries. Although some remain as capillaries, many of these newly formed vessels develop into larger vessels, in part through the addition of a variable number of concentrically disposed smooth muscle cell layers. The heterogeneity and variable complexity of blood vessels thus generated serves to ensure structural and functional diversity in the vascular tree.

Capillary blood vessels are formed by two processes: *vasculogenesis*, the in situ differentiation of mesodermal precursors into endothelial cells, which are subsequently organized into a primary capillary plexus, and *angiogenesis*, the formation of new capillary blood vessels by a process of sprouting from preexisting

Department of Morphology, University of Geneva Medical Center, 1, rue Michel-Servet, 1211 Geneva 4, Switzerland vessels (RISAU et al. 1988; PARDENAUD et al. 1989; Fig. 1). Based on our current understanding of these processes, it is believed that vasculogenesis is limited to early embryogenesis, while angiogenesis can occur throughout life. Physiological angiogenesis thus occurs during the female reproductive cycle (e.g., in the corpus luteum and regenerating endometrium), in the placenta and mammary gland during pregnancy, and during wound healing in response to tissue injury. Angiogenesis in these situations is tightly regulated and is limited by the metabolic demands of the tissues concerned. Angiogenesis also occurs in response to tissue hypoxia, which may result from increased metabolic demand (e.g. as a consequence of exercise) or from chronic tissue hypoperfusion. Although new collateral blood vessel formation in response to hypoperfusion may be beneficial in certain situations, such as myocardial ischemia or peripheral vascular insufficiency, excessive new blood vessel formation, which occurs in diabetic proliferative retinopathy in response to ischemia, frequently leads to blindness. Excessive and inappropriate angiogenesis is the central pathogenetic mechanism underlying the formation of childhood hemangiomas and, in this case, new blood vessel formation far exceeds the metabolic needs of the tissue.

Much of our interest in the angiogenic process comes from the notion that for solid tumors to grow beyond a critical size, they must recruit endothelial cells from the surrounding stroma to form their own endogenous microcirculation (FOLKMAN 1974). With respect to angiogenesis, two phases can be recognized in tumor progression: prevascular and vascular. The transition from the prevascular to the vascular phase is commonly referred to as the "angiogenic switch." The prevascular phase is characterized by an initial increase in tumor growth followed by a plateau in which the rate of tumor cell growth is balanced by an equivalent

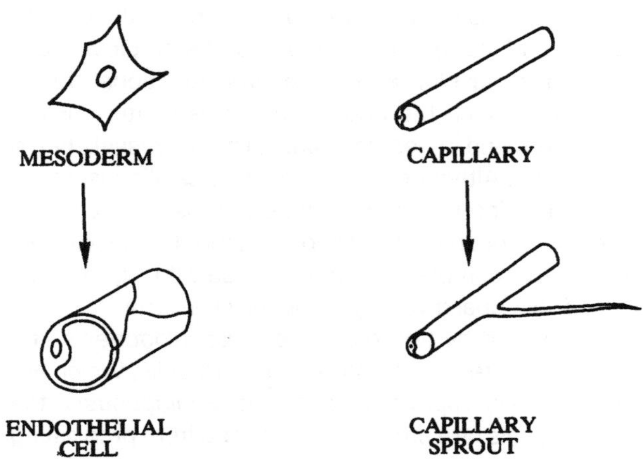

Fig. 1. Capillary blood vessels are formed by two processes: vasculogenesis (*left*), the in situ differentiation of mesodermal precursors into endothelial cells, which subsequently organize into tube-like capillaries to form a primary capillary plexus, and angiogenesis (*right*), in which new capillaries are formed by a process of sprouting from preexisting capillaries or postcapillary venules

rate of cell death (apoptosis). The prevascular phase may persist for many years and can be recognized clinically as carcinoma in situ, which is characterized by few or no metastases. During the vascular phase, which is characterized by exponential growth, tissue invasion, and the hematogenous spread of tumor cells, the rate of tumor cell proliferation is unaltered, and the increase in tumor growth is largely due to a decrease in the rate of tumor cell apoptosis (HOLMGREN et al. 1995). In a sense, tumor angiogenesis might almost be considered as "appropriate," in that newly formed vessels serve to meet the metabolic demands of the rapidly growing tumor. Although this may be beneficial to the tumor itself, it is clearly detrimental to the organism, since it allows tumor growth to continue and metastasis to occur.

The series of morphogenetic events which result in the formation of new capillary blood vessels has been well described. Angiogenesis begins with localized breakdown of the basement membrane of the parent vessel. Endothelial cells then migrate into the surrounding matrix, within which they form a capillary sprout. The sprout elongates by further migration and by endothelial cell proliferation proximal to the migrating front, and a lumen is gradually formed proximal to the region of proliferation. Contiguous tubular sprouts anastomose to form functional capillary loops, and vessel maturation is accomplished by reconstitution of the basement membrane (AUSPRUNK and FOLKMAN 1977; PAKU and PAWELETZ 1991). Alterations in at least three endothelial cell functions thus occur during this series of events: (1) modulation of interactions with the extracellular matrix, which requires alterations of cell–matrix contacts and the production of matrix-degrading proteolytic enzymes; (2) an initial increase and subsequent decrease in locomotion (migration), which allows the cells to translocate towards the angiogenic stimulus and to stop once they reach their destination; (3) an increase in proliferation, which provides new cells for the growing and elongating vessel, and a subsequent return to the quiescent state once the vessel is formed (PEPPER et al. 1995). Together, these cellular functions contribute to the process of capillary morphogenesis, i.e., the formation of three-dimensional, patent, tube-like structures.

2 Angiogenesis: A Balance Between Positive and Negative Regulators

The two most widely used assays for studying angiogenesis in vivo are the chick chorioallantoic membrane (CAM; KLAGSBRUN et al. 1976) and the rabbit corneal micropocket (GIMBRONE et al. 1974). These assays have been used for many years to describe the morphologically identifiable events which occur during angiogenesis and have also been important in the identification of positive and negative regulators (including cytokines). Subcutaneous injection or infusion of a variety of substances of interest has also been used to assess their pro- or antiangiogenic effects. Other more recently described in vivo assays involve subcutaneous implantation of polyester sponges or subcutaneous injection of Matrigel (Biomedical Products, Bedford, MA), a basement membrane-rich extracellular matrix to which angiogenesis-regulating factors can be added (Andrade et al. 1987; Passaniti et al. 1992). Based on these and other assays, a large number of both positive and negative regulators have been identified. However, it is crucial to bear in mind that even though exogenous administration of putative angiogenic or antiangiogenic factors can induce or inhibit angiogenesis in these models, it does not necessarily follow that these factors are endogenous regulators, i.e., that they are relevant to the control of angiogenesis in the intact organism. Nonetheless, for the sake of discussion we will assume that the factors mentioned in the remainder of this review are relevant, despite the fact that in most cases their precise role in the regulation of angiogenesis outside of the experimental setting still remains to be determined and that definitive studies may prove our assumption to be incorrect. It should also be remembered that although angiogenesis is regulated by a variety of factors produced by many different cell types, the ultimate target for both positive and negative regulators is the endothelial cell. This has led to the notion that angiogenesis regulators may either act directly on endothelial cells or that they may act indirectly by inducing the production of direct-acting regulators by inflammatory and other nonendothelial cells. Although this distinction may be helpful in initially sorting out the many angiogenesis regulators involved, in many cases it may be an oversimplification and should therefore be used with caution.

It is usually stated that, with the exception of angiogenesis which occurs in response to tissue injury or in female reproductive organs, endothelial cell turnover in nonpathological settings in the adult organism is very low (Engerman et al. 1967; Schwartz and Benditt 1977; Hobson and Denekamp 1984). The maintenance of endothelial quiescence may be due to the presence of endogenous negative regulators, since positive regulators are frequently detected in adult tissues in which there is apparently no angiogenesis. The converse is also true; thus positive and negative regulators often coexist in tissues in which endothelial cell turnover is increased. It is therefore currently believed that endothelial activation status is determined by a balance between positive and negative regulators: in activated (angiogenic) endothelium, positive regulators predominate, whereas endothelial quiescence is achieved and maintained by the dominance of negative regulators (Bouck 1990; Liotta et al. 1991). As indicated above, with respect to activated endothelium an important distinction needs to be made between developmental/physiological and inappropriate or excessive pathological angiogenesis. Although many of the same positive and negative regulators are operative in both settings, endothelial cell proliferation in the former is tightly controlled, whereas in the latter uncontrolled angiogenesis implies the continuous dominance of positive regulators, which results in unchecked endothelial cell growth.

3 Angiogenesis-Regulating Cytokines

Factors which affect endothelial cell activation status, either positively or negatively, include polypeptide cytokines produced by normal and tumor cells. The most thoroughly characterized of the positive regulators are vascular endothelial growth factor (VEGF) and acidic and basic fibroblast growth factors (aFGF, bFGF), which will collectively be referred to as FGF. VEGF and FGF are angiogenic when tested in the in vivo assays mentioned above (reviewed by Ferrara et al. 1992a; Klagsbrun and D'Amore 1991), and although a role for VEGF in some forms of tumor angiogenesis has been well defined (Kim et al. 1993; Millauer et al. 1994), much controversy still exists as to whether or not the FGF are relevant to the endogenous control of angiogenesis in vivo. The finding that in vitro VEGF and FGF are capable of positively regulating many endothelial cell functions relevant to angiogenesis, such as proliferation, migration, and extracellular proteolytic activity, has led to the suggestion that these factors are direct-acting positive regulators.

In contrast to VEGF and FGF, which are mitogenic for endothelial cells, transforming growth factor (TGF)-β and tumor necrosis factor (TNF)-α inhibit endothelial cell growth in vitro and have therefore been considered as direct-acting negative regulators. However, both TGF-β and TNF-α are angiogenic in vivo, which has led to the hypothesis that these cytokines induce angiogenesis indirectly by inducing the production of direct-acting positive regulators from stromal and chemoattracted inflammatory cells. In this context, then, TGF-β and TNF-α are considered to be indirect positive regulators (reviewed by Klagsbrun and D'Amore 1991).

Other cytokines which have been reported to regulate angiogenesis in vivo include interleukins (1, 6, and 8), hepatocyte growth factor, epidermal growth factor/TGF-α, platelet-derived growth factor (PDGF)-BB, interferons, and colony-stimulating factors. Finally, it should be noted that angiogenesis can be regulated by a variety of other factors, including enzymes (angiogenin, platelet-derived endothelial cell growth factor/thymidine phosphorylase), chemokines (platelet factor 4), extracellular matrix components/coagulation factors or fragments thereof (thrombospondin, angiostatin; O'Reilly et al. 1994), prostaglandins, adipocyte lipids, and copper ions (reviewed by Folkman and Klagsbrun 1987; Klagsbrun and D'Amore 1991; Leek et al. 1994; Zagzag 1995).

In attempting to further classify the activities of angiogenesis-regulating cytokines, it is useful to consider angiogenesis as occuring in two phases: a phase of *activation* and a phase of *resolution*. The phase of activation encompasses initiation and progression and includes the following steps:

1. Basement membrane degradation
2. Cell migration and extracellular matrix invasion
3. Endothelial cell proliferation
4. Capillary lumen formation

Table 1. Phases of angiogenesis

Activation: initiation and progression
Basement membrane degradation
Cell migration/matrix invasion
Cell division
Lumen formation
Resolution: termination and maturation
Cessation of migration
Inhibition of cell division
Basement membrane reconstitution
Junctional complex maturation

The phase of resolution encompasses termination and vessel maturation, comprising the following steps (PEPPER et al. 1993; Table 1):

1. Inhibition of endothelial cell proliferation
2. Cessation of cell migration
3. Basement membrane reconstitution
4. Junctional complex maturation.

While much is known about those factors which induce the activation phase, namely VEGF and bFGF, very little is known about the factors involved in the phase of resolution. A possible candidate is TGF-β1 (FLAUMENHAFT et al. 1992); in addition to its capacity in conventional two-dimensional in vitro assays to directly inhibit endothelial cell proliferation and migration and to reduce extracellular proteolysis (reviewed by KLAGSBRUN and D'AMORE 1991), all of which are required for the phase of resolution, TGF-β1 has been reported to promote the organization of single endothelial cells embedded in three-dimensional collagen gels into tube-like structures (MADRI et al. 1988; MERWIN et al. 1990), a phenomenon which could be interpreted as being representative of capillary maturation.

It is highly likely, however, that endothelial cells are rarely (if ever) exposed to a single cytokine during physiological and pathological processes. Recent studies in vivo and in vitro have pointed to the importance of interactions between positive and negative angiogenesis-regulating cytokines. The purpose of this review is not to provide an exhaustive overview of all known angiogenesis-regulating cytokines. By focusing on a limited number of cytokines, in particular VEGF, bFGF, and TGF-β1, our objectives are threefold: first, to describe some of the well-established as well as controversial issues concerning the activities of these cytokines, in an attempt to place them in an appropriate context in vivo; second, to highlight some of the interactions which have been described between different cytokines, largely by summarizing our own findings; and third, to speculate on the potential therapeutic implications of these observations. Before doing so, however, it is useful to consider some of the in vitro models which are currently used for studying the direct effects of angiogenesis regulators on endothelial cells, which, as mentioned above, are the ultimate target cells during angiogenesis.

4 In Vitro Models for the Study of Angiogenesis

The in vivo assays described above are essential to establish whether a given molecule stimulates new blood vessel formation in the intact organism. However, their interpretation is frequently complicated by the fact that the experimental conditions may inadvertently favor inflammation and that, under these conditions, the angiogenic response is elicited indirectly, at least in part, through the recruitment of inflammatory cells. Although this may be relevant to some settings in which angiogenesis occurs in vivo, it does not allow us to study the consequences of the direct interaction of angiogenesis regulators with endothelial cells. To circumvent these drawbacks, in vitro assays using populations of cultured endothelial cells have been developed for several of the cellular components of the angiogenic process; based on the geometry of the assay, these can be classified as either two-dimensional or three-dimensional. Conventional two-dimensional assays include measurement of cell proliferation, cell migration, and the production of proteolytic enzymes such as metalloproteinases and plasminogen activators (PA) (MOSCATELLI et al. 1986b; PEPPER et al. 1994b). Three-dimensional assays have as their end point the formation of capillary-like cords or tubes by endothelial cells cultured either on the surface of (planar models), or within, simplified extracellular matrices (FOLKMAN and HAUDENSCHILD 1980; MONTESANO et al. 1983; MONTESANO and ORCI 1985; KUBOTA et al. 1988; MADRI et al. 1988; MIGNATTI et al. 1989; NICOSIA and OTTINETTI 1990; IRUELA-ARISPE et al. 1991). The relevance of planar models to the invasive nature of the angiogenic process has recently been questioned (VERNON et al. 1995).

In our own studies aimed at exploring potential interactions between angiogenesis-regulating cytokines, we have used an in vitro model of angiogenesis which consists of cultivating endothelial cells on the surface of a three-dimensional collagen gel (MONTESANO and ORCI 1985) or fibrin gel (MONTESANO et al. 1987). In control cultures, the cells form a monolayer on the surface of the gel (Fig. 2A,B). When the monolayer is treated with an angiogenic factor such as bFGF (MONTESANO et al. 1986) or VEGF (PEPPER et al. 1992), the cells invade the underlying gel and, by adjusting the plane of focus beneath the surface monolayer, branching and anastomosing cell cords can be seen within the gel (Fig. 2C,D). In cross-section, the presence of tube-like structures resembling capillaries can be observed beneath the surface monolayer (Fig. 2C). Invasion can be quantitated by measuring the total additive length of all cells which have penetrated into the underlying gel to form cell cords (PEPPER et al. 1992). Unlike planar models of in vitro angiogenesis, the model we have developed has the advantage of accurately recapitulating the invasive nature of the angiogenic process; by virtue of its three dimensional nature, it also allows histotypic morphogenesis, i.e., the formation of patent capillary-like tubes whose albuminal surfaces are in direct contact with the extracellular matrix.

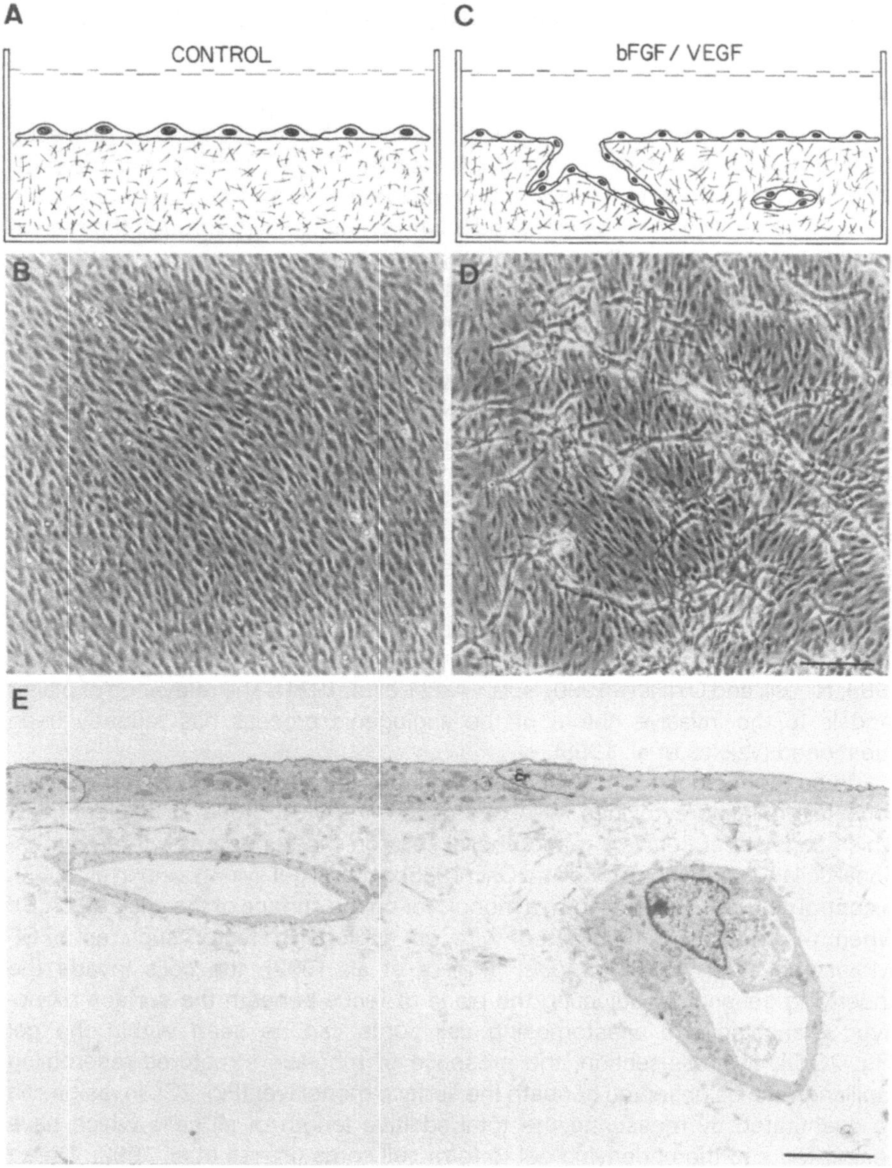

Fig. 2A–E. Three-dimensional model for the study of angiogenesis in vitro. **A,B** Endothelial cells grown on the surface of a three-dimensional collagen gel form a confluent monolayer without invading the underlying matrix (control), and when viewed from above by phase contrast microscopy the cells are seen to form a monolayer of closely apposed cells on the surface of the gel. **C,D** Addition of basic fibroblast growth factor *(bFGF)* or vascular endothelial growth factor *(VEGF)* induces the cells to invade the underlying gel and to form a network of branching capillary-like cell cords and tubes which can be seen beneath the surface monolayer. *Bar,* 150 µm. **E** When viewed in cross-section by electron microscopy, the tube-like nature of the invading cell cords, morphologically similar to capillaries seen in vivo, can be appreciated. *Bar,* 10 µm. (Fig. 2B,D is taken from MONTESANO et al. 1986)

5 Vascular Endothelial Growth Factor and Basic Fibroblast Growth Factor

VEGF (also known as vascular permeability factor, VPF) is a secreted endothelial-specific mitogen which is capable of inducing angiogenesis in the in vivo experimental models described above. Studies with neutralizing VEGF antibodies have clearly defined a role for this cytokine in experimental tumor growth (KIM et al. 1993; KONDO et al. 1993; WARREN et al. 1995). Three VEGF isoforms, which vary in their relative proportions in different tissues, are generated through alternative splicing of a single mRNA. The most abundant and most extensively studied 165-amino acid isoform ($VEGF_{165}$) has been detected in both soluble and cell/matrix-bound forms. $VEGF_{121}$ has been detected only in soluble form, while $VEGF_{181}$ appears to be localized exclusively to the cell surface and extracellular matrix. A fourth isoform, $VEGF_{206}$, has also been described, although its biological significance remains to be determined. VEGF isoforms, which exist as homodimers, have significant homology to PDGF (approximately 20% amino acid sequence similarity between $VEGF_{165}$ and PDGF; reviewed by FERRARA et al. 1992a; SENGER et al. 1993). VEGF expression by nonendothelial cells is regulated by hypoxia, glucose deprivation, prostaglandins, and estrogens as well as by a number of other cytokines (DOLECKI and CONNOLLY 1991; SHWEIKI et al. 1992; STAVRI et al. 1995; CULLINAN-BOVE and KOOS 1993; HARADA et al. 1994; STAVRI et al. 1995). Publication of genetic studies on VEGF, including targeted gene disruption in mice, is awaited.

VEGF-induced endothelial cell responses are mediated via endothelial cell-specific transmembrane tyrosine kinase receptors, which at present include VEGFR-1 (flt-1) and VEGFR-2 (KDR/flk-1) (reviewed by NEUFELD et al. 1994; MUSTONEN and ALITALO 1995). These receptors are expressed in many adult tissues, with adult brain being one important exception, and they appear to be upregulated in endothelial cells during tumor angiogenesis (reviewed by PLATE et al. 1994a); the importance of VEGFR-2 in tumor angiogenesis has been demonstrated using a dominant-negative approach (MILLAUER et al. 1994). The phenotypes of VEGFR-1- and VEGFR-2-deficient mice have recently been published. Thus, VEGFR-1-deficient mice die in utero at midsomite stages, and although homozygous deficient mice are capable of forming endothelial cells in both intra- and extra-embryonic regions, assembly of these cells into vessels is perturbed, resulting in the formation of abnormal vascular channels. The authors conclude that VEGFR-1 signaling pathways may regulate normal endothelial cell–cell or cell–matrix interactions during vascular development (FONG et al. 1995). VEGFR-2-deficient mice also die in utero between 8.5 and 9.5 days postcoitum, and this appears to be due to abortive development of endothelial cell precursors. Yolk-sac blood islands and organized embryonic blood vessels were not detectable at any stage of development. The development of hematopoietic precursors was also severely reduced (SHALABY et al. 1995). Dominant-negative and gene-targeting approaches have therefore clearly defined an essential role for VEGFR in developmental and tumor angiogenesis. Since VEGFR-expressing endothelial cells in glioblastomas are located adjacent to regions of tumor ischemia and necrosis, it is highly likely that the increase in VEGFR expression is mediated by hypoxia (PLATE et al. 1992). In this context, it has recently been demonstrated in vitro that hypoxia increases high-affinity VEGF binding to human umbilical vein endothelial cells by approximately 50% (THIEME et al. 1995). Although the molecular nature of these high-affinity binding sites was not characterized, it has recently been reported that hypoxia increases expression of VEGFR-1 and -2 ex vivo in rat lungs and that, under the same conditions, VEGFR expression is decreased by nitric oxide (NO) or an NO-related metabolite (TUDER et al. 1995). To date, with the exception of hypoxia and NO, no other factors have been described which modulate high-affinity VEGF binding to or expression of VEGFR-1 and -2.

bFGF, also known as FGF-2 (or heparin-binding growth factor-2), is a member of the FGF superfamily, which to date comprises nine distinct gene products (reviewed by BAIRD and KLAGSBRUN 1991; BASILICO and MOSCATELLI 1992; FERNIG and GALLAGHER 1994; WILKIE et al. 1995). bFGF is a cationic polypeptide (pI 9.6) which induces angiogenesis in vivo in the experimental models described above. In contrast to studies with neutralizing antibodies to VEGF, which clearly demonstrate that tumor growth is VEGF dependent, use of neutralizing antibodies to bFGF has been reported to inhibit tumor growth in some studies (REILLY et al. 1989; HORI et al. 1991), but not in others (MATSUZAKI et al. 1989; DENNIS and RIFKIN 1990). However, since bFGF mitogenicity is not endothelial cell specific, it is difficult to conclude from the former studies whether the reduction in tumor growth was mediated through inhibition of angiogenesis. Based in part on the observation that bFGF induces angiogenesis in our three-dimensional invasion model (MONTESANO et al. 1986), it has been suggested that bFGF is a direct-acting positive regulator. Although the phenotypes of mice with homozygous null mutations in FGF-3, -4 and -5 have been described (reviewed by WILKIE et al. 1995), as for VEGF, the consequences of bFGF gene inactivation have not been published.

bFGF-induced endothelial responses are mediated via transmembrane tyrosine kinase receptors (FGFR). To date, high-affinity FGFR-1 to -4 have been described, and the existence of a large number of FGFR variants (generated by alternative mRNA splicing and differential polyadenylation) further increases FGFR diversity. This diversity results in a complex pattern of overlapping binding specificities for the various FGF. It has been demonstrated that bFGF binds with high affinity to receptors 1 and 2 (reviewed by JAYE et al. 1992; GIVOL and YAYON 1992; JOHNSON and WILLIAMS 1993). The importance of FGFR-1 in postimplantation growth and mesodermal patterning has recently been demonstrated by FGFR-1-targeted gene disruption in mice, in which loss of FGFR-1 function resulted in recessive embryonic lethality during gastrulation (DENG et al. 1994; YANAGUCHI et al. 1994). Thus, although mesoderm was formed, specification of cell fate and regional patterning were several disrupted. With respect to the cardiovascular system, in a few embryos which had progressed to the appropriate stage, the heart and blood islands were present, and VEGFR-2-positive cells (endothelial precursors) were found in appropriate locations in the lateral plate and yolk sac mesoderm. This suggests that FGFR-1 is not required for differentiation of endothelial cell precursors. In man, genetic alterations in FGFR-1 to -3 have recently been linked to craniofacial developmental defects and achondroplasia (dwarfism) (reviewed by WILKIE et al. 1995).

The controversy that has arisen over the role of bFGF as an endogenous regulator of angiogenesis stems from the following observations. First, unlike VEGF, the stimulatory effects of bFGF on proliferation and migration are not restricted to endothelial cells. Second, unlike VEGF, bFGF lacks a signal peptide and therefore fails to enter the classical secretory pathway. bFGF is synthesised as both an 18-kDa and as higher molecular weight (22–25 kDa) forms, resulting from the use of alternate start codons, in which translation is initiated at CUG rather than AUG. The 18-kDA form is stored in the cytoplasm of its producer cells, while the higher molecular weight forms contain an amino-terminal nuclear localization/retention sequence which appears to mediate intranuclear accumulation. Although bFGF does not enter the classical secretory pathway, the 18-kDa form can be detected outside the cell. As a consequence of its extracellular localization, it appears that only 18-kDa bFGF, and not its high molecular weight forms, is capable of interacting with the transmembrane FGFR. The mechanisms of 18-kDa bFGF export are poorly understood, although they may involve cell injury/death and possibly an active and regulatable nonclassical export pathway (reviewed by MIGNATTI and RIFKIN 1991; BASILICO and MOSCATELLI 1992).

Although in vitro endothelial cells express FGFR-1, -2, and -4 (RUTA et al. 1988; EISEMANN et al. 1991; MIKI et al. 1992; LIAW and SCHWARTZ 1993; PLATE et al. 1992; M.S. PEPPER and S.J. MANDRIOTA, unpublished observation), which may be a consequence of culture conditions and/or serial passaging, the third and perhaps most significant issue concerning the endogenous angiogenic activity of bFGF is whether or not endothelial cells of the microvasculature express FGFR in vivo. This is important, since new capillary blood vessels arise from preexisting capillaries or postcapillary venules. A limited number of studies have reported the presence of immunoreactive FGFR-1 in endothelial cells of the microvasculature (predominantly post-capillary venules) of a wide range of normal and neoplastic adult tissues, as well as in newly formed vessels of atherosclerotic plaques and underlying adventitial vessels (HUGHES et al. 1993; HUGHES and HALL 1993; UEBA et al. 1994). However, FGFR have been undetectable in microvascular endothelial cells in virtually all settings in which there is active angiogenesis and in which expression of high affinity VEGFR has been clearly demonstrated. Thus, although FGFR mRNA are detectable in many organs during development and in the adult, this appears to be localized to neural, epithelial, and nonendothelial mesenchymal cells (reviewed by GIVOL and YAYON 1992; PARTANEN et al. 1992). Similarly, FGFR mRNA are undetectable in endothelial cells of the microvasculature in virtually all tumors studied to date. This includes highly vascular glioblastomas (MORRISON et al. 1994; PLATE et al. 1994b), although FGFR-1 immunoreactivity has been reported in endothelial cells in these tumors (UEBA et al. 1994). These observations are in striking contrast to the observations that VEGFR mRNA are expressed in a clearly defined and tightly regulated temporospatial manner in sprouting/angiogenic microvascular endothelial cells, both during development and in a variety of tumors including glioblastomas (reviewed by PLATE et al. 1994a).

The apparent discrepancy between expression of FGFR mRNA and protein in endothelial cells in vivo warrants two further comments. Thus, while FGFR mRNA has been detected in large vessel endothelial cells during development (PETERS et al. 1992) and in the adult in some (LIAW and SCHWARTZ 1993), but not other (LINDER and REIDY 1993) studies, as indicated above it appears to be absent from microvascular endothelium. Yet FGFR immunoreactivity has clearly been demonstrated in microvascular endothelium in a number of settings. How might these observations be reconciled? First, since endothelial cells are directly exposed to the circulation, soluble, circulating forms of FGFR may bind to endothelial cells or be deposited into their underlying basement membranes (HANNEKEN et al. 1995). Second, in contrast to VEGFR which appear to be abundantly expressed by endothelial cells in vivo, levels of FGFR mRNA in the microvasculature may be too low to allow detection by in situ hybridization.

bFGF immunoreactivity has clearly been demonstrated in the vascular intima (endothelial cells and underlying basement membrane) in a wide variety of settings in vivo. These include embryonic (HANNEKEN et al. 1989; GONZALEZ et al. 1990) and normal adult (CORDON-CARDO et al. 1990; HUGHES and HALL 1993) tissues, chronic inflammatory tissues (SCHULZE-OSTHOFF et al. 1990; OHTANI et al. 1993; QU et al. 1994), hemangiomas of infancy and childhood (TAKAHASHI et al. 1994), the endothelium of newly formed vessels in atherosclerotic plaques in human vessels (BROGI et al. 1993; HUGHES et al. 1993), and a variety of tumors including glioblastomas (SCHULZE-OSTHOFF et al. 1990; BREM et al. 1992; OHTANI et al. 1993; SCHULTZ-HECTOR and HAGHAYEGH 1993). However, the presence of bFGF immunoreactivity gives no information as to the origin of the molecule or the molecular nature of the structures to which it is bound. Extracellular matrix-bound bFGF is extremely stable, and since bFGF immunoreactivity but not mRNA is detectable in resting, nonhuman, aortic endothelial cells in vivo (LINDER and REIDY 1993; LIAW and SCHWARTZ 1993), it is possible that endothelium-associated bFGF may have a nonendothelial origin (circulation or medial smooth muscle cells) or that it may have been deposited in the matrix during development and postnatal growth. It has also been suggested that this may reflect the instability of bFGF mRNA relative to protein (LIAW and SCHWARTZ 1993). Finally, it has been reported that bFGF mRNA is induced during endothelial regeneration in vivo in the rat aorta (LINDER and REIDY 1993), and that bFGF-like activity, protein, and mRNA are present in cultured endothelial cells (MOSCATELLI et al. 1986a; SCHWEIGERER et al. 1987; HANNAN et al. 1988; SATO and RIFKIN 1988; LIAW and SCHWARTZ 1993; YU et al. 1993), all of which can be increased by serial passaging (SPEIR et al. 1991; LIAW and SCHWARTZ 1993). Taken together, these observations suggest that cultured endothelial cells are phenotypically closer to activated/angiogenic endothelium than to the resting endothelium from which they were derived.

If we accept that the FGF are indeed endogenous regulators of angiogenesis, despite the fact that microvascular endothelial cells may not express FGFR in vivo, a number of alternative hypotheses can be envisaged. The first two follow on from the observation that bFGF immunoreactivity can be detected in endothelium in vivo. First, bFGF-dependent, FGFR-independent signaling may occur in endothelial cells through other cell surface molecules such as heparan sulfate proteoglycans (HSPG) (QUARTO and AMALRIC 1994). Second, the observation that bFGF lacks a signal peptide and that its high molecular weight forms have nuclear localization/ retention signals raises the possibility of an autocrine/intracrine role for bFGF in vivo. A third possibility might be that bFGF is an indirect angiogenic factor which acts by stimulating the production of direct-acting cytokines by adjacent nonendothelial cells. However, it is important to note that none of these alternate hypotheses which have been proposed to explain bFGF-dependent, FGFR-independent endothelial cell activation in vivo have yet been clearly substantiated. As a general observation, it appears that providing justification for the role of bFGF in the regulation of endogenous angiogenesis requires extensive and elaborate argument when compared to VEGF. It is, however, possible that the role of bFGF in angiogenesis may be more subtle than that of VEGF.

Although VEGF is a good candidate as a pathophysiologically relevant positive regulator, a number of important questions still remain. First, both VEGF and its receptors are expressed at high levels in a variety of adult tissues in which there is apparently no angiogenesis. This might mean that VEGF has additional functions not directly related to angiogenesis. It may thus be involved in the regulation of vascular permeability (hence its alternate name, VPF; SENGER et al. 1993), it may be important for endothelial survival (trophic effect), or it may be required for the maintenance of organ-specific endothelial characteristics such as fenestrae (ROBERTS and PALADE 1995). Despite the dogma that endothelial cell turnover in the adult organism is very low, the presence of VEGF may indicate the requirement for a constant low-grade angiogenesis, which is necessary to regenerate capillaries damaged in organs with high rates of blood flow (lung and kidney). The absence of extensive angiogenesis in the face of high levels of VEGF (and its receptors) might also indicate the dominance of negative regulators which serve to prevent uncontrolled angiogenesis in these organs. Second, levels of VEGF and its receptors appear to be unchanged in certain organs and tumors in which there is extensive angiogenesis. These include, although are unlikely to be limited to, the mammary gland during pregnancy (M.S. PEPPER and S.J. MANDRIOTA, unpublished observation) and some experimental tumors in which the levels of VEGF are similar in the prevascular and vascular phases of tumor progression (CHRISTOFORI and HANAHAN 1994). One explanation might be that angiogenesis in these tissues is in fact VEGF dependent and that this is induced following the loss of a negative regulator. Alternatively, it may imply that some other positive regulator might be acting in these settings. A third hypothesis might be that VEGF-induced angiogenesis is contextual, in that it depends on the synergistic interaction with other positive regulators whose expression is regulated in an appropriate temporo-spatial manner.

Although, as mentioned above, some controversy still exists concerning its precise role as an endogenous regulator of angiogenesis, it has recently been demonstrated that bFGF is expressed in the mammary gland during early pregnancy, i.e., during the phase of active angiogenesis, and declines during late pregnancy and lactation (KRNACIK-COLEMAN and ROSEN 1994), at which time angiogenesis is markedly reduced. In addition, induction of the vascular phase in certain experimental tumors in which VEGF levels are unchanged appears to coincide with the export of bFGF from tumor cells (CHRISTOFORI and HANAHAN 1994). To test the hypothesis that VEGF-induced angiogenesis might be bFGF dependent, we used our three-dimensional invasion assay to assess the effect of simultaneous addition of bFGF and VEGF on the in vitro angiogenic response. We found that, when added separately at equimolar concentrations, bFGF was about twice as potent as VEGF (Fig. 3). However, when bFGF and VEGF were coadded, the resulting effect on invasion was far greater than additive (Fig. 3), and invasion

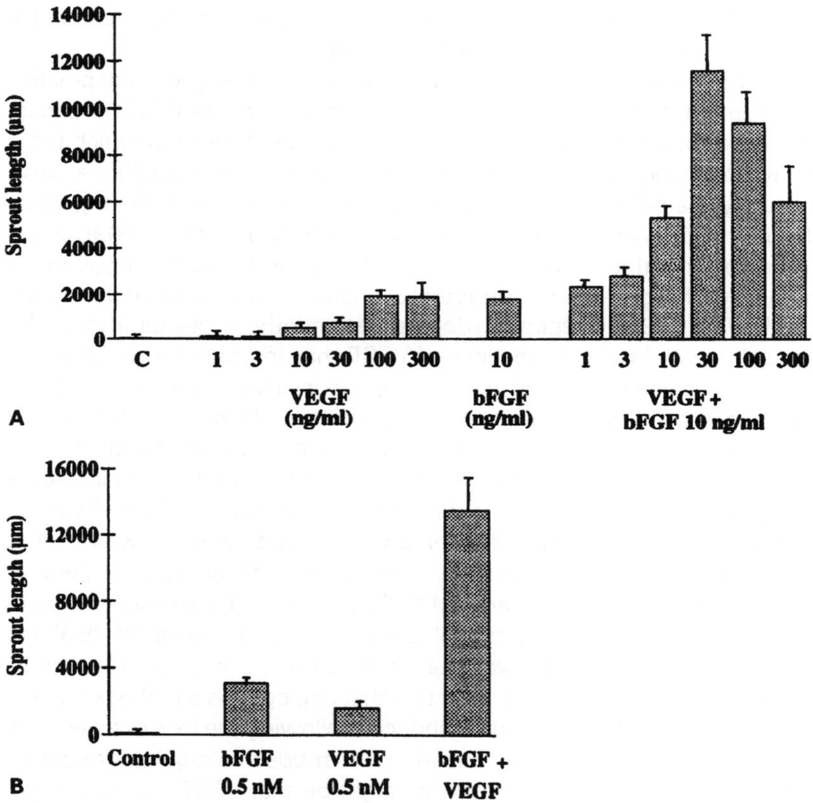

Fig. 3A,B. Synergistic effect of basic fibroblast growth factor (*bFGF*) and vascular endothelial growth factor (*VEGF*) on in vitro angiogenesis. Randomly selected fields of bovine microvascular endothelial cell monolayers treated with $VEGF_{165}$ and/or bFGF for 4 days were photographed at a single level beneath the surface monolayer. Endothelial cell invasion was quantitated by measuring the total additive length of all cell cords, and values are expressed as mean ± SEM from at least three experiments per condition. **A** VEGF dose–response and effect of coaddition of bFGF **B** Comparison of equimolar concentrations (0.5 n*M*) of bFGF (9 ng/ml) and VEGF (22.5 ng/ml and effect of coaddition. bFGF was more potent than VEGF when assessed separately; simultaneous addition of the two cytokines induced a synergistic response. From PEPPER et al. (1992) Biochem Biophys Res Commun 189: 824–831 (reproduced with permission from Academic Press, Inc.)

occurred with greater rapidity than in response to either cytokine alone (PEPPER et al. 1992). Further preliminary observations using inhibitory anti-bFGF antibodies indicate that VEGF-induced in vitro angiogenesis and VEGF-induced increase in urokinase-type plasminogen activator (uPA) activity are dependent on endogenous bFGF (M.S. PEPPER and S.J. MANDRIOTA, unpublished observation). The synergistic interaction between bFGF and VEGF was subsequently confirmed in an independent study using a related three-dimensional in vitro assay (GOTO et al. 1993), and synergism has also been observed in vivo in a rabbit model of hindlimb ischemia (ASAHARA et al. 1994) and in the rat sponge implant model (HU and FAN 1995). It is important, however, to bear in mind that results from our in vitro observations assume that endothelial cells are able to respond to exogenously added bFGF through its interaction with the cell surface (e.g., FGFR or HSPG). Findings obtained in vivo do not allow us to differentiate this possibility from the possibility that bFGF is acting indirectly through induction of a positive regulator by nonendothelial cells, which in turn contributes to the synergistic effect.

In attempting to understand the mechanisms responsible for this synergistic effect, we initially assessed the effect of coaddition of bFGF and VEGF in conventional two-dimensional in vitro assays of endothelial cell proliferation, migration, and PA-mediated extracellular proteolysis. In none of these situations was the effect of simultaneous addition of bFGF and VEGF greater than additive (MANDRIOTA et al. 1995; PEPPER et al. 1994a), although others have been able to detect a synergistic effect of bFGF and VEGF on endothelial proliferation when cells were grown in three-dimensions (GOTO et al. 1993). Our subsequent approach has been to determine whether bFGF and VEGF might modulate expression of FGFR and VEGFR in monolayer culture. Our preliminary observations suggest that, while neither cytokine, either alone or in combination, is capable of modulating expression of FGFR-1, bFGF increases expression of VEGFR-2 in bovine endothelial cells (M.S. PEPPER and S.J. MANDRIOTA, unpublished data). The importance of VEGFR-2 in angiogenesis has recently been demonstrated in mice using a dominant-negative approach (MILLAUER et al. 1994), and none of the endothelial cell lines we have used appear to express VEGFR-1 (BARLEON et al. 1994; M.S. PEPPER and S.J. MANDRIOTA, unpublished observation). It has also been demonstrated that, when expressed independently, VEGFR-2 but not VEGFR-1 can mediate chemotaxis and mitogenicity upon VEGF stimulation, although maximal responses were obtained upon coexpression of the two receptors (WALTENBERGER et al. 1994). At least in vitro therefore, VEGF-induced invasion might require the presence of a second cytokine, such as bFGF, which increases VEGFR-2 expression and hence VEGF-mediated signal transduction above a critical threshold required for mitosis, migration, and increased proteolytic activity, all of which are necessary for the formation of new capillary sprouts.

These findings are significant in a number of ways. First, VEGF-dependent angiogenesis in certain settings might require the presence of a second positive regulator. Although this hypothesis is based on findings obtained from our in vitro invasion assay, in which bFGF is clearly angiogenic, endogenous regulators of synergism might include cytokines other than bFGF as well as hypoxia. Second, our findings might provide a partial explanation for the observation that VEGF and its receptors are expressed, sometimes at relatively high levels, in adult tissues in which angiogenesis is apparently not occuring. The lack of angiogenesis in these tissues might indicate the absence of a synergistic cofactor required for VEGF-dependent angiogenesis (and/or the presence of a dominant, negative regulator). Third, modulation of new capillary blood vessel formation may serve as an alternative/adjunct to current therapeutic modalities in several angiogenesis-associated diseases. Although at first sight the redundancy of angiogenesis-regulating cytokines might suggest that therapeutic strategies based on neutralization of single angiogenic factors might be unrealistic, if the synergism which we have observed in vitro is relevant to the endogenous regulation of angiogenesis in vivo, angiogenesis would be more prominent in tumors or other pathological settings in which more than one angiogenic factor is produced. This may justify anti-angiogenesis strategies based on the neutralization of a single angiogenic factor, since this would reduce the synergistic effect. Recent work has demonstrated that administration of angiogenic factors can enhance the growth of collateral vessels in animal models of myocardial, peripheral, and cerebral arterial occlusion (reviewed by Höckel et al. 1993; Symes and Sniderman 1994). We suggest that the effect of coaddition of two cytokines whose interaction is synergistic would be greater than that derived from the addition of one of these cytokines alone. Support for this hypothesis has recently been provided by an in vivo study in which coadministered bFGF and VEGF synergized in the induction of collateral blood vessel formation in a rabbit model of hindlimb ischemia (Asahara et al. 1994).

In summary, our findings on the synergism between two positive regulators may be relevant both to understanding the biology of angiogenesis as well as to positive and negative therapeutic modulation of this process. Our observations also highlight the importance of a three-dimensional environment for the study of angiogenesis in vitro: had we relied exclusively on traditional two-dimensional assays of proliferation, migration, or proteolysis, the synergism between bFGF and VEGF would not have been detected.

6 Transforming Growth Factor-β1

TGF-β1 is a member of the TGF-β cytokine superfamily, which to date comprises more than 25 members. Three TGF-β (1–3) have been described in mammals. The prototype, TGF-β1, is a 25-kDa homodimer which is secreted as an inactive latent precursor and which can be activated by plasmin, cathepsin D, and low pH. TGF-β achieve their biological effects through binding to cell surface receptors (TGF-βR) designated types I,II, and III. TGF-β binds directly to the type II receptor. Binding is followed by recruitment of type I receptor and the formation of a stable ternary complex. Receptor II is autophosphorylated and constitutively active. Following recruitment, receptor I is phosphorylated on serine/threonine by receptor II. This is followed by receptor I-mediated intracellular signal transduction; the nature of the biological response to ligand therefore appears to be specified primarily by the type I receptor isoform engaged in the ternary complex. In nonendothelial cells, the type III receptor is beta-glycan, a transmembrane proteoglycan with a short cytoplasmic domain, containing both heparan sulfate and chondroitin sulfate glycosaminoglycans. In endothelial cells, the type III receptor is endoglin. The type III receptor does not appear to be required for signal transduction and may serve to modulate TGF-β binding to receptor type II. TGF-β also bind to the decorin core protein, which neutralizes their activity, as well as to thrombospondin, a large multifunctional glycoprotein which mediates the adhesion of both endothelial and nonendothelial cells to the extracellular matrix (reviewed by ATTISANO et al. 1994; DERYNCK 1994; KINGSLEY 1994; WRANA et al. 1994).

TGF-β is an angiogenesis-regulating cytokine that has been described as being either angiogenic or antiangiogenic depending on the nature of the assay used. Thus, in vivo, TGF-β is a potent inducer of angiogenesis when administered subcutaneously into newborn mice (ROBERTS et al. 1986; SPRUGEL et al. 1987), when tested in the rabbit cornea (PHILLIPS et al. 1992, 1993) or when applied to the chick embryo CAM (YANG and MOSES 1990). Stable transfection of TGF-β1 confers a growth advantage on Chinese hamster ovary cells in vivo but not in vitro, and this is accompanied by an increase in capillary density; local administration of neutralizing antibodies to TGF-β1 reduced both capillary density and tumor growth (UEKI et al. 1992). It should be noted, however, that overexpression of TGF-β1 in a tissue-specific manner does not necessarily induce angiogenesis in vivo. Thus, while overexpression in the heart (KOH et al. 1995) resulted in a measurable angiogenic response, overexpression of TGF-β in the vessel wall (NABEL et al. 1993), the liver (SANDERSON et al. 1995), or the insulin-producing β-cells of the endocrine pancreas (SANVITO et al. 1995) does not induce angiogenesis. Finally, TGF-β1 inhibits aFGF-induced angiogenesis in subcutaneously injected Matrigel (PASSANITI et al. 1992). One explanation for these apparently discrepant observations is that in some of the settings in which TGF-β1 is angiogenic, this is preceeded either by massive influx of inflammatory cells or chemotaxis and activation of connective tissue or epithelial cells (YANG and MOSES 1990; PHILLIPS et al 1993); the angiogenesis which follows is then mediated by positive regulators produced by TGF-β1-recruited inflammatory or connective tissue cells (WAHL et al. 1987; POSTLETHWAITE et al. 1987; WISEMAN et al 1988; MCCARTNEY-FRANCIS et al. 1990; PHILLIPS et al. 1992). An additional explanation might be that TGF-β1-mediated angiogenesis is contextual, i.e., angiogenesis is only induced in the presence of additional positive regulators which induce an additive or supra-additive angiogenic response.

Targeted disruption of the TGF-β1 gene in mice results in lethal cardiac abnormalities in mice born from a homozygous null female (LETTERIO et al. 1994). Mice born from heterozygous females show no developmental abnormalities, but at 3–4 weeks develop a rapidly progressing wasting syndrome resulting from massive inflammatory cell infiltration in many organs, which leads to multiple organ failure and ultimately death (SHULL et al. 1992; KULKARNI et al. 1993). Although these mice lack a functional copy of the TGF-β1 gene, TGF-β1 derived from the maternal circulation by transplacental transfer and from breast milk rescues homozygous null embryos from developmental abnormalities (LETTERIO et al. 1994). With the possible exception of cardiac abnormalities, no phenotype has yet been described which might indicate a role for TGF-β1 in the endogenous regulation of angiogenesis. It is possible that defining a role for TGF-β1 during angiogenesis in these mice might be complicated by redundancy in the expression of the various isoforms of the TGF-β family.

Two-dimensional in vitro assays have revealed that TGF-β1 has a pronounced effect on a number of essential components of the angiogenic process. These include endothelial cell proliferation, migration, and extracellular proteolytic activity. TGF-β1 promotes the synthesis and deposition of matrix components such as fibronectin, collagens I, IV and V, and thrombospondin and also increases integrin expression (reviewed by ROBERTS and SPORN 1989; RAYCHAUDHURY and D'AMORE 1991). Results from three-dimensional in vitro assays demonstrate that the response to TGF-β1 varies depending on the assay used. Thus it inhibits endothelial cell invasion of three-dimensional collagen gels (MÜLLER et al. 1987) or fibrin gels (PEPPER et al. 1990,1991,1993), as well as the explanted amnion (MIGNATTI et al. 1989). In addition, TGF-β1 reduces lumen size in bFGF-induced capillary-like tubes in three-dimensional fibrin gels (PEPPER et al. 1990,1993; Fig. 4). These results support the notion that TGF-β1 is a direct-acting inhibitor of extracellular matrix invasion and tube formation. However, it has also been reported that TGF-β1 promotes organization of endothelial cells into tube-like structures (MADRI et al. 1988; MERWIN et al., 1990). These findings may be reconciled by considering that when interacting directly with endothelial cells, TGF-β1 might have different functions on vessel formation at different stages of the angiogenic process. Thus it may prevent endothelial cell activation by inhibiting proliferation and invasion and thus sprout formation. Once a new vessel has formed, TGF-β1 may promote the phase of resolution by maintaining endothelial cell quiescence (inhibition of further proliferation and migration) and by inducing vessel maturation through the deposition and organization of a functional basement membrane.

An additional possibility is that the direct effect of TGF-β1 on endothelial cell function is concentration dependent, particularly since this cytokine has been described as a bifunctional regulator in a variety of other biological processes (NATHAN and SPORN 1991). To address this possibility, we have used our three-dimensional invasion model to assess the effect of a wide range of concentrations of TGF-β1 on bFGF- or VEGF-induced in vitro angiogenesis. We found that in the presence of TGF-β1, bFGF- or VEGF-induced invasion was increased at 200–500 pg TGF-β1/ml and decreased at 5–10 ng TGF-β1/ ml (Fig. 5); (PEPPER et al. 1993). Similar findings in three-dimensional collagen gels have also been reported by others (GAJDUSEK et al. 1993). This biphasic effect of TGF-β1 is in accord with previous studies in which endothelial cell wound-induced migration (HEIMARK et al. 1986; MÜLLER et al. 1987) and invasion of three-dimensional collagen gels

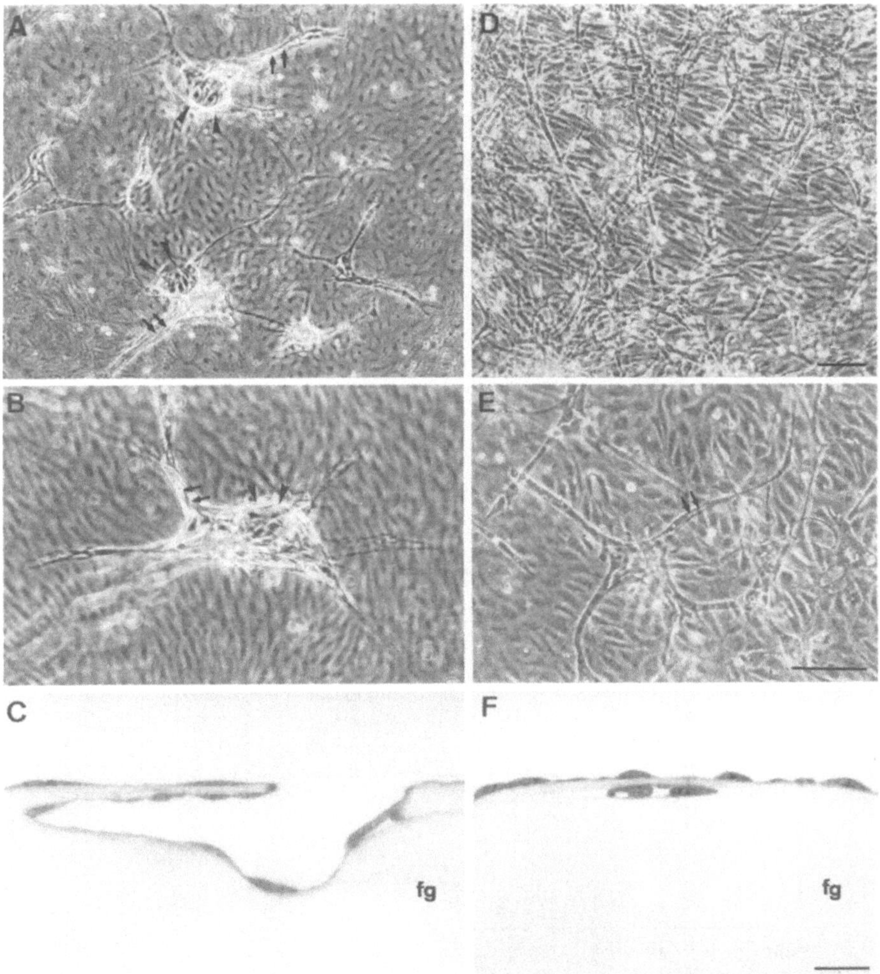

Fig. 4A–F. Transforming growth factor (TGF)-β1 modulates lumen size during in vitro angiogenesis. Basic fibroblast growth factor (bFGF) (30 ng/ml) was added without (**A, B, C**) or with 500 pg/ml TGF-β1 (**D, E, F**) to confluent monolayers of microvascular endothelial cells on fibrin gels. The resulting capillary-like tubular structures were viewed by phase-contrast microscopy (**A,B,D,E**) and semithin sections (**C,F**). bFGF induced endothelial cells to invade from a circular opening in the surface monolayer (*arrowheads* in **A** and **B**), to form well-organised cell cords with a clearly visible refringent lumen (*arrows* in **A** and **B**), which tapered down progressively in the distal part of the cords. Semithin sectioning revealed that the proximal part of the cords was often cavernous (**C**). When 500 pg/ml TGF-β1 was coadded with bFGF, the total additive length of the invading cell cords was increased (compare **A** and **D**, and see Fig. 5 for quantitation), and clearly distinguishable lumina were present beneath the surface monolayer (white refrigent line indicated by the *arrows* in **E**). Semithin sectioning revealed that lumen size was decreased to a more physiological size when compared to cultures treated with bFGF alone (compare **C** and **F**). *Bars* in **A,D,** 100 μm, in **B,E,** 50 μm, and in **C,F** 20 μm. From PEPPER et al. (1993) Exp Cell Res 204: 356–363 (reproduced with permission from Academic Press, Inc.)

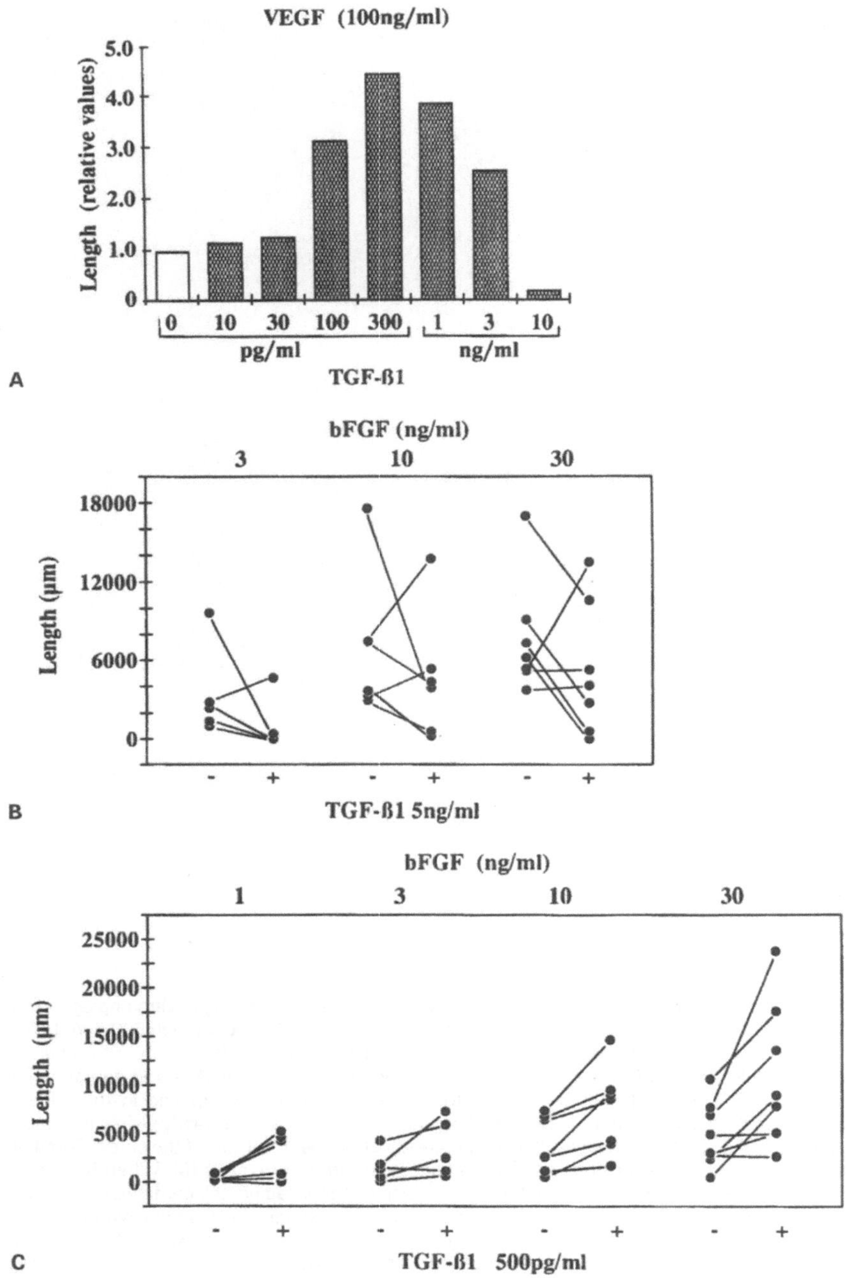
VEGF (100ng/ml)
Length (relative values)
5.0
4.0
3.0
2.0
1.0
0
0 10 30 100 300 1 3 10
pg/ml
ng/ml
TGF-ß1
A
bFGF (ng/ml)
3 10 30
18000
12000
6000
0
Length (µm)
- + - + - +
TGF-ß1 5ng/ml
B
bFGF (ng/ml)
1 3 10 30
25000
20000
15000
10000
5000
0
Length (µm)
- + - + - + - +
TGF-ß1 500pg/ml
C

(MÜLLER et al. 1987) or the explanted amnion (MIGNATTI et al. 1989) were inhibited at relatively high concentrations (1–10 ng/ml), whereas 500 pg TGF-β1/ml potentiated two-dimensional wound-induced migration (GAJDUSEK et al. 1993; M.S. PEPPER, unpublished observation). Although a limited number of studies have demonstrated a similar biphasic effect of TGF-β1 on vascular endothelial cell proliferation in subconfluent cultures (MYOKEN et al. 1990), most studies have demonstrated that TGF-β1 is inhibitory over a wide range of concentrations (FRÁTER-SCHRÖDER et al. 1986; BAIRD and DURKIN 1986; HEIMARK et al. 1986; MÜLLER et al. 1987; GAJDUSEK et al. 1993; IRUELA-ARISPE and SAGE 1993). Using a planar model of angiogenesis in postconfluent endothelial cell cultures, it has been demonstrated that TGF-β1 (0.5 pg–5.0 ng/ml) selectively stimulates proliferation of endothelial cells undergoing cord formation (IRUELA-ARISPE and SAGE 1993). These authors suggest that the specific phenotype of the cord-forming cells (including the composition of the extracellular matrix) distinguishes the way in which they respond to TGF-β1.

The mechanisms responsible for this biphasic effect are not known. One hypothesis is based on alterations in the net balance of extracellular proteolysis (PEPPER and MONTESANO 1990). Thus at the dose of TGF-β1 which potentiates bFGF- or VEGF-induced invasion, an optimal balance between proteases and protease inhibitors may be achieved at the cell surface, which allows for focal pericellular extracellular matrix degradation, while at the same time protecting the matrix against excessive degradation and inappropriate destruction (PEPPER et al. 1994a). An additional explanation might be related to our observation that $\alpha_5\beta_1$-integrin expression (protein and mRNA) is differentially affected at the potentiating and inhibitory concentrations of TGF-β1 (G. COLLO and M.S. PEPPER, in preparation). Thus basal levels of $\alpha_5 \beta_1$ are minimally increased by either bFGF or TGF-β1, wheras coaddition of the two cytokines results in a synergistic induction of this integrin. At a constant dose of bFGF, induction is greater with more prolonged kinetics at 5 ng than at 500 pg TGF-β1/ml. Since both in vivo and in vitro endothelial cells migrate on a fibronectin-rich extracellular matrix which they synthesize themselves (CLARK et al. 1982a,b; SARIOLA et al. 1984; RISAU and LEMMON 1988), we hypothesize that maximal invasion is dependent on an optimal degree of $\alpha_5\beta_1$-mediated cell adhesion to the matrix; submaximal invasion occurs when adhesion is either greater or less than that achieved with the potentiating dose of TGF-β1. The notion that 500 pg TGF-β1/ml stimulates adhesion to an extent which

◄

Fig. 5A–C. Biphasic effect of transforming growth factor (*TGF*)-β1 on vascular endothelial growth factor (*VEGF*) on basic fibroblast growth factor (*bFGF*)-induced in vitro angiogenesis. **A** Confluent monolayers of bovine microvascular endothelial cells were treated for 4 days with VEGF (100ng/ml) and/or TGF-β1, and the total additive length of all invading cell cords determined as described in Fig. 4. Median values from at least three experiments per condition are expressed relative to controls. **B,C** Cotreatment of confluent monolayers with bFGF and TGF-β1 at the indicated concentrations for 7 days. Each *point* represents the mean of three randomly selected photographic fields from a single experiment, and bFGF-treated cultures with (+) or without (–) coadded TGF-β1 in the same experiment are joined by a *single line*. bFGF- or VEGF-induced invasion was potentiated by TGF-β1 at 200–500 pg/ml and inhibited by TGF-β1 at 5–10 ng/ml. From PEPPER et al. Exp Cell Res 204: 356–363 (1993) (reproduced with permission from Academic press, Inc.)

is optimal for migration is consistent with the observation that 500 pg TGF-β1/ml potentiates whereas 5 ng TGF-β1/ml inhibits bFGF-stimulated wound-induced two-dimensional migration (GAJDUSEK et al. 1993; M.S. PEPPER, unpublished observation).

Despite rapid progress in identifying and characterizing cell surface receptors for TGF-β little is known about the repertoire and function of TGF-β R in endothelial cells, and virtually nothing is known about the role of TGF-β ligand–receptor interactions in the regulation of angiogenesis in vivo. Type I and type II receptors have been identified in cultured endothelial cells (SEGARINI et al. 1989; FAFEUR et al. 1990; MYOKEN et al. 1990; MERWIN et al. 1991a; HIRAI and KAJI 1992). The type II receptor is downregulated by bFGF, and unresponsiveness to the growth inhibitory effects of TGF-β1 has been correlated with a glycosylation defect in type II receptor in mutagenized endothelial cells (FAFEUR et al. 1990, 1993). With respect to expression in vivo, in situ hybridization studies have revealed type II receptor mRNA in epithelial, but not endothelial, cells both during development as well as in normal and psoriatic human skin (LAWLER et al. 1994; MATSUURA et al. 1994). In endothelial cells, the type III receptor is endoglin, a homodimeric transmembrane glycoprotein with subunits of 95 kDa (GOUGOS and LETARTE 1990) which binds TGF-β1 and -β3 with high affinity, but fails to bind TGF-β2 (CHEIFETZ et al. 1992). This may explain why endothelial cells respond equally to TGF-β1 and -β3, and poorly, if at all, to TGF-β2 (JENNINGS et al. 1988; CHEIFETZ et al. 1990; MERWIN et al. 1991a,b). Endoglin is constitutively expressed at high levels in the endothelium of capillaries and larger blood vessels in a wide variety of tissues; it is also abundantly expressed by syncitiotrophoblast and is present at lower levels on leukemic and normal bone marrow cells (ST.-JACQUES et al. 1994 and references therein). On the basis of chromosomal mapping and genetic linkage, endoglin has recently been assessed as a candidate gene for hereditary hemorrhagic telangectasia (HHT). HHT is an autosomal dominant disorder characterized by multisystemic vascular dysplasia and recurrent hemorrhage. The earliest detectable change in the telangiectatic lesions is dilatation of postcapillary venules in the upper dermis; endothelium and endothelial junctions appear to be normal. Mutations which would result in the synthesis of truncated proteins have been identified in the endoglin gene of patients with HHT (MCALLISTER et al. 1994). Since TGF-β induces the synthesis and assembly of the endothelial basement membrane, it is possible that defective TGF-βR signaling in endothelial cells of affected individuals may be responsible for the early dilatation of postcapillary venules through the formation of a structurally incompetent endothelial basement membrane.

To summarize the effects of TGF-β1 on the angiogenic response, it appears that the direct effect of TGF-β1 on endothelial cells not only varies at different stages of the angiogenic process, but is also concentration dependent. Thus, in addition to its indirect angiogenic effect, TGF-β1 could either promote or inhibit angiogenesis when interacting directly with endothelial cells, which in turn would depend on the context in which it is acting as well as on the repertoire of TGF-βR expressed by both endothelial and nonendothelial cells (Table 2).

Finally, although the indirect inflammatory cell-mediated hypothesis might be applicable during wound healing and tumor growth, blood vessel formation during development is unlikely to involve an inflammatory cell mediator.

7 Other Cytokines

As indicated earlier, many cytokines have been shown to modulate the angiogenic response in the experimental setting. We will not attempt to be exhaustive in our discussion of these factors. Rather, by focussing on our own findings we will discuss two families of cytokines, namely the interferons and those cytokines which signal via gp130. We do not wish to imply that these cytokines are more important than others, particularly since like many other factors, their precise role in the endogenous regulation of angiogenesis remains to be established.

7.1 Interferon-α_{2a}

The interferons (IFN), a heterogeneous family of pleiotropic cytokines that play a major role in host defense against viral and other infections, are comprised of two families, designated type I and type II. In humans, the type I family encompasses more than 20 highly related IFN-α and one IFN-β, all of which are encoded by separate intronless genes. Members of the type I family are single polypeptides composed of 165–170 amino acid residues with molecular weights of approximately 20–25 kDa. The type II family comprises only IFN-γ, which is structurally unrelated to members of the type I family. IFN-γ is a noncovalent head-to-tail homodimer composed of two basic 143 amino acid monomers, each with a molecular weight of 20–25 kDa. Type I IFN are constitutively produced by most cell types and are rapidly upregulated in response to viral infection. IFN-γ is produced by activated T lymphocytes and natural killer (NK) cells. Cellular

Table 2. Transforming growth factor (TGF)-β1 and angiogenesis

In vivo	In vitro
Indirect effect: Stimulates angiogenesis	Direct effect: Inhibits[a]/stimulates[b] proliferation Migration/invasion: biphasic
Direct effect?	Reduces extracellular proteolysis Promotes vessel maturation by promoting deposition and organization of a basement membrane and the maturation of intercellular junctional complexes

[a] 1–10 ng TGF-β1/ml.
[b] At pg/ml concentrations.

responses to the IFN require interaction of the ligand with high-affinity cell surface receptors. While all type I IFN appear to interact with the same (type I) receptor, IFN-γ binds to a separate (type II) molecule. Type I and II receptors are multisubunit structures, and in neither case does the IFN-binding component appear to possess a catalytic domain. Interaction with other components are therefore required for signaling (reviewed by VILCEK 1990; PELLEGRINI and SCHINDLER 1993; JOHNSON et al. 1994).

Interest in IFN in the regulation of angiogenesis stems from reports describing the use of IFN-α_{2a} in the treatment of childhood hemangiomatous diseases (WHITE et al. 1989,1991; ORCHARD et al. 1989; EZEKOWITZ et al. 1992; RICKETTS et al. 1994). As mentioned in the introduction, hemangiomas result essentially from uncontrolled angiogenesis, in which the extensive development of the microvascular bed far exceeds the metabolic requirements of the tissue concerned. Based on the observation that IFN-α_{2a} inhibits experimental angiogenesis in mice (SIDKEY and BORDEN 1987; MAJEWSKI et al. 1994), it has been proposed that the therapeutic effect of IFN-α is derived in part from its ability to inhibit new blood vessel formation. We used our three-dimensional in vitro angiogenesis assay to determine whether the antiangiogenic effect of IFN-α_{2a} might be mediated through a direct interaction with endothelial cells. We found that IFN-α_{2a} inhibited bFGF-induced microvascular endothelial cell invasion and tube formation with a mean effective dose (ED_{50}) of approximately 30 IU/ml (Fig. 6). We also observed that IFN-α_{2a} altered the proteolytic profile of endothelial cells by decreasing both basal and bFGF-induced levels of uPA and plasminogen activator inhibitor-1 (PAI-1) (PEPPER et al. 1994b). Paradoxically, however, we and others have found that interferon IFN-α_{2a} increases endothelial cell growth, which may be due to an

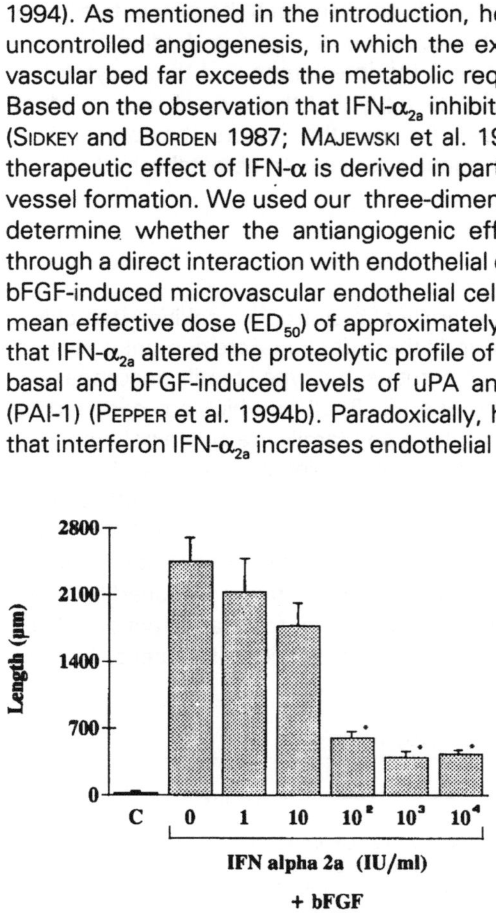

Fig. 6. Interferon (*IFN*)-α_{2a} inhibits basic fibroblast growth factor (bFGF)-induced in vitro angiogenesis. Confluent monolayers of bovine microvascular endothelial cells were treated with bFGF and IFN-α_{2a} at the indicated concentrations, and invasion was quantitated after 7 days as described in the legend to Fig. 4. Values are expressed as mean ±SEM from at least three experiments per condition. IFN α_{2a} decreased invasion in a dose-dependent manner, with a mean effective dose (ED_{50}) of approximately 30 IU/ml *$p<0.001$. IFN-α_{2a} on its own did not significantly affect basal levels of invasion (data not shown). From PEPPER et al. (1994) J Cell Biochem 55: 419–434 (reproduced with permission from Wiley-Liss, Inc.)

enhancement in endogenous bFGF synthesis and release (Cozzolino et al. 1993; Pepper et al. 1994b). These findings are in opposition to other reports in which IFN-α_{2a} has been shown to inhibit endothelial cell proliferation in the range of concentrations used in our studies (Heyns et al. 1985; Ruszczak et al. 1990). In contrast to its effect on proliferation, we observed that IFN-α_{2a} decreased two-dimensional wound-induced endothelial cell migration in a dose-dependent manner (Pepper et al. 1994b), which is consistent with the finding that human leukocyte IFN-α inhibits bovine microvascular endothelial cell migration (Brouty-Boyé and Zetter 1980).

Despite the above indications that IFN-α_{2a} might be an important negative regulator of angiogenesis, a number of questions still remain. First, virtually nothing is known about the expression or function of the type I IFN receptor in endothelial cells and its possible relation to the control of angiogenesis in vitro. Similarly, although the intracellular mechanisms which convey regulatory signals for gene regulation following IFN binding are relatively well described (reviewed by Pellegrini and Schindler 1993; Johnson et al. 1994), nothing is known in this context concerning endothelial cells. Receptors for IFN-γ, which also inhibits endothelial cell proliferation (Stolpen et al. 1986; Friesel et al. 1987), but which is not used in the treatment of hemangiomas, have been detected in endothelial cells of small- to medium-sized vessels in nonlymphoid organs as well as in hepatic sinusoidal and vascular endothelial cells (Valente et al. 1992; Volpes et al. 1991). With respect to hemangioma therapy, it will be important to determine whether IFN-α_{2a} interacts directly with endothelial cells in vivo or whether its therapeutic effect is mediated by the downregulation of direct-acting positive regulators such as VEGF or bFGF produced by stromal cells. Since VEGF and bFGF levels can be measured in body fluids and are elevated in other angiogenesis-associated diseases (Chodak et al. 1988; Nguyen et al. 1994; Yeo et al. 1993; Kondo et al. 1994; Ivanovic et al. 1995), it should be possible to determine whether serum or urinary levels of these regulators are altered during therapy with IFN-α_{2a}. Finally, since type I IFN are produced constitutively at low levels by most cell types, it will be of interest to determine whether a reduction in type I IFN production in the skin, i.e., the loss of an endogenous negative regulator, might be one of the factors that initiates hemangioma formation.

7.2 Leukemia-Inhibitory Factor, Oncostatin M, and Interleukin-6

Leukemia-inhibitory factor (LIF), oncostatin M (OSM), interleukin-6 (IL-6), interleukin-11 (IL-11), and ciliary neurotrophic factor (CNTF) are cytokines which share a common signal transduction pathway mediated via the gp130 signal converter (reviewed by Stahl and Yancopoulos 1993; Taga and Kishimoto 1993). LIF is a highly glycosylated, 180-amino acid single-chain polypeptide varying in molecular weight from 38 to 67 kDa. Although LIF was initially purified and cloned using a bioassay based on its ability to induce monocyte differentiation, it has a multitude of effects on both hemopoietic and nonhemopoietic cells and like OSM, IL-6, and IL-11, is one of a growing number of cytokines which are characterized by pleiotropy and functional redundancy (reviewed by HILTON 1992; ALEXANDER et al. 1994).

Based on the observation that LIF inhibits aortic endothelial cell proliferation in vitro (FERRARA et al. 1992b), we have assessed LIF's potential as an angiogenesis regulator using our three-dimensional in vitro model. LIF was found to inhibit invasion with a 50% inhibitory concentration (IC_{50}) of less than 1 ng/ml and to lack the concentration-dependent stimulatory effect characteristic of TGF-β1 (Fig. 7). The inhibitory effect was observed on both microvascular and aortic endothelial cells and occurred irrespective of the angiogenic stimulus, which included bFGF, VEGF, or the synergistic effect of the two factors in combination (PEPPER et al. 1995). Inhibition of invasion could be correlated with inhibition of proliferation of both microvascular and aortic endothelial cells, although this was more marked with the latter, confirming previous observations (FERRARA et al. 1992b). In addition, LIF decreased the proteolytic potential of both microvascular and aortic endothelial cells by increasing their expression of PAI-1 (PEPPER et al. 1995).

Since LIF shares a common signal transduction pathway with OSM, IL-6, and CNTF, we have assessed the effects of these cytokines in our in vitro invasion model. We found that, like LIF, both OSM and IL-6 inhibit angiogenesis, and at equimolar concentrations OSM is slightly more potent than LIF, while IL-6 is significantly less potent than either OSM or LIF; CNTF had no effect. As for LIF, the inhibitory effects of OSM and IL-6 were observed on both microvascular and aortic endothelial cells and occurred irrespective of the angiogenic stimulus (M.S. PEPPER et al., manuscript in preparation). In addition, both OSM and IL-6 inhibit aortic and microvascular endothelial cell proliferation at concentrations which are comparable to those which inhibit angiogenesis in vitro. The inhibitory effects of OSM (T.J. BROWN et al. 1990; MILES et al. 1992) and IL-6 (MAY et al. 1989; ROSEN et al. 1991) on endothelial cell proliferation have previously been reported.

Although nothing is known about the role of LIF and OSM in the regulation of angiogenesis in vivo, in situ hybridization studies have demonstrated that IL-6 expression correlates with phases of angiogenesis in the murine ovary (ROSEN et al. 1991). However, when the effect of IL-6 is assessed on bFGF-induced angiogenesis in vivo, it is either inhibitory (PASSANITI et al. 1992) or has no effect (HU et al. 1993), and when transfected into human melanoma cells, IL-6 reduces their tumorigenicity, possibly through a reduction is new vessel formation (SUN et al. 1992).

With arguments both for and against a role for IL-6 in the regulation of angiogenesis in vivo, it is not easy to reconcile existing in vivo observations with our in vitro findings. However, as has been extensively alluded to in this review, it is becoming increasingly apparent that with the existence of complex cytokine networks in vivo, the nature of the endothelial cell response elicited by a specific cytokine is contextual, i.e., depends on the presence of other regulatory molecules including other cytokines and extracellular matrix components in the pericellular environment. The notion, taken together with our finding that LIF, IL-6, and OSM, all of which utilize a common signal converter (gp130),

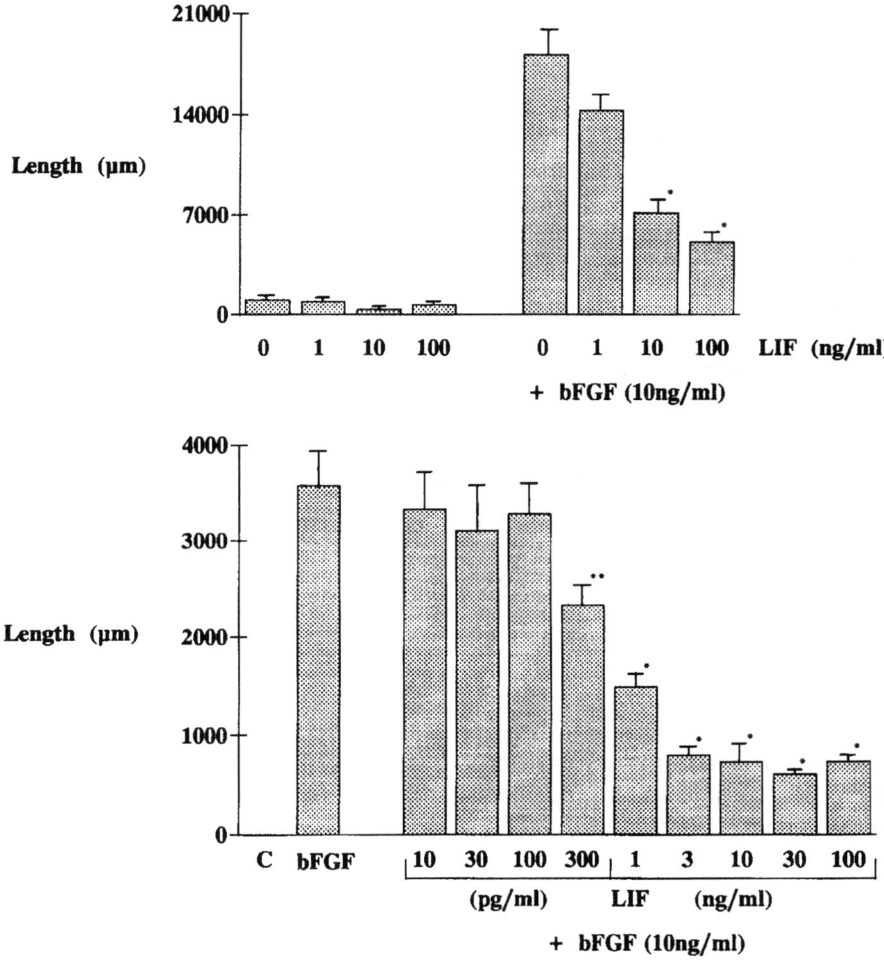

Fig. 7. Leukemia-inhibitory factor (LIF) inhibits basic fibroblast factor (bFGF)-induced in vitro angiogenesis. Confluent monolayers of bovine aortic (BAE; *top*) or microvascular (BME; *bottom*) endothelial cells were treated with bFGF and LIF at the indicated concentrations, and invasion was quantitated after 4 (BAE) or 7 (BME) days, as described in the legend to Fig. 4. Values are expressed as mean ± SEM from at least three experiments per condition. LIF decreased invasion in a dose-dependent manner, with a mean effective dose (ED_{50}) of less than 1 ng/ml in BME cells. **p* < 0.001; ***p* < 0.005. On its own LIF did not significantly affect basal levels of invasion. From PEPPER et al. (1995) J Cell Sci 108: 73–83 (reproduced with permission from The Company of Biologists Ltd.)

inhibit angiogenesis in vivo, suggests that if this signal transduction pathway is relevant to the control of angiogenesis in vivo, its negative regulatory effect may serve to temper the activity of high local concentrations of positively acting angiogenic cytokines and may thus serve to prevent uncontrolled angiogenesis. This may be particularly relevant in physiological settings such as a wound healing and inflammation (M.A. BROWN et al. 1994).

8 Conclusion and Perspectives

In the experimental setting, a large number of factors have been shown to regulate the angiogenic response, including polypeptide growth factors or cytokines. Although many of these factors have been implicated in the endogenous regulation of angiogenesis, all (with the possible exception of one, namely VEGF) still require definitive studies to demonstrate their relevance to this process.

A number of important questions also remain concerning VEGF:

1. What is the meaning of high VEGF expression in tissues in which there is no angiogenesis?
2. Why is VEGF not modulated in certain tissues in which there is extensive angiogenesis?
3. If VEGF-dependent angiogenesis is indeed contextual, what are the factors which modulate its activity?

With respect to other cytokines which have been reported to regulate angiogenesis in the experimental setting, it should be pointed out that, although many of them may in fact not be true endogenous regulators of angiogenesis, this should not detract from their potential importance as therapeutic agents.

Using a three-dimensional model of in vitro angiogenesis, we have demonstrated that important interactions exist between different cytokines. Thus synergism was observed between bFGF and VEGF, TGF-β1 had a biphasic effect, and IFN-α_{2a} and LIF inhibited in vitro angiogenesis. Based on these observations, and assuming that the factors we have studied are indeed relevant to the endogenous control of angiogenesis, we suggest that the temporally coordinated and concentration-dependent activity of a number of cytokines is necessary for the control of different phases of the angiogenic process in specific and appropriate settings. The identification of cytokines sharing a common signal transduction pathway, which inhibit angiogenesis by acting directly on endothelial cells and which lack paradoxical in vivo and in vitro effects, might facilitate our understanding of the physiological factors which regulate this process. Further investigation into the "contextual activity" of angiogenic cytokines may also allow us to uncover additional examples of synergistic interactions.

Inhibition and stimulation of new capillary blood vessel formation have been envisaged as alternatives/adjuncts in the management of angiogenesis-associated diseases. Although angiogenesis is required for the continued growth of solid tumors, under physiological conditions it is only required for wound healing and reproductive function, which suggests that inhibition of angiogenesis should be well tolerated by most adults. Inhibition of angiogenesis is also of potential benefit in the treatment of ocular neovascularization (e.g., diabetic proliferative retinopathy) and of life-threatening childhood hemangiomas. Knowledge of the factors involved in the regulation of angiogenesis, as well as their potential interactions, is likely to provide novel therapeutic targets in these settings. Stimulation of angiogenesis, on the other hand, might accelerate wound healing and promote growth of collateral vessels in ischemic tissues. Although ischemia resulting from arterial obstruction normally stimulates the development of collateral vessels, in the face of persistent ischemia this is often insufficient to maintain normal tissue perfusion and may result in myocardial infarction, stroke, and gangrene of the extremities. Considering that currently no effective drug therapy exists for many patients with unstable angina or critical leg ischemia, stimulation of angiogenesis in these settings using positive regulators of angiogenesis (including cytokines), either alone or in combination, has important therapeutic potential.

Acknowledgments. We would like to thank our colleagues D. Belin, N. Ferrara, and L. Schweigerer for their valuable contributions to our work and C. Di Sanza, M. Guisolan, M. Quayzin, J. Rial-Robert, and P.-A. Ruttimann for excellent technical assistance. Our own work reviewed in this article was supported by grants from the Swiss National Science Foundation and grants-in-aid from the Juvenile Diabetes Foundation (International) and the Sir Jules Thorn Charitable Overseas Trust.

References

Alexander HR, Billingsley KG, Block MI, Fraker DL (1994) D-factor/leukemia inhibitory factor: evidence for its role as a mediator in acute and chronic inflammatory disease. Cytokine 6: 589–596

Andrade SP, Fan T-PD, Lewis GP (1987) Quantitative in vivo studies on angiogenesis in a rat sponge model. Br J Exp Pathol 68: 755–766

Asahara T, Bauters C, Zheng LP, Takeshita S, Bunting S, Ferrara N, Symes JF, Isner JM (1994) In vivo synergistic effects of vascular endothelial growth factor and basic fibroblast growth factor on angiogenesis in rabbit ischemic hindlimb. Circulation 4: I–585

Attisano L, Wrana JL, López-Casillas F, Massagué J (1994) TGF-β receptors and actions. Biochim Biophys Acta 1222: 71–80

Ausprunk DH, Folkman J (1977) Migration and proliferation of endothelial cells in preformed and newly formed blood vessels during angiogenesis. Microvasc Res 14: 52–65

Baird A, Durkin T (1986) Inhibition of endothelial cell proliferation by type β-transforming growth factor: interactions with acidic and basic fibroblast growth factors. Biochem Biophys Res Commun 138: 476–482

Baird A, Klagsbrun M(1991) The fibroblast growth factor family. Cancer Cells 3: 239–243

Barleon B, Hauser S, Schöllmann C, Weindel K, Marmé D, Yayon A, Weich HA (1994) Differential expression of the two VEGF receptors flt and KDR in placenta and vascular endothelial cells. J Cell Biochem 54: 56–66

Basilico C, Moscatelli D (1992) The FGF family of growth factors and oncogenes. Adv Cancer Res 59: 115–165

Bouck N (1990) Tumor angiogenesis: the role of oncogenes and tumor suppressor genes. Cancer Cells 2: 179–185

Brem S, Tsanaclis AMC, Gately S, Gross JL, Herblin WF (1992) Immunolocalization of basic fibroblast growth factor to the microvasculture of human brain tumors. Cancer 70: 2673–2680

Brogi E, Winkles JA, Underwood R, Clinton SK, Alberts GF, Libby P (1993) Distinct patterns of expression of fibroblast growth factors and their receptors in human atheroma and nonatherosclerotic arteries. J Clin Invest 92: 2408–2418

Brouty-Boyé D, Zetter BR (1980) Inhibition of cell motility by interferon. Science 208: 516–518

Brown MA, Metcalf D, Gough NM (1994) Leukemia inhibitory factor and interleukin 6 are expressed at very low levels in the normal adult mouse and are induced by inflammation. Cytokine 6: 300–309

Brown TJ, Rowe JM, Shoyab M, Gladstone P (1990) Oncostatin M: a novel regulator of endothelial cell properties. In: Roberts R, Schneider D (eds) Molecular biology of the cardiovascular system. UCLA Symposia on molecular and cellular biology, New Series, vol 131, New York, pp 195–206

Cheifetz S, Hernandez H, Laiho M, ten Dijke P, Iwata KK, Massagué J (1990) Distinct transforming growth factor-β (TGF-β) receptor subsets as determinants of cellular responsiveness to three TGF-β isoforms. J Biol Chem 265: 20533–20538
Cheifetz S, Bellóin T, Calés C, Vera S, Bernabeu C, Massagué J, Letarte M (1992) Endoglin is a component of the transforming growth factor-β receptor system in human endothelial cells. J Biol Chem 267: 19027–19030
Chodak GW, Hospelhorn V, Judge SM, Mayforth R, Koeppen H, Sasse J (1988) Increased levels of fibroblast growth factor-like activity in urine from patients with bladder or kidney cancer. Cancer Res 48: 2083–2088
Christofori G, Hanahan D (1994) Molecular dissection of multi-stage tumorigenesis in transgenic mice. Semin Cancer Biol 5: 3–12
Clark RAF, DellaPelle P, Manseau E, Lanigan JM, Dvorak HF, Colvin RB (1982a) Blood vessel fibronectin increases in conjunction with endothelial cell proliferation and capillary ingrowth during wound healing. J Invest Dermatol 79: 269–276
Clark RAF, Quinn JF, Winn HJ, Lanigan JM, Dellepella P, Colvin RB (1982b) Fibronectin is produced by blood vessels in response to injury. J Exp Med 156: 646–651
Cordon-Cardo C, Vlodavsky I, Haimovitz-Friedman A, Hicklin D, Fuks Z (1990) Expression of basic fibroblast growth factor in normal human tissues. Lab Invest 63: 823–840
Cozzolino F, Torcia M, Lucibello M, Ziche M, Platt J, Fabiani S, Brett J, Stern D (1993) Interferon-α and interleukin 2 synergistically enhance basic fibroblast growth factor synthesis and induce release, promoting endothelial cell growth. J Clin Invest 91: 2504–2512
Cullinan-Bove K, Koos RD (1993) Vascular endothelial growth factor/vascular permeability factor expression in the rat uterus: rapid stimulation by estrogen correlates with estrogen-induced increases in uterine capillary permeability and growth. Endocrinology 133: 829–837
Deng C-X, Wynshaw-Boris A, Shen MM, Daugherty C, Ornitz DM, Leder P (1994) Murine FGFR-1 is required for early postimplantion growth and axial organization. Genes Dev 8: 3045–3057
Dennis PA, Rifkin DB (1990) Studies on the role of basic fibroblast growth factor in vivo: inability of neutralizing antibodies to block tumor growth. J Cell Physiol 144: 84–98
Derynck R (1994) TGF-β-receptor mediated signaling. Trends Biochem Sci 19: 548–553
Dolecki GJ, Connolly DT (1991) Effects of a variety of cytokines and inducing agents on vascular permeability factor mRNA levels in U937 cells. Biochem Biophy Res Commun 180: 572–578
Eisemann A, Ahn JA, Graziani G, Tronick SR, Ron D (1991) Alternative splicing generates at least five different isoforms of the human basic-FGF receptor. Oncogene 6: 1195–1202
Engerman RL, Pfaffenbach D, Davis MD (1967) Cell turnover of capillaries. Lab Invest 17: 738–743
Evans HM (1909) On the development of the aortae, cardinal and umbilical veins, and the other blood vessels of vertebrate embryos from capillaries. Anat Rec 3: 498–518
Ezekowitz RAB, Mulliken JB, Folkman J (1992) Interferon alfa-2a therapy for life-threatening hemangiomas of infancy. N Engl J Med 326: 1456–1463
Fafeur V, Terman BI, Blum J, Böhlen P (1990) Basic FGF treatment of endothelial cells down-regulates the 85-KDa TGFβ receptor subtype and decreases the growth inhibitory response to TGF-β1. Growth Factors 3: 237–245
Fafeur V, O'Hara B, Böhlen P (1993) A glycosylation-deficient endothelial cell mutant with modified responses to transforming growth factor-β and other growth inhibitory cytokines: evidence for multiple growth inhibitory signal transduction pathways. Mol Biol Cell 4: 135–144
Fernig DG, Gallagher JT (1994) Fibroblast growth factors and their receptors: an information network controlling tissue growth, morphogenesis and repair. Prog Growth Factor Res 5: 353–377
Ferrara N, Houck K, Jackeman L, Leung DW (1992a) Molecular and biological properties of the vascular endothelial growth factor family of proteins. Endocrine Rev 13: 18–35
Ferrara N, Winer J, Henzel WJ (1992b) Pituitary follicular cells secrete an inhibitor of aortic endothelial cell growth: identification as leukemia inhibitory factor. Proc Natl Acad Sci USA 89: 698–702
Flaumenhaft R, Abe M, Mignatti P, Rifkin DB (1992) Basic fibroblast growth factor-induced activation of latent transforming growth factor beta in endothelial cells: regulation of plasminogen activator activity. J Cell Biol 118: 901–909
Folkman J (1974) Tumor angiogenesis. Adv Cancer Res 19: 331–358
Folkman J, Haudenschild C (1980) Angiogenesis in vitro. Nature 288: 551–555
Folkman J, Klagsbrun M (1987) Angiogenic factors. Science 235: 442–447
Fong G-H, Rossant J, Gertsenstein M, Breitman ML (1995) Role of the Flt-1 receptor tyrosine kinase in regulating the assembly of vascular endothelium. Nature 376: 66–70
Fráter-Schröder M, Müller G, Birchmeier W, Böhlen P (1986) Transforming growth factor-beta inhibits endothelial cell proliferation. Biochem Biophys Res Commun 137: 295–302

Friesel R, Komoriya A, Maciag T (1987) Inhibition of endothelial cell proliferation by gamma-interferon. J Cell Biol 104: 689–696

Gajdusek CM, Luo Z, Mayberg MR (1993) Basic fibroblast growth factor and transforming growth factor beta-1: synergistic mediators of angiogenesis in vitro. J Cell Physiol 157: 133–144

Gimbrone MA Jr, Cotran RS, Leapman SB, Folkman J (1974) Tumor growth and neovascularization: an experimental model using the rabbit cornea. J Natl Cancer Inst 52: 413–427

Givol D, Yayon A (1992) Complexity of FGF receptors: genetic basis for structural diversity and functional specificity. FASEB J 6: 3362–3369

Gonzalez A-M, Buscaglia M, Ong M, Baird A (1990) Distribution of basic fibroflast growth factor in the 18-day rat fetus: localization in the basement membranes of diverse tissues. J Cell Biol 110: 753–765

Goto F, Goto K, Weindel K, Folkman J (1993) Synergistic effects of vascular endothelial growth factor and basic fibroblast growth factor on the proliferation and cord formation of bovine capillary endothelial cells within collagen gels. Lab Invest 69: 508–517

Gougos A, Letarte M (1990) Primary structure of endoglin, an RGD-containing glycoprotein of human endothelial cells. J Biol Chem 265: 8361–8364

Hannan RL, Kourembanas S, Flanders KC, Rogelj SJ, Roberts AB, Faller DV, Klagsbrun M (1988) Endothelial cells synthesize basic fibroblast growth factor and transforming growth factor beta. Growth Factors 1: 7–17

Hannekan A, Lutty GA, McLeod DS, Robey F, Harvey AK, Hjelmeland LM (1989) Localization of basic fibroblast growth factor to the developing capillaries of the bovine retina. J Cell Physiol 138: 115–120

Hanneken A, Maher PA, Baird A (1995) High affinity immunoreactive FGF receptors in the extracellular matrix of vascular endothelial cells – implications for the modulation of FGF-2. J Cell Biol 128: 1221–1228

Harada S, Nagy JA, Sullivan KA, Thomas KA, Endo N, Rodan GA, Rodan SB (1994) Induction of vascular endothelial growth factor expression by prostaglandin E2 and E1 in osteoclasts. J Clin Invest 93: 2490–2496

Heimark RL, Twardzik DR, Schwartz SM (1986) Inhibition of endothelial regeneration by type-beta transforming growth factor from platelets. Science 233: 1078–1080

Heyns AP, Eldor A, Vlodavsky I, Kaiser N, Fridman R, Panet A (1985) The antiproliferative effect of interferon and the mitogenic activity of growth factors are independent cell cycle events. J Invest Dermatol 161: 297–306

Hilton DJ (1992) LIF: lots of interesting functions. Trends Biochem Sci 17: 72–76

Hirai R, Kaji K (1992) Transforming growth factor β1-specific binding proteins on human vascular endothelial cells. Exp Cell Res 201: 119–125

Hobson B, Denekamp J (1984) Endothelial proliferation in tumors and normal tissues: continuous labelling studies. Br J Cancer 49: 405–413

Holmgren L, O'Reilly MS, Folkman J (1995) Dormancy of micrometastases: balanced proliferation and apoptosis in the presence of angiogenesis supression. Nature Med 1: 149–153

Hori A, Sasasda R, Matsutani E, Naito K, Sakura Y, Fujita T, Kozai Y (1991) Suppression of solid tumor growth by immunoneutralizing monoclonal antibody against human basic fibroblast growth factor. Cancer Res 51: 6180–6184

Hockel M, Schlenger K, Doctrow S, Kissel T, Vaupel P (1993) Therapeutic angiogenesis. Arch Surg 128: 423–429

Hu DE, Fan T-PD (1995) Suppression of VEGF-induced angiogenesis by the protein tyrosine kinase inhibitor, lavendustin A. Br J Pharmacol 114: 262–268

Hu DE, Hori Y, Fan T-PD (1993) Interleukin-8 stimulates angiogenesis in rats. Inflammation 17: 135–143

Hughes SE, Hall PA (1993) Immunolocalization of fibroblast growth factor receptor 1 and its ligands in human tissues. Lab Invest 69: 173–182

Hughes SE, Crossman D, Hall PA (1993) Expression of basic and acidic fibroblast growth factors and their receptor in normal and atherosclerotic human arteries. Cardiovasc Res 27: 1214–1219

Iruela-Arispe ML, Hasselaar P, Sage H (1991) Differential expression of extracellular proteins is correlated with angiogenesis in vitro. Lab Invest 64: 174–186

Iruela-Aripse ML, Sage EH (1993) Endothelial cells exhibiting angiogenesis in vitro proliferate in response to TGF-β1. J Cell Biochem 52: 414–430

Ivanovic V, Melman A, Davis-Joseph B, Valcic M, Geliebter J (1995) Elevated levels of TGF-β1 in patients with invasive prostate cancer. Nature Med 1: 282–284

Jaye M, Schlessinger J, Dionne CA (1992) Fibroblast growth factor receptor tyrosine kinases: molecular analysis and signal transduction. Biochim Biophys Acta 1135: 185–199

Jennings JC, Mohan S, Linkhart TA, Widstrom TA, Baylink DJ (1988) Comparison of the biological actions of TGF beta-1 and TGF beta-2: differential activity in endothelial cells. J Cell Physiol 137: 167–172
Johnson DE, Williams LT (1993) Structural and functional diversity in the FGF receptor multigene family. Adv Cancer Res 60: 1–41
Johnson HM, Bazer FW, Szente BE, Jarpe MA (1994) How interferons fight disease. Sci Am 270: 40–47
Kim KJ, Li B, Winer J, Armanini M, Gillet N, Phillips HS, Ferrara N (1993) Inhibition of vascular endothelial growth factor induced angiogenesis suppresses tumor growth in vivo. Nature 362: 841–844
Kingsley DM (1944) The TGF-β superfamily: new members, new receptors, and new genetic tests of function in different organisms. Genes Dev 8: 133–146
Klagsbrun M, D'Amore PA (1991) Regulators of angiogenesis. Annu Rev Physiol 53: 217–239
Klagsbrun M, Knighton D, Folkman J (1976) Tumor angiogenesis activity in cells grown in tissue culture. Cancer Res 36: 110–114
Koh GY, Kim S-J, Klug MG, Park K, Soonpaa MH, Field LJ (1995) Targeted expression of transforming growth factor-β1 in intracardiac grafts promotes vascular endothelial cell DNA synthesis. J Clin Invest 95: 114–121
Kondo S, Asano M, Suzuki H (1993) Significance of vascular endothelial growth factor/vascular permeability factor for solid tumor growth, and its inhibition by the antibody. Biochem Biophys Res Commun 194: 1234–1241
Kondo S, Asano M, Matsuo K, Ohmori I, Suzuki H (1994) Vascular endothelial growth factor/vascular permeability factor is detectable in the sera of tumor-bearing mice and cancer patients. Biochim Biophys Acta 1221: 221–214
Krnacik-Coleman S, Rosen JM (1994) Differential temporal and spatial gene expression of fibroblast growth factor family members during mouse mammary gland development. Mol Endocrinol 8: 218–229
Kubota Y, Kleinman HK, Martin GR, Lawley TJ (1988) Role of laminin and basement membrane in the morphological differentiation of human endothelial cells into capillary-like structures. J Cell Biol 107: 1589–1598
Kulkarni AB, Huh C-G, Becker D, Geiser A, Lyght M, Flanders KC, Roberts AB, Sporn MB, Ward JM, Karlsson S (1993) Transforming growth factor β1 null mutation in mice causes excessive inflammatory response and early death. Proc Natl Acad Sci USA 90: 770–774
Lawler S, Candia AF, Ebner R, Shum L, Lopez AR, Moses HL, Wright CVE, Derynck R (1994) The murine type II TGF-β receptor has a coincident embryonic expression and binding preference for TGF-β1. Development 120: 165–175
Leek RD, Harris AL, Lewis CE (1994) Cytokine networks in solid human tumors: regulation of angiogenesis. J Leukocyte Biol 56: 423–435
Letterio JJ, Geiser AG, Kulkarni AB, Roche NS, Sporn MB, Roberts AB (1994) Maternal rescue of transforming growth factor-β1 null mice. Science 264: 1936–1938
Liaw L, Schwartz SM (1993) Comparison of gene expression in bovine aortic endothelium in vivo versus in vitro. Differences in growth regulatory molecules. Arterioscler Thromb 13: 985–993
Linder V, Reidy MA (1993) Expression of basic fibroblast growth factor and its receptor by smooth muscle cells and endothelium in injured rat arteries. An en face study. Circ Res 73: 589–595
Liotta LA, Steeg PS, Stetler-Stevenson WG (1991) Cancer metastasis and angiogenesis: an imbalance of positive and negative regulation. Cell 64: 327–366
Madri JA, Pratt BM, Tucker AM (1988) Phenotypic modulation of endothelial cells by transforming growth factor-β depends on the composition and organization of the extracellular matrix. J Cell Biol 106: 1357–1384
Majewski S, Szmurlo NA, Marczak M, Jablonska S, Bollag W (1994) Synergistic effect of retinoids and interferon α on tumor-induced angiogenesis: anti-angiogenic effect on HPV-harboring tumor cell lines. Int J Cancer 57: 81–85
Mandriota SJ, Seghezzi G, Vassalli J-D, Ferrara N, Wasi S, Mazzieri R, Mignatti P, Pepper MS (1995) Vascular endothelial growth factor increases urokinase receptor expression in vascular endothelial cells. J Biol Chem 270: 9709–9716
Matsuura H, Myokai F, Arata J, Noji S, Taniguchi S (1994) Expression of type II transforming growth factor-beta receptor mRNA in human skin, as revealed by in stiu hybridization. J Dermatol Sci 8: 25–32
Matsuzaki K, Yoshitake Y, Matuo Y, Sasaki H, Nishikawa K (1989) Monoclonal antibodies against heparin-binding growth factor II/basic fibroblast growth factor that block its biological activity: invalidity of the antibodies for tumor angiogenesis. Proc Natl Acad Sci USA 86: 9911–9915

May LT, Torcia G, Cozzolino F, Ray A, Tatter SB, Santhanam U, Sehgal PB, Stern D (1989) Interleukin-6 gene expression in human endothelial cells: RNA start sites, multiple IL-6 proteins and inhibition of proliferation. Biochem Biophys Res Commun 159: 991–998

McAllister KA, Grogg KM, Johnson DW, Gallione CJ, Baldwin MA, Jackson CE, Helmbold EA, Markel DS, McKinnon WC, Murrell J, McCormick MK, Pericak-Vance MA, Heutink P, Oostra BA, Haitjema T, Westerman CJJ, Porteous ME, Guttmacher AE, Letarte M, Marchuk DA (1994) Endoglin, a TGF-β binding protein of endothelial cells, is the gene for hereditary haemorrhagic telangiectasia type 1. Nature Genetics 8: 345–351

McCartney-Francis N, Mizel D, Wong H, Wahl L, Wahl S (1990) TGF-β regulates production of growth factors and TGF-β by human peripheral blood monocytes. Growth Factors 4: 27–35

Merwin JR, Anderson JM, Kocher O, van Itallie CM, Madri JA (1990) Transforming growth factor-β1 modulates extracellular matrix organization and cell-cell junctional complex formation during in vitro angiogenesis. J Cell Physiol 142: 117–128

Merwin JR, Newman W, Beall LD, Tucker A, Madri J (1991a) Vascular cells respond differentially to transforming growth factors beta 1 and beta 2 in vitro. Am J Pathol 138: 37–51

Merwin JR, Roberts A, Kondaiah P, Tucker A, Madri J (1991b) Vascular cell responses to TGF–β3 mimick those of TGF-β1 in vitro. Growth Factors 5: 149–158

Mignatti P, Rifkin DB (1991) Release of basic fibroblast growth factor, an angiogenic factor devoid of secretory signal sequence; a trivial phenomenon or a novel secretion mechanism? J Cell Biochem 47: 201–207

Mignatti P, Tsuboi R, Robbins E, Rifkin DB (1989) In vitro angiogenesis on the human amniotic membrane: requirements for basic fibroblast growth factor-induced proteases. J Cell Biol 108: 671–682

Miki T, Bottaro DP, Fleming TP, Smith CL, Burgess WH, Chan AM-L, Aaronson SA (1992) Determination of ligand-binding specificity by alternative splicing: two distinct growth factor receptors enconded by a single gene. Proc Natl Acad Sci USA 89: 246–250

Miles SA, Martinez-Maza O, Rezai A, Magpantay L, Kishimoto T, Nakamura S, Radka SF, Linsley PS (1992) Oncostatin M as a potent mitogen for AIDS-Kaposi's sarcoma-derived cells. Science 255: 1432–1434

Millauer B, Shawver LK, Plate KH, Risau W, Ullrich A (1994) Glioblastoma growth inhibited in vivo by a dominant negative Flk-1 mutant. Nature 367: 576–579

Montesano R, Orci L (1985) Tumor-promoting phorbol esters induce angiogenesis in vitro. Cell 42: 469–477

Montesano R, Orci L, Vassalli P (1983) In vitro rapid organization of endothelial cells into capillary-like networks is promoted by collagen matrices. J Cell Biol 97: 1648–1652

Montesano R, Vassalli J-D, Baird A, Guillemin R, Orci L (1986) Basic fibroblast growth factor induces angiogenesis in vitro. Proc Natl Acad Sci USA 83: 7297–7301

Montesano R, Pepper MS, Vassalli J-D, Orci L (1987) Phorbol ester induces cultured endothelial cells to invade a fibrin matrix in the presence of fibrinolytic inhibitors. J Cell Physiol 132: 509–516

Morrison RS, Yamaguchi F, Bruner JM, Tang M, McKeehan W, Berger MS (1994) Fibroblast growth factor receptor gene expression and immunoreactivity are elevated in human gliblastoma multiforme. Cancer Res 54: 2794–2799

Moscatelli D, Presta M, Joseph-Silverstein J, Rifkin DB (1986a) Both normal and tumor cells produce basic fibroblast growth factor. J Cell Physiol 129: 273–276

Moscatelli D, Presta M, Rifkin DB (1986b) Purification of a factor from human placenta that stimulates capillary endothelial cell protease production, DNA synthesis and migration. Proc Natl Acad Sci USA 83: 2091–2095

Müller G, Behrens J, Nussbaumer U, Böhlen P, Birchmeier W (1987) Inhibitory action of transforming growth factor-β on endothelial cells. Proc Natl Acad Sci USA 84: 5600–5604

Mustonen T, Alitalo K (1995) Endothelial receptor tyrosine kinases involved in angiogenesis. J Cell Biol 129: 895–898

Myoken Y, Kan M, Sato GH, McKeehan WL, Sato D (1990) Bifunctional effects of transforming growth factor-β (TGF-β) on endothelial cell growth correlate with phenotypes of TGF-β binding sites. Exp Cell Res 191: 299–304

Nabel EG, Shum L, Pompili VJ,Yang Z-Y, Shu HB, Liptay S, Gold L, Gordon D, Derynck R, Nabel GJ (1993) Direct gene transfer of transforming growth factor β1 into arteries stimulates fibrocellular hyperplasia. Proc Natl Acad Sci USA 90: 10759–10763

Nathan C, Sporn M (1991) Cytokines in context. J Cell Biol 113: 981–986

Neufeld G, Tessler S, Gitay-Goren H, Cohen T, Levi BZ (1994) Vascular endothelial growth factor and its receptors. Prog Growth Factor Res 5: 89–97

Nguyen M, Watanabe H, Budson AE, Richie JP, Hayes DF, Folkman J (1994) Elevated levels of an angiogenic peptide, basic fibroblast growth factor, in the urine of patients with a wide spectrum of cancers. J Natl Cancer Inst 86: 356–361

Nicosia RF, Ottinetti A (1990) Growth of microvessels in serum-free matrix culture of rat aorta. Lab Invest 63: 115–122

Ohtani H, Nakamura S, Watanabe Y, Mizoi T, Saku T, Nagura H (1993) Immunocytochemical localization of basic fibroblast growth factor in carcinomas and inflammatory lesions of the human digestive tract. Lab Invest 68: 520–527

Orchard PJ, Smith CM, Woods WG, Day DL, Dehner LP, Shapiro R (1989) Treatment of hemangioendotheliomas with alpha interferon. Lancet 2: 565–567

O'Reilly MS, Holmgren L, Shing Y, Chen C, Rosenthal R, Moses M, Lane WS, Cao Y, Sage EH, Folkman J (1994) Angiostatin: a novel angiogenesis inhibitor that mediates the suppression of metasases by a Lewis lung carinoma. Cell 79: 315–328

Paku S, Paweletz N (1991) First steps of tumor-related angiogenesis. Lab Invest 65: 334–346

Pardenaud L, Yassine F, Dieterlen-Lièvre F (1989) Relationship between vasculogenesis, angiogenesis and haemopoiesis during avian ontogeny. Development 105: 437–485

Partanen J, Vainikka S, Korhonen J, Armstrong E, Alitalo K (1992) Diverse receptors for fibroblast growth factors. Prog Growth Factor Res 4: 69–83

Passaniti A, Taylor RM, Pili R, Guo Y, Long PV, Haney JA, Pauly RR, Grant DS, Martin GR (1992) A simple, quantitative method for assessing angiogenesis and antiangiogenic agents using reconstituted basement membrane, heparin, and fibroblast growth factor. Lab Invest 67: 519–528

Pellegrini S, Schindler C (1993) Early events in signaling by interferons. Trends Biochem Sci 18: 338–342

Pepper MS, Montesano R (1990) Proteolytic balance and capillary morphogenesis. Cell Diff Dev 32: 319–328

Pepper MS, Belin D, Montesano R, Orci L, Vassalli J-D (1990) Transforming growth factor beta1 modulates basic fibroblast growth factor-induced proteolytic and angiogenic properties of endothelial cells in vitro. J Cell Biol 111: 743–755

Pepper MS, Montesano R, Vassalli J-D, Orci L (1991) Chondrocytes inhibit endothelial sprout formation in vitro: evidence for the involvement of a transforming growth factor-beta. J Cell Physiol 146: 170–179

Pepper MS, Ferrara N, Orci L, Montesano R (1992) Potent synergism between vascular endothelial growth factor and basic fibroblast growth factor in the induction of angiogenesis in vitro. Biochem Biophys Res Commun 189: 824–831

Pepper MS, Vassalli J-D, Orci L, Montesano R (1993) Biphasic effect of transforming growth factor-beta 1 on in vitro angiogenesis. Exp Cell Res 204: 356–363

Pepper MS, Vassalli J-D, Orci L, Montesano R (1994a) Angiogenesis in vitro: cytokine interactions and balanced extracellular proteolysis. In: Maragoudakis ME, Gullino PM, Lelkes PI (eds) Angiogenesis. Molecular biology, clinical aspects. Plenum, New York, pp 149–170 (NATO ASI series A: life sciences, vol 263)

Pepper MS, Vassalli J-D, Wilks JW, Schweigerer L, Orci L, Montesano R (1994b) Modulation of bovine microvascular endothelial cell proteolytic properties by inhibitors of angiogenesis. J Cell Biochem 55: 419–434

Pepper MS, Ferrara N, Orci L, Montesano R (1995) Leukemia inhibitory factor (LIF) inhibits angiogenesis in vitro. J Cell Sci 108: 73–83

Peters KG, Werner S, Chen G, Williams LT (1992) Two FGF receptor genes are differentially expressed in epithelial and mesenchymal tissues during limb formation and organogenesis in the mouse. Development 114: 233–243

Phillips GD, Whitehead RA, Knighton DR (1992) Inhibition by methylprednisolone acetate suggests an indirect mechanism for TGF-β induced angiogenesis. Growth Factors 6: 77–84

Phillips GD, Whitehead RA, Stone AM, Ruebel MW, Goodkin ML, Knighton DR (1993) Transforming growth factor beta (TGF-B) stimulation of angiogenesis: an electron microscopic study. J Submicrosc Cytol Pathol 25: 149–155

Plate KH, Breier G, Weich HA, Risau W (1992) Vascular endothelial growth factor is a potent tumor angiogenesis factor in human gliomas in vivo. Nature 359: 845–848

Plate KH, Breier G, Risau W (1994a) Molecular mechanisms of developmental and tumor angiogenesis. Brain Pathol 4: 207–218

Plate KH, Breier G, Weich HA, Mennel HD, Risau W (1994b) Vascular endothelial growth factor and glioma angiogenesis: coordinate induction of VEGF receptors, distribution of VEGF protein and possible in vivo regulatory mechanisms. Int J Cancer 59: 520–529

Postlewaite AE, Keski-Oja J, Moses HL, Kang AH (1987) Stimulation of the chemotactic migration of human fibroblasts by transforming growth factor β. J Exp Med 165: 251–256

Qu Z, Picou M, Dang TT, Angell E, Planck SR, Hart CE, Rosenbaum JT (1994) Immunolocalization of basic fibroblast growth factor and platelet-derived growth factor-A during adjuvent arthritis in the Lewis rat. Am J Pathol 145: 1127–1139

Quarto N, Amarlic (1994) Heparan sulfate proteoglycans as transducers of FGF-2 signaling. J Cell Sci 107: 3201–3212

RayChaudhury A, D'Amore PA (1991) Endothelial cell regulation by transforming growth-factor-beta. J Cell Biochem 47: 224–229

Reilly TM, Taylor DS, Herblin WF, Thoolen MJ, Chiu AT, Watson DW, Timmermans PBMWM (1989) Monoclonal antibodies directed against basic fibroblast growth factor which inhibit its biological activity in vitro and in vivo. Biochem Biophys Res Commun 164: 736–743

Ricketts RR, Hatley RM, Corden BJ, Sabio H, Howell CG (1994) Interferon-alpha-2a for the treatment of complex hemangiomas of infancy and childhood. Ann Surg 219: 605–614

Risau W, Lemmon V (1988) Changes in the vascular extracellular matrix during embryonic vasculogenesis and angiogenesis. Dev Biol 125: 441–450

Risau W, Sariola H, Zerwes H-G, Sasse J, Ekblom P, Kemler R, Doetschman T (1988) Vasculogenesis and angiogenesis in embryonic stem cell derived embryoid bodies. Development 102: 471–478

Roberts AB, Sporn MB, Assoian RK, Smith JM, Roche NS, Wakefield LM, Heine UI, Liotta LA, Falanga V, Kehrl JH, Fauci AS (1986) Transforming growth factor type β: rapid induction of fibrosis and angiogenesis in vivo and stimulation of collagen formation in vitro. Proc Natl Acad Sci USA 83: 4167–4171

Roberts AB, Sporn AB (1989) Regulation of endothelial cell growth, architecture, and matrix synthesis by TGF-β. Am Rev Respir Dis 140: 1126–1128

Roberts WG, Palade GE (1995) Increased microvascular permeability and endothelial fenestration induced by vascular endothelial growth factor. Cell Sci 108: 2369–2379

Rosen EM, Liu D, Setter E, Bhargava M, Goldberg ID (1991) Interleukin-6 stimulates motility of vascular endothelium. Experientia Suppl 59: 194–205

Ruszczak Z, Detmar M, Imcke E, Orfanos CE (1990) Effects of rIFN alpha, beta and gamma on the morphology, proliferation, and cell surface antigen expression of human dermal microvascular endothelial cells in vitro. J Invest Dermatol 95: 693–699

Ruta M, Howk R, Ricca G, Drohan W, Zabelshansky M, Laureys G, Barton DE, Francke U, Schlessinger J, Givol D (1988) A novel protein tyrosine kinase gene whose expression is modulated during endothelial cell differentiation. Oncogene 3: 9–15

Sanderson N, Factor V, Nagy P, Kopp J, Kondai P, Wakefield L, Roberts AB, Sporn MB, Thorgeirsson SS (1995) Hepatic expression of mature transforming growth factor β1 in transgenic mice results in multiple tissue lesions. Proc Natl Acad Sci USA 92: 2572–2576

Sanvito F, Nichols A, Herrera P-L, Huarte J, Wohlwend A, Vassalli J-D, Orci L (1995) TGF-β1 overexpression in murine pancreas induces chronic pancreatitis and, together with TNF-α, triggers insulin-dependent diabetes. Biochem Biophys Res Commun 217: 1279–1286

Sariola H, Peault B, LeDouarin N, Buck C, Dieterlen-Lièvre F, Saxén L (1984) Extracellular matrix and capillary ingrowth in interspecies chimeric kidneys. Cell Differ 15: 43–51

Sato Y, Rifkin DB (1988) Autocrine activities of basic fibroblast growth factor: regulation of endothelial cell movement, plasminogen activator synthesis, and DNA synthesis. J Cell Biol 107: 119–1205

Schultz-Hector S, Haghayegh S (1993) β-fibroblast growth factor expression in human and murine squamous cell carcinomas and its relationship to regional endothelial cell proliferation. Cancer Res 53: 1444–1449

Schulze-Osthoff K, Risau W, Vollmer E, Sorg C (1990) In situ detection of basic fibroblast growth factor by highly specific antibodies. Am J Pathol 137: 85–92

Schwartz SM, Benditt EP (1977) Aortic endothelial cell replication, I. Effects of age and hypertension in the rat. Circ Res 41: 248–255

Schweigerer L, Neufeld G, Friedman J, Abraham JA, Fiddes JC, Gospodarowicz D (1987) Capillary endothelial cells express basic fibroblast growth factor, a mitogen that promotes their own growth. Nature 325: 257–259

Segarini PR, Rosen DM, Seydin SM (1989) Binding of transforming growth factor-β to cell surface proteins varies with cell type. Mol Endocrinol 3: 261–272

Senger DR, Van De Water L, Brown LF, Nagy JA, Yeo K-T, Yeo T-K, Berse B, Jackman RW, Dvorak AM, Dvorak HF (1993) Vascular permeability factor (VPF, VEGF) in tumor biology. Cancer Metastasis Rev 12: 303–324

Shalaby F, Rossant J, Yamaguchi TP, Gertsenstein M, Wu X-F, Breitman ML, Schuch AC (1995) Failure of blood-island formation and vasculogenesis in Flk-1-deficient mice. Nature 376: 62–66
Shull MM, Ormsby I, Kier AB, Pawlowski S, Diebold RJ, Yin M, Allen R, Sidman C, Proetzel G, Calvin D, Annunziata N, Doetschman T (1992) Targeted disruption of the mouse transforming growth factor-β1 gene results in multifocal inflammatory disease. Nature 359: 693–699
Shweiki D, Itin A, Soffer D, Keshet E (1992) Vascular endothelial growth factor induced by hypoxia may mediate hypoxia-initiated angiogenesis. Nature 359: 843–845
Sidkey YA, Borden EC (1987) Inhibition of angiogenesis by interferons: effects of tumor- and lymphocyte-induced vascular responses. Cancer Res 47: 5155–5161
Speir E, Sasse J, Shrivastav S, Casscells W (1991) Culture-induced increase in acidic and basic fibroblast growth factor activities and their association with the nuclei of vascular endothelial and smooth muscle cells. J Cell Physiol 147: 362–373
Sprugel KH, McPherson JM, Clowes AW, Ross R (1987) Effects of growth factors in vivo I. Cell ingrowth into porous subcutaneous chambers. Am J Pathol 129: 601–613
St-Jacques S, Cymerman U, Pece N, Letarte M (1994) Molecular characterization and in situ localization of murine endoglin reveal that it is a transforming growth factor-β binding protein of endothelial and stromal cells. Endocrinology 134: 2645–2657
Stahl N, Yancopoulos GD (1993) The alphas, betas and kinases of cytokine receptor complexes. Cell 74: 587–590
Stavri GT, Hong Y, Zachary IC, Breier G, Baskerville PA, Ylä-Herttuala S, Risau W, Martin JF, Erusalimsky JD (1995) Hypoxia and platelet-derived growth factor-BB synergistically upregulate the expression of vascular endothelial growth factor in vascular smooth muscle cells. FEBS Lett 358: 311–315
Stoplen AH, Guinan EC, Fiers W, Pober JS (1986) Recombinant tumor necrosis factor and immune interferon act singly and combination to reorganize human vascular endothelial cell monolayers. Am J Pathol 123: 16–24
Sun WH, Kreisle RA, Phillips AW, Ershler WB (1992) In vivo and in vitro characteristics of interleukin 6-transfected B16 melanoma cells. Cancer Res 52: 5412–5415
Symes JF, Sniderman AD (1994) Angiogenesis: potential therapy for ischemic disease. Curr Opin Lipidology 5: 305–312
Taga T, Kishimoto T (1993) Cytokine receptors and signal transduction. FASEB J 7: 3387–3396
Takahashi K, Mulliken JB, Kozakewich HPW, Rogers RA, Folkman J, Ezekowitz RAB (1994) Cellular markers that distinguish the phases of hemangioma during infancy and childhood. J Clin Invest 93: 2357–2364
Thieme H, Aiello LP, Takagi H, Ferrara N, King GL (1995) Comparative analysis of vascular endothelial growth factor receptors on retinal and aortic vascular endothelial cells. Diabetes 44: 98–103
Tuder RM, Flook BE, Voelkel NF (1995) Increased gene expression for VEGF and the VEGF receptors KDR/flk and flt in lungs exposed to acute or chronic hypoxia. J Clin Invest 95: 1798–1807
Ueba T, Takahashi JA, Fukumoto M, Ohata M, Ito N, Oda Y, Kikuchi H, Hatanaka M (1994) Expression of fibroblast growth factor receptor-1 in human glioma and meningioma tissues. Neurosurgery 34: 221–226
Ueki N, Nakazato M, Ohkawa T, Ikeda T, Amuro Y, Hada T, Higashino K (1992) Excessive production of transforming growth factor-β1 can play an important role in the development of tumorigenesis by its action for angiogenesis: validity of neutralizing antibodies to block tumor growth. Biochim Acta 1137: 189–196
Valente G, Ozmen L, Novelli F, Geuna M, Palestro G, Forni G, Garotta G (1992) Distribution of interferon gamma receptor in human tissues. Eur J Immunol 22: 2403–2412
Vernon RB, Lara SL, Drake CJ, Iruela-Arispe ML, Angello JC, Litttle CD, Wight TN, Sage EH (1995) Organized type I collagen influences endothelial patterns during "spontaneous angiogenesis in vitro": planar cultures as models of vascular development. In Vitro Cell Dev Biol 31: 120–131
Vilcek J (1990) Interferons. In: Sporn MB, Roberts AB (eds) Peptide growth factors and their receptors II. Springer, Berlin Heidelberg New York, pp 3–38
Volpes R, van den Oord JJ, De Vos R, Depla E, De Ley M, Desmet VJ (1991) Expression of interferon gamma receptor in normal and pathological human liver tissue. J Hepatol 12: 195–202
Wahl SM, Hunt DA, Wakefield LA, McCartney-Francis N, Wahl LM, Roberts AB, Sporn MB (1987) Transforming growth factor type β induces monocyte chamotaxis and growth factor production. Proc Natl Acad Sci USA 84: 5788–5792
Waltenberger J, Claesson-Welsh L, Siegbahn A, Shibuya M, Heldin C-H (1994) Different signal transduction properties of KDR and flt-1, two receptors for vascular endothelial growth factor. J Biol Chem 269: 26988–26995

Warren RS, Yuan H, Matli MR, Gillett NA, Ferrara N (1995) Regulation by vascular endothelial growth factor of human colon cancer tumorigenesis in a mouse model of experimental liver metastasis. J Clin Invest 95: 1789–1797
White CW, Sondheimer HM, Crouch EC, Wilson H, Fan LL (1989) Treatment of pulmonary hemangiomatosis with recombinant interferon alfa-2a. N Engl J Med 320: 1197–1200
White CW, Wolf SJ, Korones DN, Sondheimer HM, Tosi MF, Yu A (1991) Treatment of childhood angiomatous diseases with recombinant interferon alfa-2a. J Pediatr 118: 59–66
Wilkie AOM, Morris-Kay GM, Jones EY, Health JK (1995) Functions of fibroblast growth factors and their receptors. Curr Biol 5: 500–507
Wiseman DM, Polverini PJ, Kamp DW, Leibovich SJ (1988) Transforming growth factor beta (TGFβ) is chemotactic for human monocytes and induces their expression of angiogenic activity. Biochem Biophys Res Commun 157: 793–800
Wrana JL, Attisano L, Wieser R, Ventura F, Massagué J (1994) Mechanism of activation of the TGF-β receptor. Nature 370: 341–347
Yamaguchi TP, Harpal K, Hankemeyer M, Rossant J (1994) fgfr-1 is required for embryonic growth and mesodermal patterning during mouse gastrulation. Genes Dev 8: 3032–3044
Yang EY, Moses HL (1990) Transforming growth factor β1-induced changes in cell migration, proliferation, and angiogenesis in the chicken chorioallantoic membrane. J Cell Biol 111: 731–741
Yeo K-T, Wang HH, Nagy JA, Sioussat TM, Ledbetter SR, Hoogewerf AJ, Zhou Y, Masse EM, Senger DM, Dvorak HF, Yeo T-K (1993) Vascular permeability factor (vascular endothelial growth factor) in guinea pig and human tumor and inflammatory effusions. Cancer Res 53: 2912–2918
Yu Z-X, Biro S, Fu Y-M, Sanchez J, Smale G, Sasse J, Ferrans VJ, Casscells W (1993) Localization of basic fibroblast growth factor in endothelial cells: immunohistochemical and biochemical studies. Exp Cell Res 204: 247–259
Zagzag D (1995) Angiogenic growth factors in neural embryogenesis and neoplasia. Am J Pathol 146: 293–309

Adhesion Molecules and Tumor Cell-Vasculature Interactions: Modulation by Bioactive Lipid Molecules

D.G. TANG[1] and K.V. HONN[2,3]

1 Introduction 69

2 12(*S*)-HETE Regulates Tumor Cell-Matrix Interactions 75

3 12(*S*)-HETE Modulates Endothelial Cell Phenotypes, Expression/Function of Endothelial Cell Adhesion Molecules and Tumor Cell-Endothelial Cell Interactions 77

4 Molecular Mechanisms of 12(*S*)-HETE on Regulating the Adhesiveness of Tumor Cells and Endothelial Cells 80

References 84

1 Introduction

Adhesive interactions between disseminating tumor cells and the microvasculature (endothelial cells, leukocytes, platelets, extracellular matrix components, etc.) represent an essential molecular event in the metastatic process. As in the case of other metastatic determinants such as angiogenesis, proteolysis, motility and invasion, the adhesive interactions also are mediated by distinct groups of molecules, termed adhesion receptors, which are represented by four major families, i.e., integrins, the immunoglobulin supergene family, cadherins, and selectins. Many other molecules such as cell surface carbohydrates and lectins, proteoglycans, and some growth factors also possess adhesive functions under special circumstances. The relationship between adhesion and metastasis is simply exemplified by an in vitro adhesion assay utilizing tumor cells of differential metastatic potential. As illustrated in Fig.. 1, B16 F10 murine melanoma cells (a high metastatic subline) demonstrated a more rapid and quantitatively greater adhesion to microvascular endothelium that F1 cells (a low metastatic subline), thus suggesting a positive correlation between the metastatic ability and adhesive ability of tumor cells. The literature is replete with examples demonstrating

[1]Department of Radiation Oncology, Wayne State University, 431 Chemistry, Detroit, MI 48202, USA
[2]Departments of Radiation Oncology, Pathology, and Chemistry, Wayne State University, Detroit, MI 48202, USA
[3]Gershenson Radiation Oncology Center, Harper Hospital, Detroit, MI 48201, USA

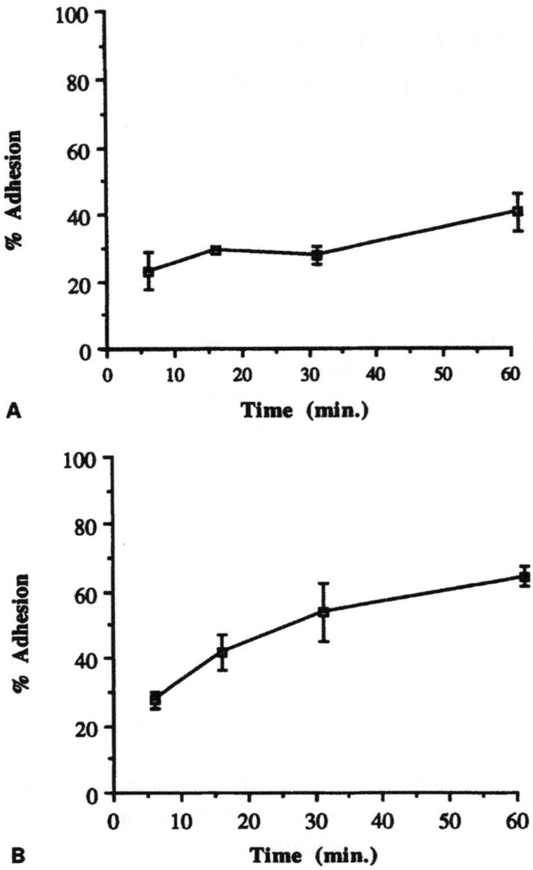

Fig. 1A,B. Murine B16 F1 (**A**) and F10 (**B**) melanoma cells exhibited differential abilities to adhere to murine microvascular endothelial cells. (^{3}H) thymidine labeled F1 or F10 cells were seeded onto confluent CD clone 3 murine microvascular endothelial cells and adhesion was allowed to proceed for various intervals as indicated. The adhesion was determined by counting radioactivity (Tang et al. 1994). Quadriplicate wells were run for each time point and the results presented as % of the total number of cells adhered. The *bars* represent mean ± S.D. derived from three independent experiments

a cause-and-effect relationship between adhesion molecules and tumor cell metastasis. Quite convincing data have been presented for three adhesion molecules, i.e., α5β1 integrin, E-cadherin and C-CAM, which possess metastasis-suppresive functions. Tumor cells are known to exhibit decreased adhesion to matrix compared to their normal counterparts. Only recently was it realized that this altered phenotype is largely caused by a decreased expression of integrin α5β1 (Zutter et al. 1990; Feldman et al. 1991), the prototypical fibronectin receptor. Chinese hamster ovary cells deficient in α5β1 expression demonstrate enhanced tumorigenecity (Schreiner et al. 1991b) which could be suppressed by gene-driven expression of this receptor (Giancotti and Ruoslahti 1990). Human colonic epithelial cells gradually lose the expression of integrin α5β1 during progression from the benign to the malignant phenotype (Stallmach et al. 1992). Ectoptic expression of α5β1 prevents the malignant transformation process as well as the tumorigenecity (Varner et al. 1992). E-cadherin normally plays a key role in maintaining the integrity of the epithelial sheet. Decreased expression of E-cadherin levels has been documented in many different types of human cancers including those of breast, colon, esophagus, lung, prostate, and head and neck (UMBAS et al. 1994; SOMMERS et al. 1991; SCHIPPER et al. 1991). Restoration of E-cadherin expression in human carcinoma cells prevented their invasion and reduced their metastasis (FRIXEN et al. 1991; VLEMINCKX et al. 1991). C-CAM is a member of the immunoglobulin superfamily and is homologous to carcino-embryonic antigen (CEA). This adhesion molecule appears to be expressed exclusively in prostate epithelial cells and its expression is regulated by androgen hormones (HSIEH et al. 1995). Using immunocytochemical staining for C-CAM, KLEINERMAN et al. (1995) observed a typical plasma membrane staining pattern in the basal cell layer of the normal prostate gland obtained from fetal, juvenile, and adult prostates. In sharp contrast to this expression pattern, human prostate cancers demonstrated a complete loss of expression of C-CAM (KLEINERMAN et al. 1995). Another immunoglobulin superfamily member, DCC (deleted in colon cancer; FEARON et al. 1990), also may possess a tumor-suppresive role; however, its role in cell adhesion has not been characterized definitely.

In contrast to α5β1, E-cadherin, and C-CAM, other adhesion molecules may positively contribute to tumorigenesis and metastasis. These molecules, according to their expression patterns and modes of actions, can be classified into three groups. The first group includes molecules which are quantitatively overexpressed in tumor cells compared to normal cells (reviewed in HONN and TANG 1992). Typical examples include expression of integrins α6β1 in colon carcinoma cells (SCHREIBER et al.1991a) and α6β4 in 3LL lung carcinoma cells (PERROTTI et al. 1990), CEA in colon carcinoma (JOHNSON 1991), and a 67 KDa non-integrin laminin receptor in multiple cancer cells (breast, colon, and lung; (SATOH et al. 1992). The second group involves adhesion molecules whose alternatively spliced isoforms are utilized by tumor cells for their progression and dissemination. The prototypical example for this group is CD44, a transmembrane hyaluronate receptor which is found on a variety of different cell types including endothelial cells, epithelial cells, fibroblasts, leukocytes, and many solid tumor cells (HAYNES et al. 1991). CD44 plays a crucial physiological role in lymphocyte homing. Intriguingly, many tumor cells express special isoforms of alternatively spliced CD44 molecules and the expression levels are correlated positively with the metastatic potential of tumor cells (GUNTHERT et al. 1991). Transfection of variant CD44 isoform genes into normal cells confers on them a metastatic phenotype when injected into syngeneic animals (GUNTHERT et al. 1991; HOFFMANN et al. 1991). A monoclonal antibody to this variant CD44 protein retards metastasis formation by the transfected cells (GUNTHERT et al. 1991; HOFFMANN et al. 1991). The third group includes a large number of adhesion molecules whose "abnormal," ectopic expression contributes to tumor cell-matrix interactions and tumor metastasis (reviewed in TANG and HONN 1995). Generally, these ectopically expressed adhesion molecules are those normally expressed on blood-borne cells or normal migratory cells such as neural crest cells, thus strengthening the similarity between the metastatic process and the migration processes of select normal cells. Integrin α4β1 is expressed normally on a subset of lymphocytes and functions in lymphocyte homing. However, human melanoma cells have been found to utilize this receptor for their adhesion to matrix proteins (GEHLSEN et al. 1992). Another adhesion molecule frequently utilized by melanoma cells is ICAM-1 (intercellular adhesion molecule1; JOHNSON et al. 1989) which is normally expressed on cytokine activated endothelial cells, which may originate from the same stem cells (i.e., the hemangioblastic precursor cell) as blood cells (AUGUSTIN et al. 1994). Numerous adhesion molecules, once thought to be expressed exclusively on blood cells and vascular endothelial cells, are also found on various tumor cells. Typical examples include integrin αIIbβ3 (GROSSI et al. 1988; CHEN et al. 1992), PECAM-1 (platelet endothelial cell adhesion molecule-1; TANG et al. 1993f; NICKLOFF 1993), and LFA-1 (lymphocyte function-associated molecule-1; $\alpha_L\beta2$; (ROOSIEN et al. 1989). NCAM, an immunoglobulin supergene family member normally expressed on migrating neural crest cells and neuronal cells, is expressed on many human solid tumor cells where it clearly plays a role in mediating tumor cell adhesion to matrix and endothelium (SOLER et al. 1993; ZOCCHI et al. 1993).

The expression and function of various adhesion molecules are spatiotemporally regulated. For example, integrin αvβ3, which is expressed on endothelial cells and tumor cells, mediates cell-cell adhesion as well as cell adhesion to a variety of matrix proteins (e.g., fibronectin, vitronectin, fibrinogen, von Willebrand factor, thrombospondin, and denatured collagen). The expression/function of this adhesion receptor has been shown to be modulated by divalent cations (Ca^{2+}, Mg^{2+}, and Mn^{2+}; SMITH et al. 1994), integrin-associated proteins (SCHWARTZ et al. 1993; BARTFELD et al. 1993), bioactive mediators including transforming growth factor-β (TGF-β; IGNOTZ et al. 1989), tumor necrosis factor-α (TNF-α) and interferon-γ (IFN-γ; DEFILIPPI et al. 1991) interleukin-1 (IL-1; LAFRENIE et al. 1992), GM- and M-CSF (colony-stimulating factor; DENICHILO and BURNS 1993), 1α, 25-dihydroxyvitamin D_3 (MEDHORA et al. 1993), retinoic acid (ROSSINO et al. 1991), and basic fibroblast growth factor (bFGF; KLEIN et al. 1991). Another physiologically important class of regulators of cell adhesion and adhesion molecules are various bioactive lipid molecules. The binding capacity of integrin αvβ3 has been shown to depend on the membrane phospholipid composition (CONFORTI et al. 1990). Similarly, an uncharacterized lipid molecule (IMF-1) can modulate the binding affinity and function of leukocyte integrin LFA-1 (HERMANOWSKI-VOSATKA et al. 1992). Another lipid molecule, 13-HODE (13-hydroxyoctadecaenoic acid) is observed to regulate the surface expression of integrin αvβ3 (BUCHANAN et al. 1993). IMF-1, 13-HODE, and membrane phospholipid molecules most likely modulate the integrin functions via direct lipid-protein interactions (CONFORTI et al. 1990; HERMANOWSKI-VOSATKA et al. 1992; BUCHANAN et al. 1993). In contrast, a large array of bioactive lipid molecules may modulate the expression/function of adhesion receptors (especially integrins) through distinct intracellular signaling mechanisms. These include prostaglandins(MILAM et al. 1991), 14, 15-EET (14, 15-epoxyeicosatetraenoic acids; JONES and HONN 1992), and 12(*S*)-HETE ([12(*S*)-hydroxyeicosatetraenoic acid; see discussion below). In addition, lipid molecules may regulate the expression/function of adhesion receptors via both direct binding and indirect signaling mechanisms. A prototypical example is PAF (platelet activating factor) which modulates leukocyte-endothelial cell interactions by binding to the PAF receptor and activating β2 integrin expression (ZIMMERMAN et al. 1990; LORANT et al. 1991).

12(*S*)-HETE is derived from arachidonic acid metabolism via the 12 lipoxygenase pathway. Many cell types including platelets, leukocytes, endothelial cells, smooth muscle cells, and neuronal cells can synthesize 12(*S*)-HETE. Various solid tumor cells including murine B16a melanoma, Lewis lung carcinoma, rat W256 carcinosarcoma and Dunning prostate adenocarcinoma, and human colon, breast, and prostate cancer cells also can synthesize 12(*S*)-HETE (CHEN et al. 1994; LIU et al. 1994b; reviewed in HONN et al. 1994b and TANG and HONN 1994). The widespread expression of 12-lipoxygenase mRNA was demonstrated by northern blotting and RNase protection assays in solid tumor cells derived from mouse, rat, and human (Fig. 2). In many experimental systems, there is a positive correlation among 12-lipoxygenase mRNA expression, 12(*S*)-HETE biosynthesis, tumor cell adhesion (to either endothelium or matrix), and the metastatic potential of tumor cells. Murine B16a melanoma cells maintained in syngeneic C57BL/6J mice were isolated with centrifugal elutriation (HONN et al. 1992) into four distinct subpopulations which possess increasing metastatic capacities (from 100 to 340). These subpopulations express increasing amounts of 12-lipoxygenase mRNA (Fig. 2, left panel). Correspondingly, HM(high metastatic) 340 cells produced three- to four-fold more 12-HETE than the LM (low metastatic) 180 cells from exogenously added arachidonic acid. Also, the level of endogenous 12(*S*)-HETE (as determined by RIA) in HM340 cells is twice as much as that in LM180 cells (LIU et al. 1994b; TANG and HONN 1994). In vitro, adhesion of B16a cells to vascular endothelium is accompanied or immediately followed by a surge of 12(*S*)-HETE biosynthesis by tumor cells, which was correlated with tumor cell-induced endothelial cell retraction (HONN et al. 1994a). LM180 B16a cells generated little 12(*S*)HETE upon adhesion and did not induce endothelial cell retraction. In contrast, HM340 B16a cells adhering to endothelium biosynthesized large amounts of 12(*S*)-HETE and induced prominent retraction of endothelial cell monolayers (HONN et al. 1994a). To confirm that the greater capacity of adhesion to endothelium by HM340 cells was due to their higher level of biosynthesis of 12-HETE, HM340 cells were pretreated with BHPP (*N*-benzyl-*N*-hydroxy-5-phenylpentanamide), a select 12-lipoxygenase inhibtor, and then incubated with 5(*S*)-, 11(*S*)-, 12(*S*)-, 12(*R*)-, or 15(*S*)-HETE (0.01–1 μ*M*) and assayed for adhesion to microvessel endothelium. As expected, cells pretreated with BHPP and then incubated with 12(*S*)-HETE demonstrated an enhanced ability to adhere to endothelium. No other monohydroxy fatty acid tested produced this effect (TANG and HONN 1994).

A positive correlation between 12-HETE production and metastatic potential also was observed in low (AT 2.1 and GP 9F3) and high (MAT Lu and MLL) metastatic Dunning rat prostatic carcinoma cells, murine B16F1 (low metastatic) and F10 (high metastatic) cell lines, and murine K-1735 melanoma-derived sub-clones, C1-11 (low metastatic) and M1 (high metastatic). These observations suggest that 12(*S*)-HETE may represent an important regulator of cell adhesion and tumor cell-microvasculature interactions. Physiologically, 12(*S*)-HETE has been shown to induce platelet aggregation, stimulate insulin secretion, suppress

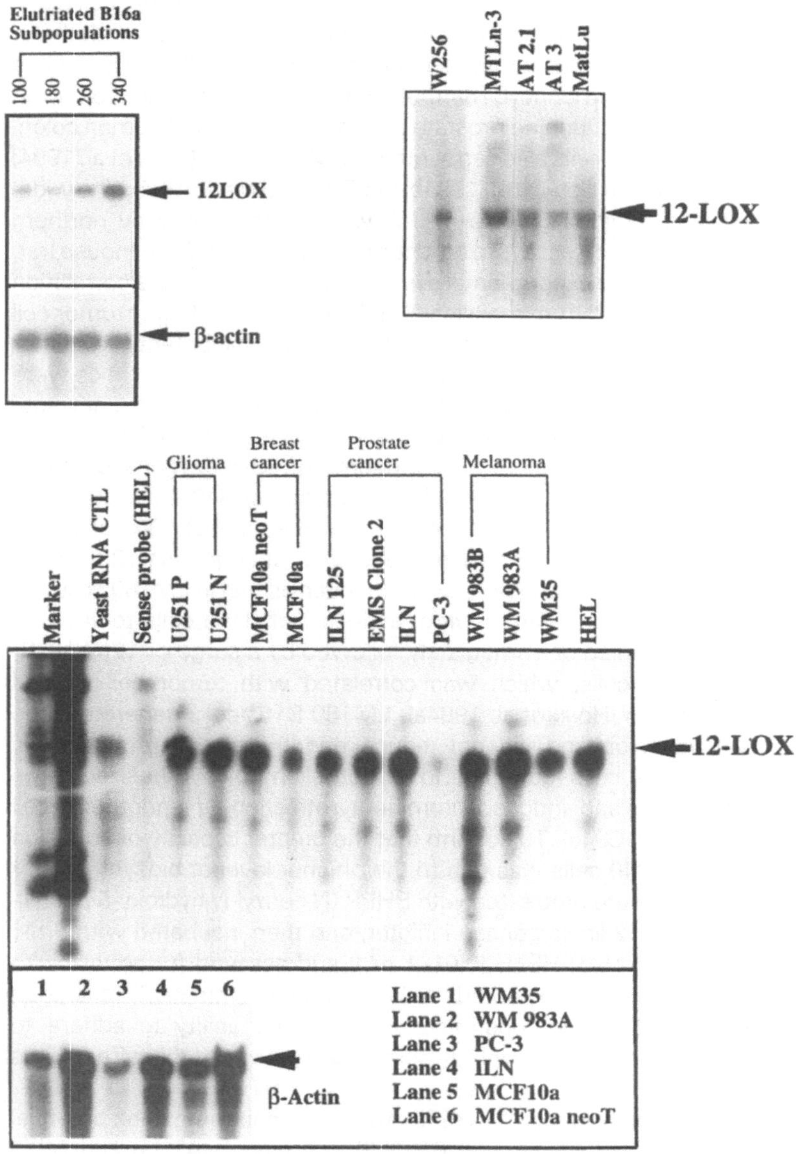

renin production, chemoattract leukocytes, facilitate macrophage adhesion, and inhibit prostacyclin biosynthesis by vascular endothelial cells (SPECTOR et al. 1988). In addition, more and more signaling functions have been ascribed to 12(*S*)-HETE (see below). Experiments performed in our laboratory and others have demonstrated that 12(*S*)-HETE possesses versatile biological activities in modulating the phenotypic properties of tumor cells and the metastatic processes such as tumor cell-platelet-endothelial cell interactions (HONN et al. 1994b; TANG and HONN 1994). In this chapter we will focus on the regulatory effect of 12(*S*)-HETE on the expression/function of adhesion receptors and adhesive interactions among tumor cells, endothelial cells, and matrix.

2 12(*S*)-HETE Regulates Tumor Cell-Matrix Interactions

In our early studies we observed that 12(*S*)-HETE stimulated metastatic murine tumor cell adhesion to fibronectin and subendothelial matrix in a dose- and time-dependent fashion, with a maximal effect observed at 0.1 μM of 12(*S*)-HETE 15 min after stimulation (GROSSI et al. 1989; TIMAR et al. 1995). The 12(*S*)-HETE enhanced tumor cell adhesion appeared to result from an increased surface expression of integrin receptors $\alpha IIb\beta 3$, which is a consequence of cytoskeleton-dependent receptor translocation to the cell surface from the intracellular pool without involving up-regulated gene transcription (TIMAR et al. 1995). Not surprisingly, 12(*S*)-HETE possesses a prominent regulatory effect on the tumor cell cytoskeleton (TIMAR et al. 1993). The earliest event post-12(*S*)-HETE treatment is a decreased labeling of cytoplasmic stress fibers with a concomitantly enhanced cortical actin labeling. These alterations in microfilaments are accompanied or immediately followed by microtubule polymerization and vimentin intermediate

Fig. 2. 12-lipoxygenase mRNA is widely expressed in solid tumor cells derived from various species. *Top left panel*: Murine B16a melanoma cells maintained in syngeneic C57BL/6J mice were isolated with centrifugal elutriation (HONN et al. 1992) into four distinct subpopulations which possess increasing metastatic capacities (from 100 to 340). A total of 5 µg of mRNA was separated on a denaturing formaldehyde/agarose gel and transferred to the nylon membrane, which was subsequently probed with ^{32}P-labeled full-length cDNA probe of platelet type 12-lipoxygenase (kindly provided by Dr. C. FUNK, University of Pennsylvania). The same membrane was then deprobed and reprobed with labeled β-actin cDNA probe. The 12-lipoxygenase mRNA (~3.0 kb) was indicated. The expression levels of 12-lipoxygenase mRNA were positively correlated with the metastatic ability. *Top right panel*: 12 lipoxygenase mRNA expression in rat tumor cells. W256, Walker carcinosarcoma; MTLn-3, rat mammary adenocarcinoma; AT2.1 and AT3, rat prostate adenocarcinoma; MatLu, rat prostate cancer metastasis to the lung. A total of 20 µg RNA was used in the northern blotting using ^{32}P-labeled full-length cDNA probe of platelet type 12-lipoxygenase, as described above. The 12-lipoxygenase mRNA band was indicated on the *right*. *Bottom panel*: Detection of 12-lipoxygenase mRNA expression in human cancer cells by RNase protection assays. A total of 20 µg of total RNA isolated from human melanoma, prostate cancer, breast cancer, and glioma was used in a protection assay using a 250 bp ^{32}P-labeled riboprobe of 12-lipoxygenase (indicated on the *right*). *Inset* β-actin protection assays for six tumor cell lines. *HEL*, human erythroleukemia cell line

filament bundling and also by an increase in vinculin-containing focal adhesions (TIMAR et al. 1992, 1993), which mediates increased tumor cell spreading on fibronectin (TIMAR et al. 1992; TANG et al. 1995). B16a cell spreading on fibronectin is mediated, at least in part, by integrin αIIbβ3 (TIMAR et al. 1992). 12(*S*)-HETE also promotes B16a cell spreading in collagen gel (Fig. 3). The integrin receptor(s) involved in 12(*S*)-HETE-promoted B16a cell spreading in collagen gel is unknown. The 12(*S*)-HETE effect on tumor cell cytoskeleton and focal adhesions is protein kinase C (PKC)-dependent since pretreatment of tumor cells with a selective PKC inhibitor, i.e., calphostin C abolished the 12(*S*)-HETE effects (TIMAR et al. 1992, 1993; TANG et al. 1995b). Recently, it was demonstrated that 12(*S*)-HETE-promoted B16a cell spreading on fibronectin may also involve increased tyrosine phosphorylation of $pp125^{FAK}$ (TANG et al. 1995b). In B16a cells tyrosine phosphorylated proteins are colocalized with vinculin and PKC in focal adhesion plaques (TANG et al. 1995b), important cell-matrix contact sites where signals across the plasma membrane are transduced through integrin receptors. Stimulation of B16a cells plated on fibronectin with 12(*S*)-HETE resulted, in a time- and dose-dependent manner, in increased tyrosine phosphorylation of several proteins including a focal adhesion-specific pp150, p120–130 protein cluster, p76, and a p42/44 doublet (TANG et al. 1995b). One protein in the p120–130 protein cluster was shown to be $pp125^{FAK}$ (TANG et al. 1995b). The 150 kDa protein may represent the αIIb or α5 subunit although the experimental evidence for this has yet to be provided. Likewise, the molecular identities of p76 and p42/44 are not known, although based on their relative mobilities on SDS-PAGE they may represent paxillin and $pp41/43^{FRNK}$ or p42/44 MAP kinase, respectively.

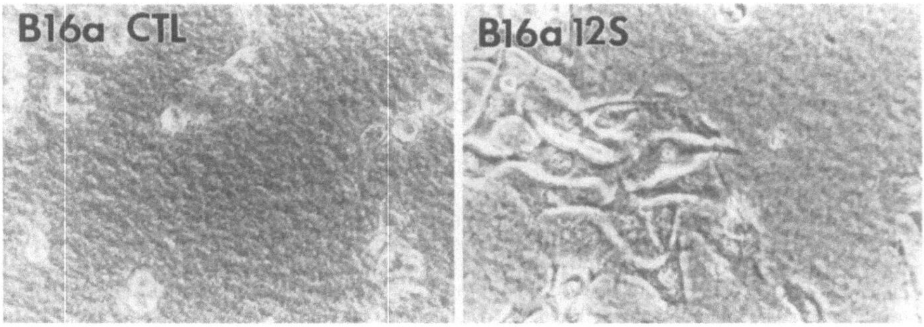

Fig. 3. 12(*S*)-HETE promotes B16a melanoma cell spreading in three dimensional collagen gel. B16a cells grown in reconstituted collagen gel (Vitrogen) were treated with either solvent control (*CTL*) ethanol (*left panel*) or 0.1 μ*M* 12 (*S*)-HETE (the *right panel*) for 24 h. As could be clearly seen, 12(*S*)-HETE treated B16a cells demonstrated a well-spread morphology while control cells remained rounded. x200

3 12(*S*)-HETE Modulates Endothelial Cell Phenotypes, Expression/Function of Endothelial Cell Adhesion Molecules and Tumor Cell-Endothelial Cell Interactions

Tumor cell-endothelial cell interactions, which are spatiotemporally regulated by a large array of bioactive mediators (growth factors, circulating hormones, cytokines, chemokines, lipid mediators, etc.) and can be analyzed at the molecular level with the "docking and locking" hypothesis (reviewed in Honn and Tang 1992 and Tang and Honn 1995), constitute one of the most important factors in the organ preference of metastasis. 12(*S*)-HETE plays an essential modulatory role ranging from regulating tumor cell locking onto endothelial cells by modulating integrin expression to promoting tumor cell extravasation by inducing endothelial cell retraction.

Most tumor cells exit from the microvasculature by inducing endothelial cell retraction. The biological mediators responsible for tumor cell induced endothelial cell retraction largely remain unknown. Our laboratory presented the first experimental evidence that 12(*S*)-HETE is capable of inducing a nondestructive reversible retraction of cultured large vessel endothelial cell monolayers (Honn et al. 1989). Subsequent studies established that 12(*S*)-HETE also induces retraction of microvessel endothelial cells (Honn et al. 1994a; Tang et al. 1993b,e). More importantly, tumor cell adhesion to microvessel endothelium induced a non-denuding retraction of endothelial cell monolayers (Tang et al. 1993b), which, in many aspects, mimicked the 12(*S*)-HETE-induced endothelial cell retraction. Tumor cell induced endothelial cell retraction can be potentiated by homologous platelets (Honn et al. 1994c), thus providing an experimental rationale for the platelet contribution to tumor cell metastasis. Further investigations revealed that both platelet-augmented and tumor cell induced endothelial cell retraction were mediated through 12(*S*)-HETE (Tang et al. 1993b; Honn et al. 1994c). Mechanistic studies provided evidence that 12(*S*)-HETE activates PKC in endothelial cells, which subsequently phosphorylates several major cytoskeletal elements including myosin light chain, actin and vimentin, leading to a reorganization of the cytoskeleton network (microfilament, intermediate filament, and focal adhesions) and apparent endothelial cell retraction (Tang et al. 1993b,e).

12(*S*)-HETE induced endothelial cell retraction also may involve alterations in adhesion molecules that normally are concentrated at cell-cell junctions. Integrin $\alpha v\beta 3$ and PECAM-1 are localized to the cell junctional areas in confluent endothelial cell cultures (Tang et al. 1993a; Fig. 4). 12(*S*)-HETE treatment results in a series of well-defined changes in $\alpha v\beta 3$-containing focal adhesions, leading to an eventual decrease in $\alpha v\beta 3$ localization to focal adhesion plaques at both the cell body and the cell-cell borders (Tang et al. 1993a). In contrast to its effect on integrin $\alpha v\beta 3$, 12(*S*)-HETE does not appear to alter the normal cellular distribution patterns of another integrin receptor, i.e., $\alpha 5\beta 1$, the classic fibronectin receptor. The 12(*S*)-HETE induced reorganization of $\alpha v\beta 3$-enriched focal adhesions resembles its effect on vinculin and again appears to be dependent on PKC activation (Tang

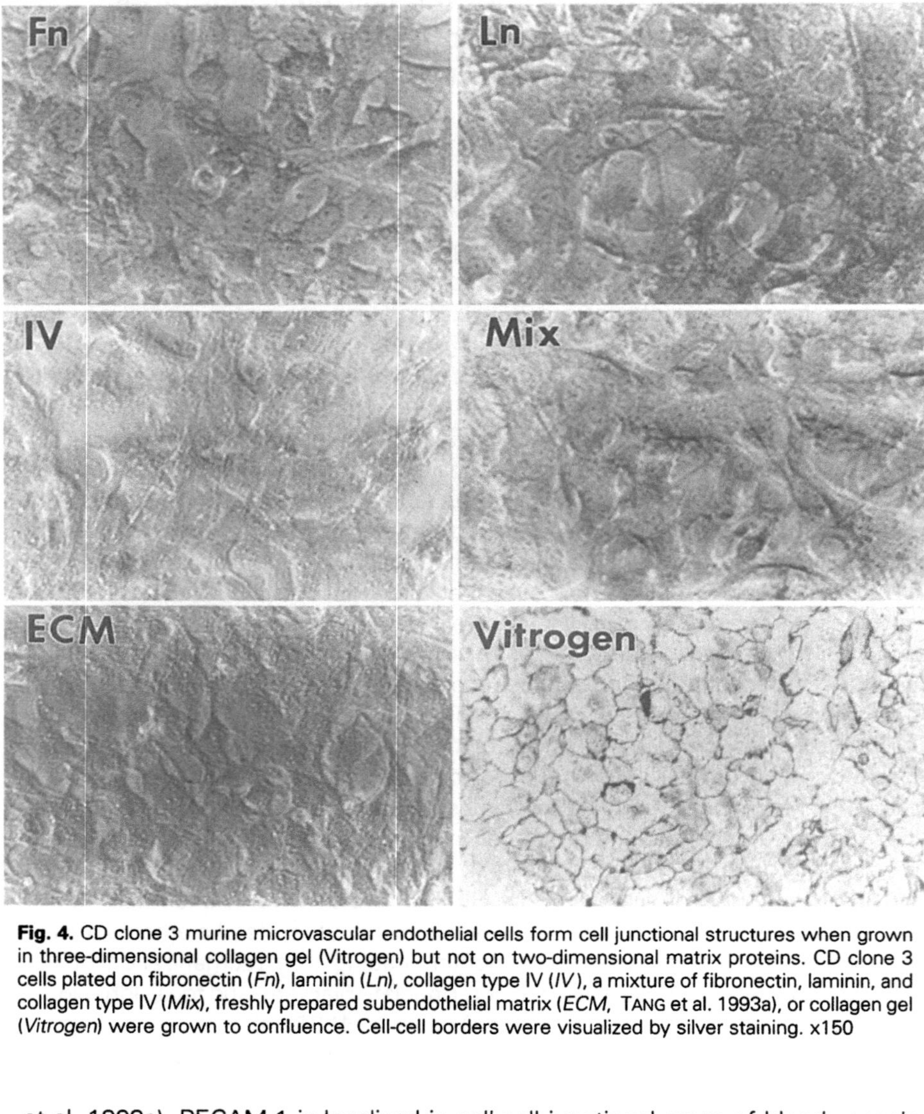

Fig. 4. CD clone 3 murine microvascular endothelial cells form cell junctional structures when grown in three-dimensional collagen gel (Vitrogen) but not on two-dimensional matrix proteins. CD clone 3 cells plated on fibronectin (*Fn*), laminin (*Ln*), collagen type IV (*IV*), a mixture of fibronectin, laminin, and collagen type IV (*Mix*), freshly prepared subendothelial matrix (*ECM*, TANG et al. 1993a), or collagen gel (*Vitrogen*) were grown to confluence. Cell-cell borders were visualized by silver staining. x150

et al. 1993a). PECAM-1 is localized in cell-cell junctional areas of blood vessels in vivo and has been proposed to be involved in maintaining the integrity of the endothelial cell monolayers (NEWMAN et al. 1990; ALBELDA et al. 1991). Culture of murine microvascular endothelial cells (CD clone 3) in three-dimensional collagen gel induced formation of typical cell junctional structures which could be visualized by silver staining (Fig. 4). No cell border formation was observed when these microvascular endothelial cells were grown on two dimensional matrix proteins (i.e., fibronectin, laminin, collagen type IV, or a mixture of these proteins) or freshly prepared subendothelial matrix (Fig. 4). Interestingly, only CD clone 3 cells

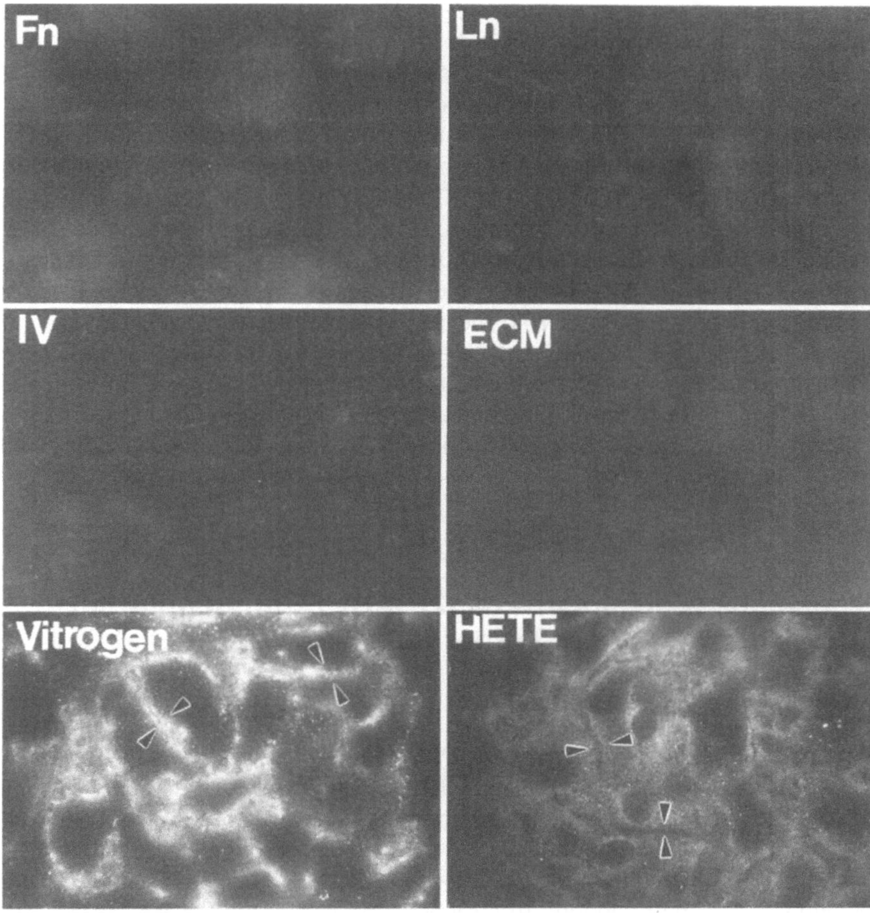

Fig. 5. CD clone 3 cells grown in collagen gel express PECAM-1 in cell junctional areas and 12(*S*)-HETE disrupts the junctional localization of PECAM-1. The *arrowheads* point to the cell-cell border localization of PECAM-1. CD clone 3 cells were grown to confluence as described in Fig. 4 and then processed for PECAM-1 staining using polyclonal anti-PECAM-1 (SEW-3, kindly provided by Dr. P.J. Newman) followed by fluorescein-labled goat anti-rabbit IgG. In one set of samples, cultures were treated with 0.1 μ*M* 12(*S*)-HETE (*HETE*) and then processed for immunolabeling. *Fn*, Fibronectin; *Ln*, laminin; *IV*, collagen type IV; *ECM*, extracellular matrix; x400

grown in a collagen gel (Vitrogen) expressed PECAM-1 to cell-cell borders (Fig. 5). Treatment of confluent CD clone 3 cultures dispersed cell border-localized PECAM-1 molecules to the whole cell body (Fig. 5). It is not clear at present whether this alteration of PECAM-1 is a trigger or a consequence of 12(*S*)-HETE induced endothelial cell retraction.

12(*S*)-HETE also modulates the surface expression of integrin receptors on endothelial cells. 12(*S*)-HETE stimulation up-regulates the surface expression of integrin αvβ3 (but not α5β1) in both large vessel and CD clone 3 microvascular endothelial cells (Tang et al. 1993a,c, 1994). The increased αvβ3 surface expression appears to be concomitant with a decrease in αvβ3-enriched focal adhesions and is a posttranscriptional, PKC- and cytoskeleton-dependent process in CD clone 3 cells (Tang et al. 1994). Interestingly, when another murine microvascular endothelial cell line (i.e., CD clone 4) was stimulated with exogenous 12(*S*)-HETE, it was found that the mRNA level of integrin αv was enhanced two- to-five-fold (Tang et al. 1995a), suggesting that 12(*S*)-HETE also may activate the gene transcription of some integrin receptors. How 12(*S*)-HETE triggers a posttranscriptional response in CD clone 3 cells but a transcriptional response in CD clone 4 cells with respect to αv expression is unknown. At any rate, the increased αv appears to primarily associate with the β3 subunit (Tang et al. 1995a). The end result of 12(*S*)-HETE enhanced surface expression of αvβ3 integrin receptors is increased tumor cell adhesion to the activated endothelium (Tang et al. 1993a,c, 1994), possibly through linkage molecules such as fibrinogen binding to integrin αIIbβ3 in tumor cells and activated integrin αvβ3 in vascular endothelial cells (Tang et al. 1993d).

4 Molecular Mechanisms of 12(*S*)-HETE on Regulating the Adhesiveness of Tumor Cells and Endothelial Cells

In our early experiments we observed that 12(*S*)-HETE mimicked the tumor promoter phorbol myristate acetate (PMA) in enhancing tumor cell integrin expression and adhesion (Grossi et al. 1989), suggesting that 12(*S*)-HETE may work via the activation of PKC. This hypothesis was supported by subsequent experiments demonstrating that, in rat W256 carcinosarcoma cells, 12(*S*)-HETE induced a 100% increase in membrane-associated PKC activity and that down-regulation of PKC by prolonged treatment with PMA abolished the 12(*S*)-HETE enhanced W256 cell adhesion to endothelium (Liu et al. 1991). The mechanism by which 12(*S*)-HETE activates PKC has not been clearly established. The physiological activators of PKC are diacylglycerol (DAG) and Ca^{2+}. It appears that 12(*S*)-HETE in different cell types activates different isoform(s) of PKC. In murine B16a melanoma cells, PKCα is the only isoform that can be detected by immunoblotting using isoform-specific anti-peptide antibodies (Honn et al. 1994b). 12(*S*)-HETE stimulation leads to an increased translocation of PKC from the diffuse cytoplasmic pool to the plasma membrane (Honn et al. 1994b). When B16a cells plated on fibronectin were stimulated with 12(*S*)-HETE, increased membrane-associated PKC was observed to colocalize with vinculin to the focal adhesion plaques (Honn et al. 1994b). In rat AT2.1 prostate carcinoma cells, two isoforms of PKC, α and δ, were identified (Liu et al. 1994a). Interestingly, 12(*S*)-HETE preferentially activates PKCα by increasing the membrane association of this isoform, resulting in an up-regulated tumor cell motility and invasion (Liu et al. 1994a). Treating tumor cells with thymelea toxin, a selective activator of Ca^{2+}-dependent

PKCα over Ca^{2+}-independent PKCδ, also promoted tumor cell motility and invasion (LIU et al. 1994), thus confirming the preferential activation of PKCα by 12(*S*)-HETE in this tumor cell line. Further, BAPTA, a Ca^{2+} chelator, dose-dependently inhibited 12(*S*)-HETE-promoted tumor cell motility and invasion (LIU et al. 1994a), suggesting that the PKC isoform activated by 12(*S*)-HETE is Ca^{2+}-dependent. Stimulation of B16a melanoma cells with 12(*S*)-HETE was observed to be followed by a rapid accumulation of DAG and inositol trisphosphate (IP3) (LIU et al. 1995), therefore suggesting phosphatidylinositol 4,5 bisphosphate (PIP_2) metabolism through the action of phospholipase C (PLC). To support this, blocking the functions of G proteins by pertussis toxin as well as inhibiting PLC activity by specific antagonists also suppress the effects of exogenous 12(*S*)-HETE (LIU et al. 1995), indicating that 12(*S*)-HETE may function via activating an upstream G protein. Another alternative for the accumulation of DAG following 12(*S*)-HETE treatment is through down-regulating the DAG kinase activity (SETTY et al. 1987). In the phosphoinositide cycle, DAG derived from various sources undergoes rapid phosphorylation by DAG kinase to phosphatidic acid. In bovine aortic endothelial cells, both 12(*S*)-HETE and 15(*S*)-HETE were found to inhibit the DAG kinase activity, leading to an accumulation of DAG mass in these cells (SETTY et al. 1987).

In murine microvessel endothelial cells (i.e., CD clone 3 cells), at least two isoforms of PKC are expressed (TANG, LIU, and HONN, unpublished observations). Immunostaining in CD clone 3 cells revealed a colocalization of PKC with the filamentous cytoskeleton, primarily microfilaments and intermediate filament vimentin (Fig. 6a,b), which was confirmed by western blotting (Fig. 7). 12(*S*)-HETE stimulation of CD clone 3 cells in a time-dependent fashion reorganized the filamentous pattern of PKC localization (Fig. 6c-f). Most dramatically, a time-related redistribution of PKC from cell body to extending cell filapodia was observed (6e,f). This pattern of change was very similar to the cytoskeletal alterations during endothelial cell retraction induced by 12(*S*)-HETE (TANG et al. 1993e). PKC isoforms which are endogenously associated with the filamentous cytoskeletal elements may play an important role in regulating their dynamics. The 12(*S*)-HETE effect on vascular endothelial cells (retraction, cytoskeletal rearrangement, adhesion receptor expression) may largely result from its impact on PKC. Quite different from CD clone 3 endothelial cells, B16a melanoma cells, under unstimulated conditions, express PKC mostly in a diffuse, homogeneous pattern (Fig. 8a) without apparent cytoskeletal association. 12(*S*)-HETE stimulation resulted in the localization of PKC to focal adhesions (HONN et al. 1994b). Some of the 12(*S*)-HETE effects on melanoma cells (e.g., stimulation of B16a cell spreading on fibronectin; TANG et al. 1995) also may result from, in addition to PKC activation, activated protein tyrosine kinase activity. Unstimulated CD clone 3 endothelial cells express phosphotyrosines diffusely in the cytoplasm (Fig. 8b) and 12(*S*)-HETE stimulation does not bring about any significant alterations in this distribution pattern (TANG and HONN, unpublished observations). In contrast, B16a cells normally enrich phosphotyrosines to focal adhesions and filamentous structures (Fig. 8c). 12(*S*)-HETE stimulation of B16a cells results in increased numbers of phosphotyrosine-containing focal adhesions and increased tyrosine phosphorylation

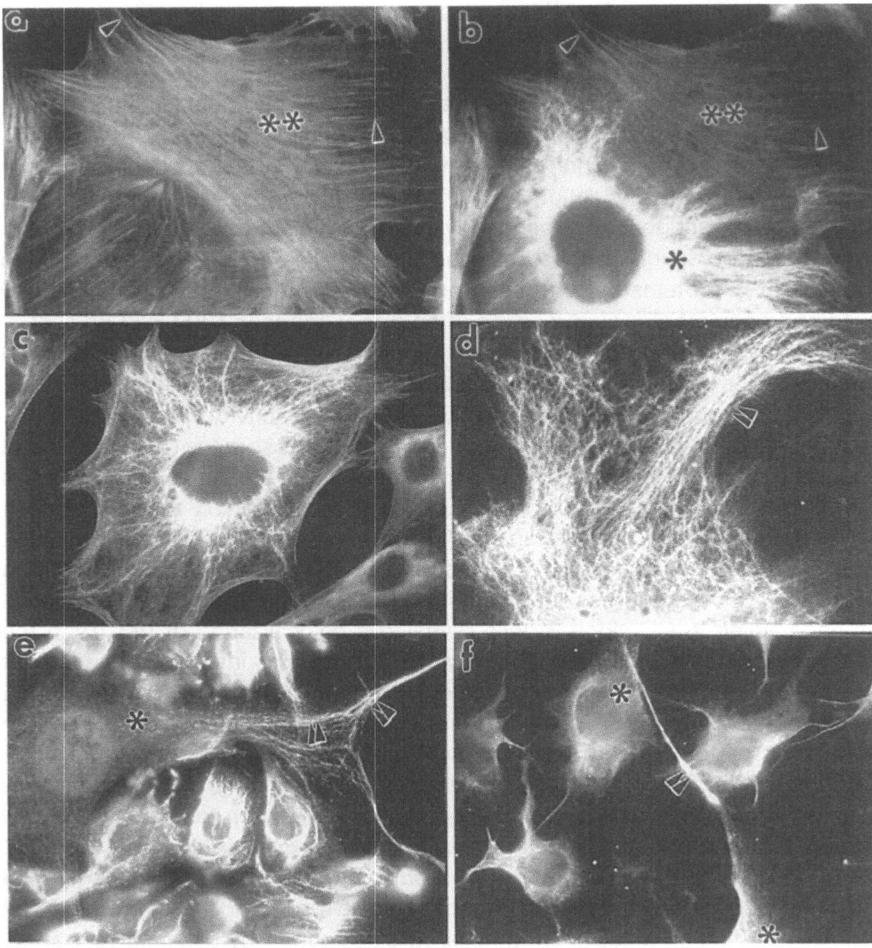

Fig. 6a–f. PKC distribution in CD clone 3 microvascular endothelial cells and the 12(*S*)-HETE effects. Subconfluent CD clone 3 cells were double labeled with rhodamine-phalloidin (**a**) and a monoclonal antibody to protein Kinase C (PKC) (mAb 1.9) (**b**). Note the colocalization of PKC with microfilaments (*double asterisks*), microfilament endings (*arrowheads*), and perinuclear vimentin intermediate filaments (*single asterisk*). **c–f** PKC labeling in control cells (**c**) and in cells treated with 0.1 μ*M* 12(*S*)-HETE for 5 (**d**), 15 (**e**), and 60 min (**f**). Note the appearance of PKC-associated filament bundles 5 min post 12(*S*)-HETE stimulation (**d**, *double arrowheads*) followed by a dramatic reorganization of PKC from cell body (*asterisk*) to the extending filapodia (*double arrowheads*). x600

of $pp125^{FAK}$, p42/44, and other proteins, leading to enhanced cell spreading on matrix proteins (Tang et al. 1995b). The differential involvement of PKC and phosphotyrosines in normal cells and tumor cells with regard to regulating cytoskeletal integrity and cell-matrix interactions (i.e., focal adhesions) may underlie the differential effects of 12(*S*)-HETE which sometimes are observed in these two different types of cells.

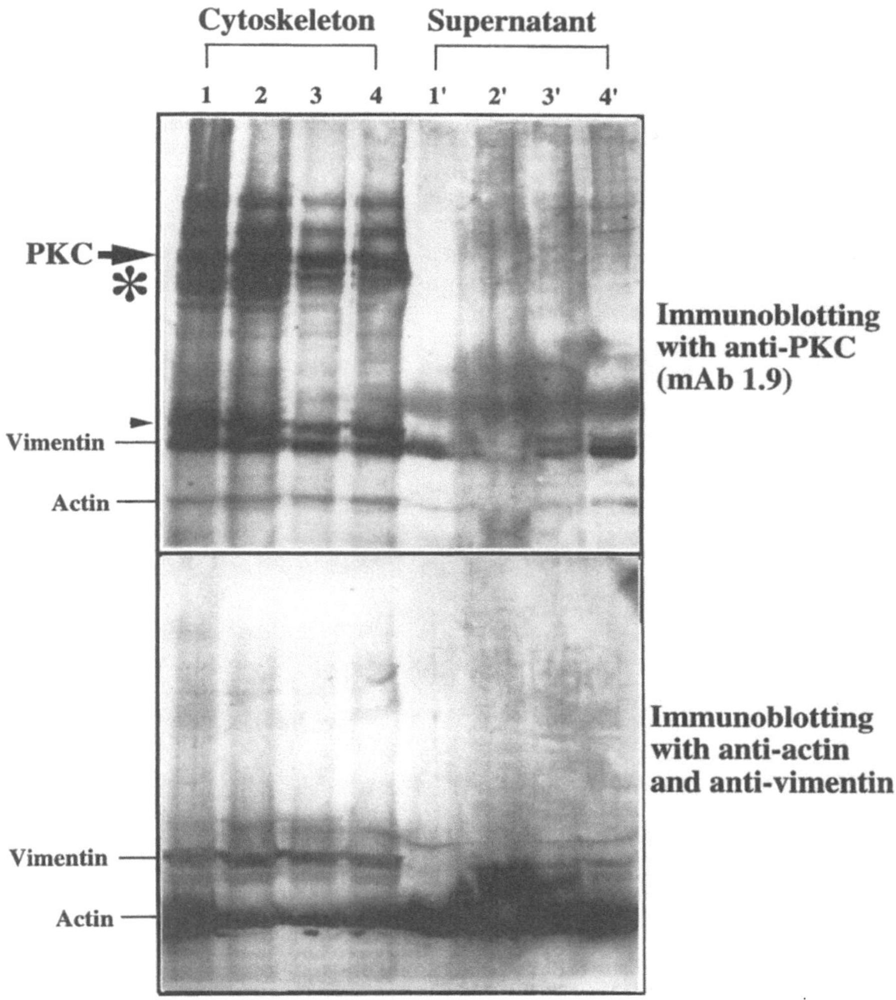

Fig. 7. Protein Kinase C (*PKC*) is endogenously associated with Triton resistant cytoskeleton in CD clone 3 endothelial cells. Endothelial cells were extracted with Triton-containing cytoskeleton extraction buffer (TANG et al. 1993e) for the number of rounds indicated. Equal amounts of cytoskeletal and cytosolic (supernatant) proteins were loaded onto a 12% denaturing SDS-PAGE and the immunoblotting performed with a monoclonal antibody (mAb) to PKC (mAb 1.9; *upper panel*). The same membrane was reprobed with a mixture of anti-actin and anti-vimentin antibodies (*lower panel*). Note that the expected PKC band (~80 kDa; *arrow*) was not extracted off from cytoskeleton although some proteins (*upper panel, asterisk*) below the PKC band were removed from the cytoskeleton after two rounds of extraction. Interestingly, mAb 1.9 also picked up vimentin and, less prominently, actin in the cytoskeletal preparation. Several protein bands above vimentin were also identified with mAb 1.9 which might represent hyperphosphorylated vimentin isoforms or some other cytoskeleton-associated proteins

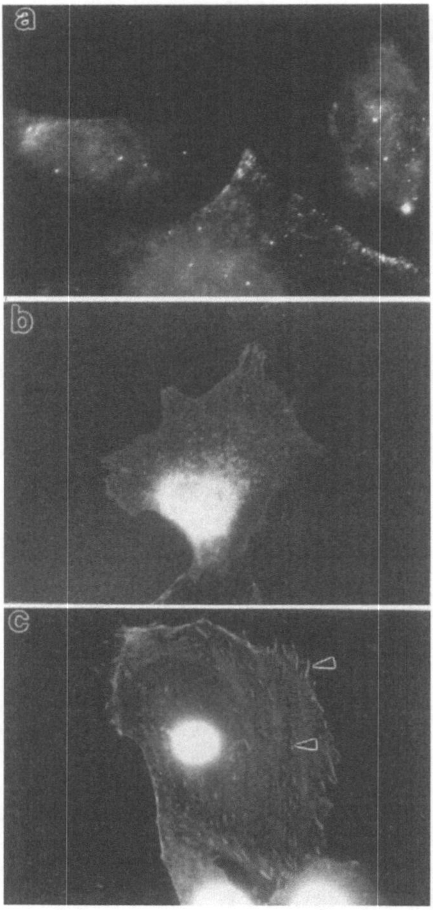

Fig. 8a–c. Differential expression of protein kinase C (PKC) and phosphotyrosine in B16a melanoma cells and CD clone 3 microvascular endothelial cells. **a** PKC staining with monoclonal antibody (mAb) 1.9 in B16a melanoma cells and **b, c** Phosphotyrosine labeling with mAb PT66 in CD clone 3 cells (**b**) and B16a cells (**c**). x600

Acknowledgments. We are indebted to Dr. C. Funk (University of Pennsylvania) for the geneous gifts of the platelet-type 12-LOX cDNA clone and Dr. P.J. Newman (The Blood Center of Southeastern Wisconsin) for antibodies against PECAM-1. Acknowledgement also goes to Dr. C. Diglio for providing us various endothelial cell lines. The work presented in this chapter was supported by the National Institute of Health grantts CA 47115 and CA 29997, the Fogarty International Research grant, and grants from the Development office of Harper Hospital and Wayne State University awarded to K.V.H.

References

Albelda SM, Mueller WA,Buck CA, Newman PJ (1991) Molecular and cellular properties of PECAM-1 (endoCAM/CD31): a novel vascular cell-cell adhesion molecule. J Cell Biol 114: 105–118

Augustin HG, Kozian DH, Johnson RC (1994) Differentiation of endothelial cells: analysis of the constitutive and activated endothelial cell phenotypes. Bioessays 16: 901–906

Bartfeld NS, Pasquale EB, Geltosky JE, Languino LR (1993) The αvβ3 integrin associates with a 190-kDa protein that is phosphorylated on tyrosine in response to platelet-derived growth factor. J Biol Chem 268: 17270–17276

Buchanan MR, Bertomeu MC, Haas TA, Orr FW, Eltringham-Smith L (1993) Localization of 13-HODE and the vitronectin receptor in human endothelial cells and endothelial cell/platelet interactions in vitro. Blood 81: 3303–3312

Chen YQ, Gao X, Timar J, Tang DG, Grossi IM, Chelladurai M, Kunicki TJ, Fligiel SEG, Taylor JD, Honn KV (1992) Identification of the αIIbβ3 integrin in murine tumor cells. J Biol Chem 267: 17314–17320

Chen YQ, Duniec ZM, Liu B, Hagmann W, Gao X, Shimoji K, Marnett LJ, Johnson CR, Honn KV (1994) Endogenous 12(S)-HETE production by tumor cell and its role in metastasis. Cancer Res 54: 1574–1579

Conforti G, Zanetti A, Pasquali-Ronchetti I, Quaglino D Jr, Neyroz P, Marchisio PC, Dejana E (1990) Modulation of vitronectin receptor binding by membrane lipid composition. J Biol Chem 265: 4011–4019

Defilippi P, Truffa G, Stefanuto G, Altruda F, Silengo L, Tarone G (1991) Tumor necrosis factor α and interferon γ modulate the expression of vitronectin receptor (integrin β3) in human endothelial cells. J Biol Chem 266: 7638–7645

De Nichilo MO, Burns GF (1993) Granulocyte-macrophage and macrophage colony-stimulating factors differentially regulate αv integrin expression on cultured human macrophages. Proc Natl Acad Sci USA 90: 2517–2521

Fearon ER, Cho KR, Nigro JM, Kern SE, Simons JW, Ruppert JM, Hamilton SR, Preisinger AC, Thomas G, Kinzler KW, Vogelstein B (1990) Identification of chromosome 18q gene that is altered in colorectal cancer. Science 247: 49–56

Feldman LE, Shin KC, Natale RB, Tòd R (1991) β1 integrin expression on human small cell lung cancers. Cancer Res 51: 1065–1070

Frixen UH, Behrens J, Saches M, Eberle G, Voss B, Warda A, Lochner D, Birchmeier W (1991) E Cadherin-mediated cell-cell adhesion prevents invasiveness of human carcinoma cells. J Cell Biol 113: 173–185

Gehlsen KR, Davis GE, Sriramarao P (1992) Integrin expression in human melanoma cells with differing invasive and metastatic properties. Clin Exp Metastasis 10: 111–120

Giancotti FG, Ruoslahti E (1990) Elevated levels of the α5β1 fibronectin receptor suppress the transformed phenotype of Chinese hamster ovary cells. Cell 60: 849–859

Grossi IM, Hatfield JS, Fitzgerald LA, Newcombe M, Taylor JD, Honn KV (1988) Role of tumor cell glycoproteins immunologically related to glycoproteins Ib and IIb/IIIa in tumor cell-platelet and tumor cell-matrix proteins. FASEB J 2: 2385–2395

Grossi IM, Fitzgerald LA, Umbarger LA, Nelson KK, Diglio CA, Taylor JD, Honn KV (1989) Bidirectional control of membrane expression and/or activation of the tumor cell IRGpIIb/IIIa receptor and tumor cell adhesion by lipoxygenase products of arachidonic and linoleic acid. Cancer Res 49:1029–1037

Günthert U, Hofmann M, Rudy W, Reber S, Zoller M, Hausmann I, Matzkus A, Wenzel A, Ponta H, Herrlich P (1991) A new variant of glycoprotein CD44 confers metastatic potential to rat carcinoma cells. Cell 65: 13–24

Haynes BF, Liao HX, Patton KL (1991) The transmembrane hyaluronate receptor (CD44): multiple function, multiple forms. Cancer Cells 3: 347–350

Hermanowski-Vosatka A, VanStrip JAG, Swiggard WJ, Wright SD (1992) Integrin modulating factor-1, a lipid that alters the function of leukocyte integrins. Cell 68:341–352

Hofmann M, Rudy W, Zoller M, Tolg C, Ponta H, Herrlich P, Günthert U (1991) CD44 splice variants confer metastatic behavior in rats: homologous sequences are expressed in human tumor cell lines. Cancer Res 51: 5292–5297.

Honn KV, Tang DG (1992) Adhesion molecules and cancer cell interaction with endothelium and subendothelial matrix. Cancer Metastasis Rev 11: 353–375

Honn KV, Grossi IM, Diglio CA, Wojtukiewicz M, Taylor JD (1989) Enhanced tumor cell adhesion to the subendothelial matrix resulting from 12(*S*)-HETE-induced endothelial cell retraction. FASEB J 3: 2285–2293

Honn KV, Nelson KK, Renaud C, Baza R, Diglio CA, Timar J (1992) Fatty acid modulation of tumor cell adhesion to microvessel endothelium and experimental metastasis. Prostaglandins 44: 413–429

Honn KV, Tang DG, Duniec ZM, Grossi IM, Timar J, Renaud C, Leithauser M, Blair I, Diglio CA, Kimler VA Taylor JD, Marnett LJ (1994a) Tumor cell-derived 12(*S*)-hydroxyeicosatetraenoic acid induces microvascular endothelial cell retraction. Cancer Res 54: 565–574

Honn KV, Tang DG, Gao X, Butovich IA, Liu B, Timar J, Hagmann W (1994) 12-Lipoxygenases and 12(*S*)-HETE: role in cancer metastasis. Cancer Metastasis Rev 13: 365–396

Honn KV, Tang DG, Grossi IM, Renaud C, Diglio CA (1994) Enhanced endothelial cell retraction mediated by 12(*S*)-HETE: a proposed mechanism for the role of platelets in tumor cell metastasis. Exp Cell Res 210: 1–9

Hsieh JT, Luo W, Song W, Wang Y, Kleinerman DI, Van NT, Lin SH (1995) Tumor suppressor role of an androgen-regulated epithelial cell adhesion molecule (C-CAM) in prostate carcinoma cell revealed by sense and antisense approaches. Cancer Res 55: 190–197

Ignotz RA, Heino J, Massague J (1989) Regulation of cell adhesion receptors by transforming growth factor-β. Regulation of vitronectin receptor and LFA-1. J Biol Chem 264: 389–392

Johnson JP (1991) Cell adhesion molecules of the immunoglobulin supergene family and their role in malignant transformation and progression to metastatic disease. Cancer Metastasis Rev 10: 11–22

Johnson JP, Stade BG, Holmann B, Schwable W, Riethmuller G (1989) De novo expression of intercellular adhesion molecule-1 in melanoma correlates with increased risk of metastasis. Proc Natl Acad Sci USA 86: 641–644

Jones CL, Honn KV (1992) Enhanced membrane expression of adhesion αIIbβ3 in Lewis lung carcinoma cells by epoxyeicosatetraenoic acids. In: Nigam S, Honn KV, Marnett LJ, Walden T (eds) Eicosanoids and other bioactive lipids in cancer, inflammation and radiation injury. Kluwer Academic, Boston, pp 671–678

Klein S, Giancotti FG, Presta M, Albelda SM, Buck CA, Rifkin DB (1993) Basic fibroblast growth factor modulates integrin expression in microvascular endothelial cells. Mol Biol Cell 4: 973–982

Kleinerman DI, Troncoso P, Lin SH, Pisters LL, Sherwood ER, Brooks T, von Eschenbach AC, Hsieh JT (1995) Consistent expression of an epithelial cell adhesion molecule (C-CAM) during human prostate development and loss of expression in prostate cancer: implication as a tumor suppressor. Cancer Res 55: 1215–1220

Lafrenie RM, Podor TJ, Buchanan MR, Orr FW (1992) Up-regulated biosynthesis and expression of endothelial cell vitronectin receptor enhances cancer cell adhesion. Cancer Res 52: 2202–2208

Liu B, Timar J, Howlett J, Diglio CA, Honn KV (1991) Lipoxygenase metabolites of arachidonic and linoleic acids modulate the adhesion of tumor cells to endothelium via regulation of protein kinase C. Cell Regul 2: 1045–1055

Liu B, Maher RJ, Hannun Y, Porter AT, Honn KV (1994a) 12(*S*)-HETE enhancement of prostate tumor cell invasion: selective role of PKCα. J Natl Cancer Inst 86: 1145–1151

Liu B, Marnett LJ, Chaudhary A, Ji C, Blair IA, Johnson CR, Diglio CA, Honn KV (1994b) Biosynthesis of 12(*S*)-hydroxyeicosatetraenoic acid by B16 amelanotic melanoma cells is a determinant of their metastatic potential. Lab Invest 70: 314–323

Liu B, Khan WA, Hannun YA, Timar J, Taylor JD, Lundy S, Butovich I, Honn KV (1995) 12(*S*)-HETE and 13(*S*)-HODE regulation of protein kinase C alpha in melanoma cells: role of receptor mediated hydrolysis of inositol phospholipids. Proc Natl Acad Sci USA 92: 9323–9327

Lorant DE, Patel KD, McIntyre TM, McEver RP, Prescott SM, Zimmerman GA (1991) Coexpression of GMP-140 and PAF by endothelium stimulated by histamine or thrombin: a juxtacrine system for adhesion and activation of neutrophils. J Cell Biol 115: 223–234

Medhora MM, Teitelbaum S, Chappel J, Alvarez J, Mimura H, Ross F, Hruska K (1993). 1α, 25-dihydroxyvitamin D3 up-regulates expression of the osteoclast integrin αvβ3. J Biol Chem 268: 1456–1461

Milam SB, Magnuson VL, Steffensen B, Chen DL, Klebe RJ (1991) IL-1β and prostaglandins regulate integrin mRNA expression. J Cell Physiol 149: 173–183

Newman PJ, Bernt MC, Gorski J, White GC, Lyman S, Paddock C, Muller WA (1990) PECAM-1 (CD31) cloning and relation to adhesion molecules of the immunoglobulin gene superfamily. Science 247: 1219–1222

Nickloff BJ (1993) PECAM-1 (CD31) is expressed on proliferating endothelial cells, stromal spindle-shaped cells, and dermal dendrocytes in Karposi's sarcoma. Arch Dermatol 129: 250–251.

Perrotti D, Cimino L, Falcioni R, Tibursi G, Gentileschi MP, Sacchi A (1990) Metastatic phenotype: growth factor dependence and integrin expression. Anticancer Res 10: 587–598

Roosien FF, DeRijik D, Birkker A, Roos E (1989) Involvement of LFA-1 in lymphoma invasion and metastasis demonstrated with LFA-1 deficient mutants. J Cell Biol 108: 1979–1985

Rossino P, Defilippi R, Silengo L, Tarone G (1991) Up-regulation of the integrin α1/β1 in human neuroblastoma cells differentiated by retinoic acid: correlation with increased neurite outgrowth reponses to laminin. Cell Regul 2: 1021–1033

Satoh K, Narumi K, Isemura M, Sakai T, Abe T, Matsushima K, Okuda K, Motomiya M (1992) Increased expression of the 67kd laminin receptor gene in human small cell lung cancer. Biochem Biophys Res Commun 182: 746–752

Schipper JH, Frixen UH, Behrens J, Unger A, Jahnke K, Birchmeier W (1991) E-cadherin expression in squamous cell carcinoma of head and neck: inverse correlation with tumor dedifferentiation and lymph node metastasis. Cancer Res 51: 6328–6337

Schreiber C, Bauer J, Margolis M, Juliano RL (1991a) Expression and role of integrins in adhesion of human colonic carcinoma cells to extracellular matrix components. Clin Exp Metastasis 2: 163–178

Schreiber C, Fisher M, Hussein S, Juliano RL (1991b) Increased tumorigenicity of fibronectin receptor deficient Chinese hamster ovary cell variants. Cancer Res 51: 1738–1740

Schwartz MA, Brown EJ, Fazeli B (1993) A 50-kDa integrin-associated protein is required for integrin-regulated calcium entry in endothelial cells. J Biol Chem 268: 19931–19934

Setty BNY, Graeber JE, Stuart MJ (1987) The mitogenic effect of 15- and 12- hydroxyeicosatetraenoic acid on endothelial cells may be mediated via diacylglycerol kinase inhibition. J Biol Chem 262: 17613–17622

Smith JW, Piotrowicz RS, Mathis D (1994) A mechanism for divalent cation regulation of β3-integrins. J Biol Chem 269: 960–967

Soler AP, Johnson KR, Wheelock MJ, Knudsen KA (1993) Rhabdomyosarcoma-derived cell lines exhibit abberant expression of the cell-cell adhesion molecules N-CAM, N-cadherin, and cadherin-associated proteins. Exp Cell Res 208: 84–93

Sommers CL, Thompson EW, Torri JA, Kemler R, Gelmann EP, Byers SW (1991) Cell adhesion molecule uvomorulin expression in human breast cancer cell lines: relationship to morphology and invasive capacities. Cell Growth Differ 2: 365–372

Spector AA, Gordon JA, Moore SA (1988) Hydroxyeicosatetraenoic acids (HETEs). Prog Lipid Res 27: 271–323

Stallmach A, Von Lampe BV, Matthes H, Bornhoft G, Riecken EO (1992) Diminished expression of integrin adhesion molecules on human colonic epithelial cells during the benign to malignant tumor transformation. Gut 33: 342–346

Tang DG, Honn KV (1994) 12-Lipoxygenase, 12(*S*)-HETE, and cancer metastasis. Ann NY Acad Sci 722: 199–215

Tang DG, Honn KV (1995) Adhesion molecules and tumor metastasis. Invasion Metastasis 14: 109–122

Tang DG, Chen Y, Diglio CA, Honn KV (1993a) PKC-dependent effects of 12(*S*)-HETE on endothelial cell vitronectin and fibronectin receptor. J Cell Biol 121: 689–704

Tang DG, Diglio CA, Honn KV (1993b) 12(*S*)-HETE-induced microvascular endothelial cell retraction results from PKC-dependent rearrangement of cytoskeletal elements and αvβ3 integrin. Prostaglandins 45: 249–268

Tang DG, Grossi IM, Diglio CA, Honn KV (1993c) 12(*S*)-HETE promotes tumor cell adhesion by increasing surface expression of αvβ3 integrins on endothelial cells. Int J Cancer 54: 102–111

Tang DG, Onoda JM, Grossi IM, Nelson KK, Umbarger L, Honn KV (1993d) Phenotypic properties of tumor cells in culture: αIIbβ3 integrin expression, tumor cell induced platelet aggregation, and tumor cell adhesion to endothelium as important parameters of metastatic potential. Int J Cancer 54: 338–347

Tang DG, Timar J, Grossi IM, Renaud C, Kimler V, Diglio CA, Taylor JD, Honn KV (1993e) The lipoxygenase metabolite, 12(*S*)-HETE, induces a protein kinase C dependent cytoskeletal re-arrangement and retraction of microvascular endothelial cell. Exp Cell Res 207: 361–375

Tang DG, Chen Y, Newman PJ, Shi L, Gao X, Diglio CA, Honn KV (1993f) Identification of PECAM-1 in solid tumor cells and its potential involvement in tumor cell adhesion to endothelium. J Biol Chem 268: 22883–22894

Tang DG, Diglio CA, Honn KV (1994) Activation of microvascular endothelium by 12(*S*)-HETE leads to enhanced tumor cell adhesion via upregulation of surface expression of αvβ3 integrin: a post-transcriptional, PKC- and cytoskeleton-dependent process. Cancer Res 54: 1119–1129

Tang DG, Diglio CA, Bazaz R, Honn KV (1995a) Transcriptional activation of integrin αv in microvascular endothelial cells by PKC activator 12(*S*)-HETE. J Cell Sci 108: 2629–2644

Tang DG, Tarrien M, Dobrzyski P, Honn KV (1995b) Melanoma cell spreading on fibronectin induced by 12(*S*)-HETE involves both PKC- and protein tyrosine kinase-dependent focal adhesion formation and tyrosine phosphorylation of focal adhesion kinase ($pp125^{FAK}$) J Cell Physiol 165: 291–306

Timar J, Chen Y, Liu B, Bazaz R, Taylor JD, Honn KV (1992) The lipoxygenase metabolite 12(*S*)-HETE promotes αIIbβ3 integrin-dependent tumor cell spreading on fibronectin. Int J Cancer 52: 594–603

Timar J, Tang DG, Bazaz R, Rong X, Kimler V, Taylor JD, Honn KV (1993) 12(*S*)-HETE induces the rearrangement of tumor cell cytoskeleton by PKC-dependent mechanism. Cell Motil Cytoskel 26: 49–65

Timar J, Bazaz R, Kimler V, Haddad M, Tang DG, Robertson D, Taylor JD, Honn KV (1995) Immunological characterization and effects of 12(*S*)-HETE on a dynamic intracellular pool of the αIIbβ3-integrin in melanoma cells. J Cell Sci 108: 2175–2186

Umbas R, Isaacs WB, Bringuier PP, Schaafsma EHE, Karthaus HFM, Oosterhof GON, Debruyne FMJ, Schalken JA (1994) Decreased E-cadherin expression is associated with poor prognosis in patients with prostate cancer. Cancer Res 54: 3929–3933

Varner JA, Fisher MH, Juliano RL (1992) Ectopic expression of integrin α5β1 suppresses in vitro growth and tumorigenicity of human colon carcinoma cells. Mol Biol Cell 3: 232a

Vleminckx K, Vakaek L, Mareel M, Fiers W, Van Roy F (1991) Genetic manipulation of E-cadherin expression reveals an invasion suppressor role. Cell 66: 107–119

Zimmerman GA, McIntyre TM, Mehra M, Prescott SM (1990) Endothelial cell-associated platelet factor: a novel mechanism for signaling intercellular adhesion. J Cell Biol 110: 529–540

Zocchi MR, Vidal M, Poggi A (1993) Involvement of CD56/N-CAM molecule in the adhesion of human solid tumor cell lines to endothelial cells. Exp Cell Res 204: 130–135

Zutter MM, Mazoujian G, Santoro SA (1990) Decreased expression of integrin adhesive protein receptors in adenocarcinoma of the breast. Am J Pathol 137: 863–870

Brain Metastasis: Role of Trophic, Autocrine, and Paracrine Factors in Tumor Invasion and Colonization of the Central Nervous System

G.L. NICOLSON, D.G. MENTER, J.L. HERRMANN, Z. YUN, P. CAVANAUGH, and D. MARCHETTI

1	Introduction	89
1.1	Malignant Tumors and Brain Metastasis	90
2	Brain-Trophic Factors and Metastasis	91
3	Neurotrophins and Neurotrophin Receptors	91
4	Neurotrophin Signaling Mechanisms	93
4.1	Cooperation in Neurotrophin Signaling Pathways	94
4.2	Neurotrophin Signal Transduction Pathways	95
5	Neurotrophin Receptors, Tumor Invasion, and Metastasis	98
5.1	Overexpression of $P75^{NTR}$ on Brain-Metastatic Human Melanoma Cells	98
5.2	Neurotrophins Enhance Invasion and Degradative Enzyme Production of Brain-Metastatic Melanoma Cells	99
6	Neurotrophins and Tumor Cell Survival and Death	99
7	Autocrine Growth Factors and Brain Metastasis	101
8	Neurotrophin Production and Brain Invasion	101
9	Paracrine Growth Factors and Brain Metastasis	102
9.1	Reciprocal Cytokine Regulation of Brain-Metastatic Cell Growth	104
10	Brain Environment and Tumor Metastasis	105
10.1	Cellular Responses to Brain Tissue Injury as a Paradigm for Brain Metastasis	106
10.2	Melanotropic Peptides May Modulate Astroglial and Tumor Cell Behavior	108
11	Conclusions	109
References		110

1 Introduction

The brain is a unique target for tumor invasion and metastasis formation (STECK and NICOLSON 1993; NICOLSON 1993a; MENTER et al. 1995a; NICOLSON et al.1994b. The central nervous system (CNS) is confined by the skull, and the brain is highly sensitive to the slightest change in the local microenvironment. The brain is also surrounded by a formidable blood-brain barrier (BBB), which does not allow penetration of most cell

Department of Tumor Biology, University of Texas, MD Anderson Cancer Center, Houston, TX 77030, USA types. Because the brain lacks lymphatic drainage, cerebral edema is a major complication resulting from tumors in the CNS. When tumors form in the brain, either from endogenous brain cells or from metastases invading the BBB, they are very difficult to successfully treat. Therapy of brain tumors and metastases are often only palliative and are often accompanied by additional complications. A relatively small tumor in the brain cavity can cause severe symptoms, including impaired cognition, headaches, seizures, and eventually paralysis.

1.1 Malignant Tumors and Brain Metastasis

To metastasize to the CNS, blood-borne malignant tumor cells must attach to microvessel endothelial cells, respond to brain-derived invasion factors, invade the BBB, and respond to paracrine survival and growth factors (for a review see Nicolson 1993a). Although metastasis formation in other organs may be tolerated or remain asymptomatic, once metastatic cells colonize the CNS, tumor growth often results in a rapid decline in the quality of life and ensuing death. In addition, as mentioned above, once metastases form in the CNS, they are extremely difficult to treat. Problems in treatment include difficulties in drug delivery to the CNS as well as problems such as cerebral edema and other conditions detrimental to patient survival (Wright and Delaney 1989).

Malignant melanoma metastasizes to the brain with one of the highest frequencies of any cancer that is capable of colonizing the CNS. Patients with disseminated malignant melanoma frequently develop metastatic lesions in the brain and spinal cord that can result in severe and debilitating neurological complications (Wright and Delaney 1989; Steck and Nicolson 1993). Approximately 13% of cancer patients will present with symptomatic complications related to brain metastases, but almost 40% of melanoma patients will be treated for complications due to brain metastases. At autopsy, 70%–80% of malignant melanoma patients have CNS lesions (Stehlin et al. 1967; Sugarbaker 1981).

Melanomas undergo progressive changes during their malignant progression. Of the phenotypic changes that occur during metastatic melanoma progression of tumors capable of forming brain metastases, differences in the expression of receptors for paracrine growth factors and the production of various autocrine growth factors are important (Herlyn et al. 1985; Albino et al. 1991). Differences in growth factor production between cultured melanocytes and malignant melanoma cell lines include increased autocrine production of transforming growth factor (TGF)-β, TGF-α basic fibroblast growth factor (bFGF), keratinocyte growth factor, and the A chain of platelet-derived growth factor (PDGF; Albino et al. 1991). The significance of these autocrine factors in modulating the malignant properties exhibited by melanoma cells remains largely unknown, but they are thought to be important in allowing malignant cells to survive in unusual compartments such as the brain, where paracrine growth factors differ from most other organs.

2 Brain-Trophic Factors and Metastasis

The growth of tumors depends on cell proliferation, and this usually occurs in response to specific growth factors from either autocrine or paracrine sources (for reviews see NICOLSON 1993b,c). Another important mechanism leading to tumor growth is the failure to initiate programmed cell death (apoptosis, RAFF 1992). During their growth, tumors frequently contain highly malignant cells that become clonally selected from the proliferating growth fraction of cells and eventually become the dominant cell subpopulations (NOWELL 1976; NICOLSON 1987). When conditions within the tumor microenvironment become growth limiting, however, malignant cells may be forced to utilize other mechanisms in order to survive. Under growth-limiting conditions, trophic factors can have a profound effect on cell survival. Trophic factors can support malignant cell survival in a state of growth arrest or dormancy, even when the absence of growth factors usually signals cell death. Consequently, tumor cells that are responsive to trophic factors may survive host selective pressures on tumor cells to proliferate, further diversify, and become clonally dominant. A heavily investigated area of developmental biology is the role of trophic factors in organ development. Research on brain-trophic factors has resulted in important hypotheses on the development and differentiation of neural cells (RAFF et al. 1993), but the role of trophism as a support mechanism for tumor cell survival is only now beginning to be examined.

One of the best examples for the role of organ-trophic substances in cell survival can be demonstrated by the actions of neurotrophins. Neurotrophins can promote targeted tissue invasion and survival of certain neuronal cells (RAFF et al.1993), and such observations resulted in the 'neurotrophic theory' to account for the massive neuronal cell death and innervation in developing embryos (for a review, see SNIDER 1994). Neuronal cells that are overproduced in the embryo must compete for limited supplies of target tissue-derived neurotrophins. In this environment, the overwhelming majority of neurons die by apoptosis, leaving only small numbers of surviving neurons, and these are the neurons that successfully innervate the target tissue. Applying this concept of trophism, death and survival to rapidly growing tumors or large numbers of metastatic cells in the blood circulation or at implantation sites may account, in part, for the survival of only certain tumor cells under growth-limiting conditions. This may be particularly true of neurotrophic effects on tumor cells that have neuroectodermal origins, such as malignant melanoma cells (MENTER et al. 1995a,b).

3 Neurotrophins and Neurotrophin Receptors

Neurotrophins are small highly basic proteins (pI 9–10.5) that are synthesized as prepropeptides that are processed to proteins containing three interchain disulfide bonds (for a review see BRADSHAW et al. 1993). The circulating forms of neurotrophins are nonglycosylated proteins approximately 26 kDa in size. Each protein monomer contains an elongate central axis made of an antiparallel β-sheet structure with a flattened hydrophobic face that is involved in dimer formation. At one end of each monomer there are three β-hairpin loops, and there is one at the opposite end comprising regions that vary between homologous neurotrophin family members.

Nerve growth factor (NGF) is the prototype neurotrophin, but other members of this homologous protein family exhibit neurotrophic properties. Brain-derived neurotrophic factor (BDNF), isolated from brain tissue, shows significant amino acid homology approximately (50%) with NGF. Similarly, neurotrophin (NT)-3 and NT-4, isolated originally from Xenopus, and its mammalian homologue NT-5, are all highly conserved in the amino acid sequence in the region of the central axis of the molecule. These molecules derive their unique functional properties from the four variable β-hairpin loops which are involved in binding to various neurotrophin receptors (BRADSHAW et al. 1993).

Neurotrophin receptors (Fig. 1) can be divided into two affinity classes, a low-affinity receptor class (K_d=approximately 2×10^{-9}) and a high-affinity receptor class (K_d=approximately 2×10^{-11}).The gene encoding the human low-affinity NGF receptor (NGFR or $p75^{NTR}$) was cloned by JOHNSON et al. (1986). This human gene encodes a 75-kDa cell surface glycoprotein made up of 399 amino acids, including a 222-amino acid extracellular domain, a 22-amino acid transmembrane domain, and a 155-amino acid cytoplasmic segment. The molecule contains four cysteine-rich extracellular domains and a G protein-binding consensus sequence in the cytoplasmic domain.

The biologic effects of NGF involve tyrosine kinase activation (MAHER 1988; MIYASAKA et al. 1990). Sequence analysis of $p75^{NTR}$, however, indicates that this molecule lacks a tyrosine kinase consensus sequence (JOHNSON et al. 1986). Despite the absence of a tyrosine kinase domain, transfection of $p75^{NTR}$ into cells enhanced tyrosine kinase phosphorylation following NGF stimulation (OHMICHI et al. 1992a), suggesting that $p75^{NTR}$ can mediate neurotrophin signaling.

The search for a high-affinity NGF receptor with tyrosine kinase activity resulted in the discovery of the TRK family of neurotrophin receptors. The TRK family of tyrosine receptor protein kinases consists of several receptor molecules (TrkA, TrkB, TrkC, TrkD) with varying degrees of specificity for the different members of the neurotrophin family (for reviews see CHAO 1992; MEAKIN and SHOOTER 1992; BARBACID 1993). These receptors constitute the high-affinity neurotrophin family members. Each mature $p140^{trkA}$ receptor contains a 375-amino acid extracellular domain, a 26-amino acid transmembrane span, and a large cytoplasmic domain of 357 amino acids. The trkB and trkC genes encode molecules of approximately 145kDa which are known to also exist as truncated forms or contain inserts in their tyrosine kinase domains. All of the TRK family members share distinct structural motifs in the glycosylated extracellular domains (Fig.1). The TRK receptors are widely distributed in neuronal and some nonneuronal tissues (for reviews see CHAO 1992; MEAKIN and SHOOTER 1992; BARBACID 1993).

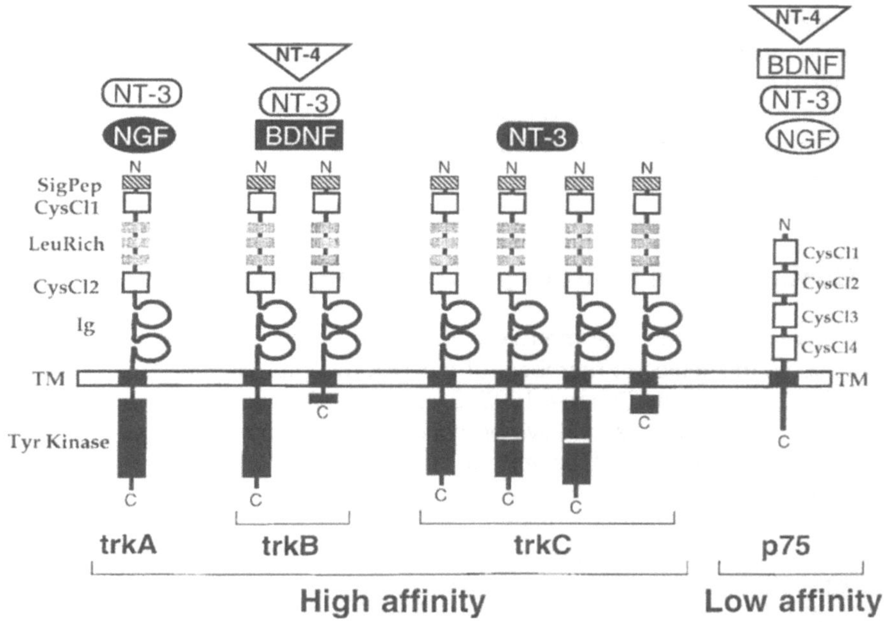

Fig. 1. Neurotrophins and the two different affinity class receptors that bind them. The neurotrophins are a family of small (approximately 13 kDa) proteins that are highly basic (p19–10.5). The family consists of nerve growth factor (*NGF*), brain-derived neurotrophic factor (*BDNF*), neurotophin (*NT*)-3, and NT-4/5. The high-affinity class receptors (approximately $K_d = 2 \times 10^{-11}$) consist of a family of tropomyosin receptor kinase (TRK) molecules. First cloned from colon carcinoma, *trkA* has been histochemically localized to nervous tissues. Hybridization cloning led to the discovery of the other family members, *trkB* and *trkC*, as well as their isoforms. These isoforms contain both deletions and insertions in their cytoplasmic domains. The characteristic extracellular domain structures shared by these receptors include a signal peptide (*SigPep*) and a leucine rich domain (*LeuRich*) flanked by two cysteine cluster regions (*CysCl1*, Cyscl *2*) as well as immunoglobulin C2 domains (*Ig*). The cytoplasmic region contains the TRK tyrosine kinase domain, which is highly conserved among full-length receptors. The primary ligands for each TRK receptors are NGF, BDNF, and NT-3, whch bind to *trkA*, *trkB*, and *trkC* respectively. There is also some cross-reactivity of NT-3 for 140^{trkA} as well as NT-3 and NT-4 for 145^{trkB}. The low-affinity receptor class consists of only $p75^{NTR}$, which is characterized by a series of cysteine clusters (CysCl1–4) in the extracellular domain. The $p75^{NTR}$ cytoplasmic domain is characterized by a G protein consensus sequence. The $p75^{NTR}$ receptor binds all of the neurotrophins with low affinity (K_d approximately 2×10^{-9})

4 Neurotrophin Signaling Mechanisms

The role of the low- and high-affinity neurotrophin receptor classes in neurotrophin signaling remains a puzzle that is just beginning to be understood. A number of studies suggest that cooperativity can occur between $p140^{trkA}$ and $p75^{NTR}$ receptors to establish functional receptor complexes in the presence of NGF. Some of the most illustrative experiments have utilized knockout mice. Genetic ablation of trkA or NGF production decreases responsiveness to painful stimuli in null allele mice. Most of these animals die by 3 weeks of age (Crowley et al. 1994; Smeyne et al. 1994). In contrast, $p75^{NTR}$ (-/-) receptor mice exhibit similar decreases in responsiveness to pain, but many of these animals live to be adults (Lee et al. 1992, 1994). The differences in responsiveness to pain are explained by differential survival of various sensory neurons. The *trkA* (-/-) animals show more massive losses of peripheral neurons than seen in the $p75^{NTR}$ (-/-) animals.

4.1 Cooperation in Neurotrophin Signaling Pathways

Some investigators feel that the role of $p75^{NTR}$ is to procure and present bound neurotrophin molecules to members of the TRK receptor tyrosine kinase family (Chao 1992; Meakin and Shooter 1992; Barbacid 1993). The $p75^{NTR}$ gene can be transfected into NGF-unresponsive or-responsive cells to establish cooperation between $p140^{trKA}$ and $p75^{NTR}$ receptors (Hempstead et al 1991). Similarly, transfection of trkA into melanoma cells that express high levels of $p75^{NTR}$ results in the expression of high-affinity NGF receptors, greater than expected for $p140^{trkA}$ alone (Hempstead et al. 1991). Using immortalized cells that are unresponsive to NGF, Verdi et al. (1994) obtained evidence that supports the notion that the introduction of the $p75^{NTR}$ gene can enhance catalytic activation of coexpressed $p140^{trKA}$ receptors. Relative to cells expressing only TrKA, coexpression of $p75^{NTR}$ and $p140^{trKA}$ resulted in increases in downstream signaling and neurotrophin responses (Hempstead et al. 1989; Berg et al. 1991).

The ability of $p75^{NTR}$ to undergo cooperative interactions with the other neurotrophin receptors appears to have been accepted by most investigators. For example, recent evidence based on anti-$p75^{NTR}$ antibody injections into chick embryos suggests that neurotrophins cannot interact with antibody-blocked $p75^{NTR}$ but $p75^{NTR}$ can still cooperate with $p140^{trKA}$ to form functional signaling pathways (von Bartheld et al. 1994). Collectively, these data illustrate the importance of cooperativity between the TRK family of receptors and $p75^{NTR}$ for enhancing the neurotrophin responses.

It seems likely that $p140^{trKA}$ and $p75^{NTR}$ interactions or cooperativity occurs after ligand receptor interaction and does not involve direct receptor-receptor contact. This notion follows from the failure of various antibodies to co-immunoprecipitate $p140^{trKA}$ and $p75^{NTR}$ from mild detergent cell lysates (Meakin and Shooter 1991), even though anti-$p75^{NTR}$ can prevent NGF at high concentrations from binding to $p75^{NTR}$ receptors on cells that express functional $p140^{trKA}$. In addition, antibody treatment was not able to prevent neurite outgrowth, suggesting that distal cytoplasmic interactions may be involved in the cooperative effects (Weskanp and Reichardt 1991). Similarly, at low concentrations of NGF, BDNF or NGF binding to $p75^{NTR}$ was blocked by anti-$p75^{NTR}$ antibodies, but the Trk signaling pathway was still functional, as evidenced by c-fos activation (Barker and Shooter 1994). Moreover, c-fos activation could be attenuated by the addition of anti-$p75^{NTR}$ antibodies in the presence of low concentrations of NGF ligand, but not by $p75^{NTR}$ saturated with BDNF (Barker and Shooter 1994). Although the authors concluded that $p75^{NTR}$ increases the local concentrations of NGF and stimulates $p140^{trKA}$ activation, they could not rule out direct signaling mediated by $p75^{NTR}$. Other evidence against the formation of heterodimeric receptor complexes is that the cross-linking of NGF to TrKA receptors yields only $p140^{trKA}$/NGF complexes, but not $p140^{trKA}$/$p75^{NTR}$/NGF complexes (MEAKIN and SHOOTER 1991). Thus $p140^{trKA}$ and $p75^{NTR}$ receptors may not be in close physical proximity in NGF signaling complexes, and their interactions may occur indirectly in cytoplasmic signaling pathways.

Some studies have implicated the direct involvement of $p75^{NTR}$ in signal transduction independent of Trk receptors. The $p75^{NTR}$ cytoplasmic tail contains a 14 amino acid mastoparan (wasptoxin)-like domain, but it lacks a tyrosine kinase domain (JOHNSON et al. 1986; FEINSTEIN and LARHAMMAR 1990). Activation of a G-stimulatory protein complex (Gs) in the presence or absence of NGF may lead to the production of cyclic adenosine monophosphate (cAMP) by adenylate cyclase and activation of protein kinase A (PKA), followed by transcription factor activation. This is supported by recent studies in which $p75^{NTR}$ gene transfection into TrkA-deficient cells resulted in increased cAMP production following NGF stimulation (KNIPPER et al 1993). Elevated cAMP activation of a protein kinase pathway may influence neurotrophin binding or downstream signaling via a TRK pathway. These pathways may have other as yet undiscovered functions in the presence or absence of ligand, which may explain some of the properties of cells expressing $p75^{NTR}$ in the absence of TRK receptors (RABIZADEH et al. 1993).

Transfection studies have shown that the cytoplasmic tail of $p75^{NTR}$ is essential for high-affinity NGF binding to $p140^{trKA}$ (HEMPSTEAD et al. 1990; HANTZOPOULOS et al. 1994). In immunoprecipitation assays following NGF stimulation, a 45- to 47-KDa serine/threonine protein kinase that is sensitive to purine analogues (protein kinase N, PKN) has been found to be bound to $p75^{NTR}$ (VOLONTE and GREENE 1992; VOLONTE et al. 1993a). The activation of this PKN in association with stimulation of ornithine decarboxylase activity may play an important role in the signaling pathways associated with $p75^{NTR}$ (VOLONTE and GREENE 1990).

Downstream signals from neurotrophin/$p75^{NTR}$/PKN complexes may cooperate in amplifying $p140^{trKA}$ signals. Thus when neurotrophin concentrations are high, the low-affinity activation of $p75^{NTR}$/PKN may amplify the $p140^{trKA}$ response pathway. In contrase, when neurotrophin levels are low, $p75^{NTR}$ signals may be routed along an alternative pathway, allowing $p75^{NTR}$ with its low affinity for neurotrophins to act as a sensitive molecular switch (MENTER et al. 1995a). Neurotrophins can signal a wide variety of effects, such as cell invasion, differentiation, survival, or apoptosis, and not much is known about the specificity of these signals and how these signaling pathways diverge, interact, or become aberrant in malignant cells.

4.2 Neurotrophin Signal Transduction Pathways

Like other receptor tyrosine kinase (RTK) receptors, neurotrophin RTK are involved in a sequence of events that include ligand binding leading to receptor dimer formation, transactivation resulting in Tyr phosphorylation, and eventually activation of ser/thr phosphorylation cascades (Fig. 2). Active signaling complexes can be formed by interactions between receptor phosphotyrosines and proteins containing SH2 (src homology-2) Tyr-binding domains. For example, an activated p140[trkA] receptor can bind to SH2-bearing phospholipase-Cγ1 (PLCγ1) (OHMICHI et al. 1991) along with a 38-kDa phosphoprotein of unknown function (OHMICHI et al 1992a). In a contrast to other RTK, this occurs in the absence of detectable association with GTPase-activating proteins. Furthermore, unlike other RTK, phosphotidylinositol-3 kinase (P13-K) is transiently activated, and accumulation of

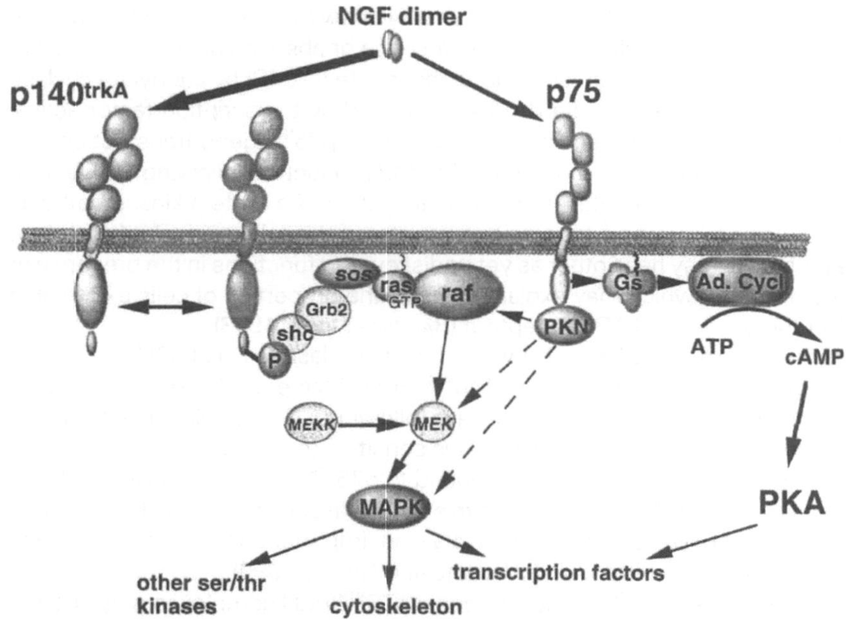

Fig. 2. Possible mechanisms involved in neurotrophic signal transduction. The dimerized form of the nerve growth factor (*NGF*) molecule, as an example, binds to either p140[trkA] or p75[NTR]. Homodimerization followed by phosphorylation of p140[trkA] and possibile recruitment of p75[NTR] leads to the cytosolic binding of src homology-2 (SH2) phosphotyrosine-binding proteins. The SH2 domain of Shc, an SH2-containing protein, binds to the phosphotyrosine at position 490 on the Trk cytoplasmic domain. Phosphorylation of Shc promotes the SH2-driven binding and activation of Grb2, which binds to and activates son of sevenless (*SOS*)-1, a nucleotide exchange factor. The action of SOS-1 elicits the stabilization of guanosine triphosphate (GTP) association with p21ras. The activated form of p21ras-GTP is stably associated with the plasma membrane during its recruitment to the signaling complex leading to Raf-1 binding to p21ras-GTP. Raf-1 activation accurs by an unknown mechanism that may utilize as yet unidentified factors and causes the Ser/Thr phosphorylation of *MEK*-(mitrogen-activated protein kinase, *MAPK*/extracellular signal-regulated kinase, *ERK*), which in turn phosphorylates MAPK on Ser/Thr. MAPK activation leads to further activation of other Ser/Thr kinases, cytoskeletal elements, and transcription factors (e.g. fos/jun). The binding of NGF to p75[NTR] can initiate G-stimulatory protein complexes (G_s) to activate cyclic adenosine monophosphate (*cAMP*) production by adenylate cyclase and activation of protein kinase A (*PKA*) followed by transcriptionaly factor activation. The p75[NTR] receptor is associated with the Ser/Thr phosphorylate protein kinase N (*PKN*), which may interact directly with Raf-1, MEK, or MAPK to amplify the NGF response. Other MEK kinases (*MEKK*) may act on the Ser/Thr kinase cascade independently of Raf-1

inositol phosphate-3 (IP3), calcium mobilization, or association of p140^{trKA} with p85 does not occur (OHMICHI et al. 1992b).

Activation of TRK signal pathways may involve Shc, an SH2-containing protein (ROZAKIS-ADCOCK et al. 1992). Several reports have demonstrated complex formation between pp140^{trKA} and the SH2 domain of Shc. Formation of this complex leads to tyrosine phosphorylation on Shc and the association of Shc with Grb2, another SH2-containing protein (OBERMEIER et al. 1993; BORELLO et al. 1994; STEPHENS et al. 1994). Signaling complexes for PC12 cell neurite extension may rely on cooperative interactions between PLCγ1 and Shc, but P13-K does not seem to be essential (OBERMEIER et al. 1994). The association of Shc with Grb2 can lead to further complex formation with the p21^{ras} nucleotide exchange factor son of sevenless-1 (SOS-1).

p21^{ras} can affect the NGF-mediated, phosphorylation-dependent activation of several key growth and differentiation molecules. These include: (a) ser/thr kinase c-Raf-1, (b) mitogen-activated protein kinase/extracellular signal-regulated kinase (MAPK/ERK kinase, MEK), and (c) mitogen-activated protein kinase (MAPK) (AVRUCH et al. 1994). The activation of MAPK can transiently induce the expression of a number of primary response genes that encode transcription factors (BATISTATOU et al. 1992). The MAPK activity may also affect other ser/thr kinases and/or cytoskeletal elements TAYLOR et al. 1994). Recently, MEK kinase (MEKK), a ser/thr kinase that can activate MEK independently of Raf-1, was shown to phosphorylate MEK in PC12 cells as they respond to NGF (LANGE-CARTER and JOHNSON 1994).

The association of Ras with Raf may help explain the coupling of different signal transduction pathways that result in cell differentiation or division (AVRUCH et al. 1994). If the Ras-derived CAAX farnesylation motif is added to the carboxyl terminal of Raf-1, it can associate with the plasma membrane and become constitutively activated (LEEVERS et al. 1994; STOKOE et al. 1994). Activated Raf-1 phosphorylates MEK, but the mechanism of Raf activation is uncertain. Other ligands, phosphorylation reactions by downstream ERK, or other kinases or zinc finger interactions may be essential for Raf-1 activation (BATISTATOU et al. 1992). Thus there may be convergent signaling pathways between the Trk pathway and the cooperative interaction of p75^{NTR} via the activation of the Raf-1 pathway by p75^{NTR}/PKN or direct activation of MEK or MAPK by p75^{NTR}/PKN (Fig. 2). Support for this hypothesis is that MEK (ERK1) and MAPK (ERK2) were found to coimmunoprecipitate with p75^{NTR} (VOLENTE et al. 1993b). Cells transfected with p75^{NTR} alone, however, did not result in activation of downstream effector molecules such as Raf-1 or MAPK, indicating that p75^{NTR} alone is probably insufficient to activate this pathway (OHMICHI et al. 1992c).

An alternative signaling pathway for p75^{NTR} may result in the activation of the sphyngomyelin cycle. By addition of cell-permeable ceramide analogues to p75^{NTR}-expressing glioma cells, growth inhibition and the formation of dendritic cell processes occurred (DOBROWSKY et al. 1994). The sphyngomyelin pathway may also be important in signaling by tumor necrosis factor (TNF)-α receptors, and this pathway appears to involve a ceramide-activated protein phosphatase

(Wolff et al. 1994). This alternate form of signal transduction by p75^{NTR} may be important in brain invasion. Brain tissue injured by tumor cell invasion may provide a source of ceramide that might influence invading tumor cells.

5 Neurotrophin Receptors, Tumor Invasion, and Metastasis

During malignant progression, tumor cells undergo genomic changes and show differences in the expression of particular gene products (Nicolson 1987,1988, 1993a). For example, malignant melanoma cells from patient brain metastases exhibit characteristic chromosomal alterations, such as a high frequency of translocation or deletion breakpoints at 11q23, terminal translocations at 17q25, or an isochromosome for the long arm of chromosome 17 (Morse et al. 1992). Human melanoma cells also show progression-associated increases in the expression of p75^{NTR} (Ross et al. 1984; Herlyn et al.1985; Herrmann et al.1993), whose gene is located at 17q21–22. In addition, the neural cell adhesion molecule (NCAM) locus at 11q23 may also be important in melanoma brain metastasis.

5.1 Overexpression of p75^{NTR} on Brain-Metastatic Human Melanoma Cells

We have examined the role of neurotrophin receptors in human melanoma invasion and brain colonization. Using a human melanoma 70W variant cells that have the capacity to form brain colonies in nude mice, parental MeWo cells that exhibit intermediate metastatic potential and nonmetastatic 3S5 cells, we studied the effects of neurotrophins and growth factors on the malignant properties of these cells. In these melanoma cells, overexpression of p75^{NTR} is associated with brain colonization and neurotrophin-mediated enhancement of extracellular matrix invasion (Herrmann et al. 1993; Marchetti et al. 1993; Menter et al. 1995b). The expression of p75^{NTR} on brain-metastatic 70W cells was determined by p75^{NTR} immunoprecipitation analysis of radioiodinated cell surface proteins. Anti-p75^{NTR} monoclonal antibody specifically precipitated higher amounts of an appropriately sized, radioiodinated p75^{NTR} component in lysates of the 70W and A875 melanoma cells. The expression of p75^{NTR} on the surface of the MeWo parental cell line was low but detectable, and we could not detect p75^{NTR} on nonmetastatic 3S5 cells. In addition, we could not detect the expression of the trkA gene or p140^{trkA} on the surfaces of any of the human MeWo melanoma cells (Herrmann et al. 1993).

We next examined whether NGF binding to p75^{NTR} leads to the formation of NGF receptor complexes in MeWo cells. Immunoprecipitation was performed in the presence of excess exogenous NGF. Addition of excess NGF caused a significant increase in the amount of immunoprecipitate formed and an increase in high molecular weight immunocomplexes. For example, an approximately 200-KDa complex was increased in amount by prior treatment with NGF or in the presence of excess exogenous NGF (HERRMANN et al. 1993). Similar complexes of approximately 200 kDa with a high affinity for NGF have been reported to be formed on A875 melanoma cells after NGF treatment (BUXSER et al. 1985). In addition, we observed by immunofluorescence using anti-$p75^{NTR}$, that receptor complexes were rapidly aggregated and probably endocytosed following NGF treatment (D. MENTER, J. HERMANN, and G.L. NICOLSON, submitted).

5.2 Neurotrophins Enhance Invasion and Degradative Enzyme Production of Brain-Metastatic Melanoma Cells

Neurotrophins can enhance the invasive properties of certain melanoma cells (HERRMANN et al. 1993; MARCHETTI et al. 1993; NICOLSON et al. 1994a). This may be required to penetrate the BBB. We have examined the effects of NGF on invasion of brain-metastatic melanoma cells using filters coated with Matrigel (Biomedical products, Bedford, MA) in a Transwell apparatus (Costar,Cambridge, MA). As a chemoattractant we placed brain microvessel endothelial cell-conditioned medium in the lower chamber; (endothelial cell motility factors from lung or liver did not substitute for brain endothelial cell motility factors). NGF treatment resulted in a 7.9-fold increase in the extent of matrix invasion of the 70W cells, corresponding with increases in MMP-2 type IV collagenase/gelatinase A and heparanase, but not MMP-9 type IV collagenase/gelatinase B activities. NGF caused less matrix invasion by MeWo parental cells, and there was no increase in invasion of nonmetastatic 3S5 cells (HERRMANN et al. 1993; MARCHETTI et al. 1993; NICOLSON et al. 1994).

The ability to invade a reconstituted basement membrane was only apparent if the human melanoma cells were grown on extracellular matrix and placed on an invasion substrate in the presence of brain microvessel endothelial cell-conditioned medium. This suggested that, in addition to their response to neurotrophic factors, melanoma cells must interact with the appropriate matrix and receive paracrine motility signals to be highly invasive. In fact, adhesive contact with RGD-containing substratum may be essential for the proper expression and function of $p75^{NTR}$ (HERRMANN et al. 1993).

6 Neurotrophins and Tumor Cell Survival and Death

Homozygous knockout mice lacking neurotrophins or their receptors have been very useful in elucidating the function of these regulatory molecules (SNIDER 1994). Differences in the properties of knockout mice vary, depending on which neurotrophic gene has been eliminated and the type of neuronal cell under observation. For example, both NT-3 and its putative receptor, $p145^{trkC}$, regulate the proliferation and survival of neuronal precursors, in addition to the branching of axons into target fields (KALCHEIM et al. 1992; BIRREN et al. 1993; DICICCO et al. 1993; SCHNELL et al. 1994). This is consistent with the lack of proprioceptor production in *trkC* (-/-) or NT-3 (-/-) mice (EERNFORS et al. 1994a; KLEIN et al. 1994). In contrast, BDNF or *trkB* gene targeting seems to effect vestibular ganglia to the greatest degree (KLEIN et al. 1993; ERNFORS et al. 1994b; JONES et al. 1994), whereas targeted disruption of NGF or *trkA* genes yields mice with defects in the superior cervical ganglia (CROWLEY et al. 1994).

Gene targeting or knockout experiments performed with each of the neurotrophins or their various receptors demonstrate profound effects on the survival of dorsal root ganglia neurons that have the same neuroectodermal origins as melanocytes. Although in these gene targeting studies the effects on melanocytes were not described, there is evidence that many neuroectodermally derived sensory neurons switch their neurotrophin dependence from BDNF or NT-3 during early embryonic development to NGF al later stages (BUCHMAN and DAVIES 1993). Therefore, the targeting of more than one neurotrophin or neurotrophin receptor in homologous recombination experiments may be required to observe an effect on melanocytes in null allele mice.

Changes in neurotrophin dependence may reflect the progressive increase in $p75^{NTR}$ production that occurs during the progression of melanocytes to malignant melanoma cells. Phorbol 12-tetra decanoate 13-acetate (TPA) was previously reported to induce $p75^{NTR}$ receptor production, and this agent also induces synthesis of Trk receptors (PEACOCKE et al. 1988; YAAR and GILCHREST 1991). Primary melanocyte cultures express low levels of *trkC* that can be upregulated by TPA stimulation (YAAR et al. 1994). Although we did not find *trkA* expression in malignant human melanoma cells, we have observed *trkC* expression (HERRMANN et al. 1993). During progression, melanoma cells may be predisposed to switching expression of neurotrophin receptor genes to those most likely to support their survival in new tissue compartments.

The $p75^{NTR}$ receptor may have alternative functions in different cell types. In addition to receiving differentiation or survival signals in neuronal cells, $p75^{NTR}$ may provide retrograde transport in certain neuronal cell types (VERDI et al. 1994), trigger apoptosis in certain transformed cells (RABIZADEH et al. 1993), or signal survival when expressed in other cell types (KANNAN et al. 1992). Certain properties of $p75^{NTR}$ may allow it to function in regulating survival and death of melanoma cells. Thus $p75^{NTR}$ may be analogous to members of the tumor necrosis factor receptor superfamily, which regulate programmed cell death (BEUTLER and VAN HUFFEL 1994; SMITH et al. 1994).

Support for a role for $p75^{NTR}$-dependent signaling in apoptosis was obtained by introducing $p75^{NTR}$ into a SV40 large T antigen-immortalized neuronal cell line deficient in $p75^{NTR}$ and $p140^{trk}$ receptors. In the absence of NGF, the $p75^{NTR}$ transfectants died; however, incubation with exogenous NGF or addition of monoclonal antibody to $p75^{NTR}$ suppressed neural cell death (RABIZADEH et al. 1993). In the embryo, these effects are dependent on developmental stage. For example, $p75^{NTR}$ supports sensory neuron survival from embryonic day 13 to at least postnatal day 2, but $p75^{NTR}$ can initiate apoptosis at later developmental stages (Barrett and Bartlett 1994). Therefore, $p75^{NTR}$ may play a bifunctional role as a molecular switch that signals either cell survival or death.

The normal role of neurotrophins is to promote neuronal cell survival, but they may also promote the survival of brain-metastatic cells. This has beeen shows recently using brain-metastatic 70W melanoma cells. Melanoma cells that expressed high numbers of $p75^{NTR}$ on their cell surface survived under limiting growth conditions, whereas cells with fewer numbers of cell surface $p75^{NTR}$ underwent rapid cell death (Menter et al. 1994).

7 Autocrine Growth Factors and Brain Metastasis

The production of autocrine and paracrine growth factors by melanoma cells can influence their survival and growth in the brain. The role of autocrine growth factors in tumor progression and metastasis has been considered elsewhere (Aaronson 1991; Nicolson 1993). A general finding is that cells established from early tumor lesions have strict growth factor requirements, whereas cells established from highly progressed, metastatic lesions have reduced requirements for growth factors (Herlyn et al. 1985, 1989). Highly progressed tumor cells or metastatic cells produce a variety of autocrine growth factors that aid their growth in new tissue compartment (Rodeck and Herlyn 1991). In addition, there is a tendency for highly progressed or metastatic cells to lose growth inhibitor responsiveness (Lu et al. 1992).

There is very limited data on the production of autocrine growth factors by brain-metastatic tumor cells. To examine the possible growth factors that could be important in brain metastasis, we examined brain-metastatic 70W melanoma cells for synthesis of various growth factor transcripts by reverse transcriptase-polymerase chain reaction (RT-PCR). RT-PCR revealed the production of TGF-β1, bFGF, TGF-α, and interleukin (IL)-1β (Menter et al. 1995b). There was no transcript observed using PDGF primers. All RT-PCR primers were intron spanning to ensure that the PCR products were derived from RNA and not from genomic DNA contamination. Thus we concluded that brain-metastatic 70W cells expressed certain cytokines that might be important in conditioning the brain microenvironment or, alternatively, these factors could operate to stimulate autocrine growth of brain metastases.

8 Neurotrophin Production and Brain Invasion

If neurotrophins are important in brain invasion and colonization by metastatic tumor cells, then there should be some evidence that they are present at the invasion front of brain metastases. Since it was established that brain-metastatic human 70W melanoma cells could produce various growth factors, including TGF-β, TGF-α, bFGF, and IL-1β, we reasoned that these factors might act as paracrine factors that regulate neurotrophin production in the brain. Indeed, many of these factors can stimulate brain astrocytes or oligodendrocytes to produce neurotrophins. Therefore, we examined whether brain-invading melanoma cells can induce changes in NGF concentrations or distribution at the invading edge of melanoma tumors in vivo (MARCHETTI et al. 1995; MENTER et al. 1995b). Brain tissue samples from human melanoma metastases and uninvolved brain tissues progressively distant from the melanoma cells were examined immunohistochemically for the presence of NGF and other neurotrophins. Hematoxylin-eosin staining confirmed the presence of brain-invasive melanoma and adjoining brain tissue with extensive gliosis. For example, staining of serial sections with anti-NGF monoclonal antibody revealed increased concentrations of NGF in the tumor-adjacent tissue at the invasive front of the metastatic lesions. Staining was highest at the interface between melanoma tumor and adjacent normal brain tissue and gradually decreased in concentration until NGF was undetectable at more distant sites. Controls without primary antibody or uninvolved brain tissue distant from the melanoma lesion possessed very low or undetectable concentrations of NGF (MENTER et al. 1995b). This preliminary study has now been expanded to include larger numbers of patients, other neurotrophins, and melanoma metastases at sites other than brain (MARCHETTI et al. 1995). The results were essentially the same. Neurotrophins such as NT-3 were expressed at highest levels at the invasion front of melanoma brain metastases. The invading melanoma cells in the brain appeared to induce high concentrations of NGF and NT-3 in normal brain tissue near the invading melanoma cells. In contrast, invading melanoma cells in tissues other than the CNS were not associated with the presence of neurotrophins at the invasion front of the tumors (MARCHETTI et al. 1995).

9 Paracrine Growth Factors and Brain Metastasis

To survive and grow, tumor cells metastatic to brain must respond to local or paracrine growth factors in the brain environment (for a review, see NICOLSON 1993a,b). A major tissue-derived paracrine growth factor for metastatic melanoma cells has been purified to homogeneity (CAVANAUGH and NICOLSON 1989) and found to be a transferrin (Tf) (CAVANAUGH and NICOLSON 1991). Tf-like factors (TfLF) are probably used as paracrine growth stimulators at organ sites such as lung, bone, and brain. TfLF and Tf may utilize the same receptor on melanoma cells, the approximately 180-kDa dimeric Tf receptor. We examined the ^{125}I-labeled Tf-binding properties and growth response to Tf of tumor cell sublines of different metastatic properties. In a murine melanoma system, we found that brain-colonizing sublines exhibited the greatest growth response to Tf and bound the most ^{125}I-labeled Tf, followed in order by ovary-colonizing, highly lung-colonizing, and finally poorly lung-colonizing sublines (NICOLSON et al. 1990). We also found a close relationship between the binding of ^{125}I-labeled Tf, growth responses to Tf, and spontaneous metastatic potential in a rat mammary adenocarcinoma metastatic system. The results indicated that Tf receptor numbers increased as spontaneous metastatic properties increased in the following order: high brain-metastasizing ability > high lung-metastasizing ability > poor metastatic capability (INOUE et al. 1993). Examination of the responses of human melanoma cell lines to Tf in the absence of serum indicated that the brain-metastatic sublines responded best to Tf and expressed the highest numbers of Tf receptors (NICOLSON et al. 1994a), suggesting that Tf response may be an important property of brain-metastatic melanoma cells.

Overexpression of particular growth factor receptors may be important in metastatic cell growth response at certain sites. Similarly, the overexpression of growth factors is also important in stimulating the growth responses of normal cells at sites of wounding or inflammation. Tumor cells that express high numbers of Tf receptors should be able to respond to low, limiting concentrations of Tf that exist in some tissue compartments, such as the brain.

In the brain, Tf, or more likely TfLF, are probably used as paracrine growth factors during fetal development (MESCHER and MUNIAM 1988). With the possible exception of the choroid plexus, uninjured adult brain does not synthesize large amounts of Tf, and Tf is normally present in limited quantities in the brain, probably due to its poor penetration through the BBB. When malignant cells metastasize to brain, it may be advantageous for them to express high numbers of Tf receptors and to respond to low concentrations of Tf. Alternatively, brain synthesizes TfLF that may not posses the same efficiency in binding to the Tf receptor. We have recently found that fetal brain synthesizes relatively large amounts of a TfLF that we have called TfLF-3 (JIA et al. 1994). It is likely that TfLF-3 is only one of several growth and inhibitory factors important in the organ preference of metastatic cells to the brain (NICOLSON 1993a–c).

In the normal brain Tf/TfLF are produced primarily by oligodendrocytes and astrocytes in the choroid plexus, cerebral plexus, amygdala, hippocampus, brain stem, and cerebellar Purkinje cells (CONNER et al. 1990; CONSTAM et al. 1992; MORRIS et al. 1992). The production of Tf (and probably TfLF) by brain astrocytes can be induced by both IL-1 and TNF-α (OH et al. 1993). Injury to the brain leads to a significant increase in both Tf/TfLF and TfR expression (ORITA et al. 1990). Also during brain injury astrocytes can respond to bFGF, endothelial growth factor (EGF), IL-1α, and IL-2 (HUNTER et al. 1993).

In the brain, certain cell types, such as oligodendrocytes, type 2 astrocytes (O-2A) progenitor cells, can respond to trauma-associated mitogenic signals from bFGF and PDGF (FRESSINAUD et al. 1993). Microglial cells also synthesize cytokines in response to trauma, such as bFGF, IL-1, and TNF-α (MERRIL 1992; FISCHER et al. 1993). Brain trauma that occurs during the pathogenesis of glioma brain tumors often leads to the production of both bFGF and vascular endothelial cell growth factor (VEGF). The levels of bFGF and VEGF are the highest in the anaplastic astrocytes that surround abnormal blood vessels, usually areas of endothelial cell proliferation (ALVAREZ et al. 1992). Brain microvascular endothelial cells are also important sources of cytokines (FABRY et al. 1993). Furthermore, since endothelial cells respond to angiogenesis factors released by tumor cells, a reciprocal relationship probably exists between tumor cells and specific organ-derived endothelial cells at the secondary site (Fig. 3).

9.1 Reciprocal Cytokine Regulation of Brain-Metastatic Cell Growth

Coculture and conditioned medium experiments have provided important information about the reciprocal cytokine relationships between tumor cells and their parenchymal counterparts (NICOLSON 1993b,c; NICOLSON et al. 1994b). This

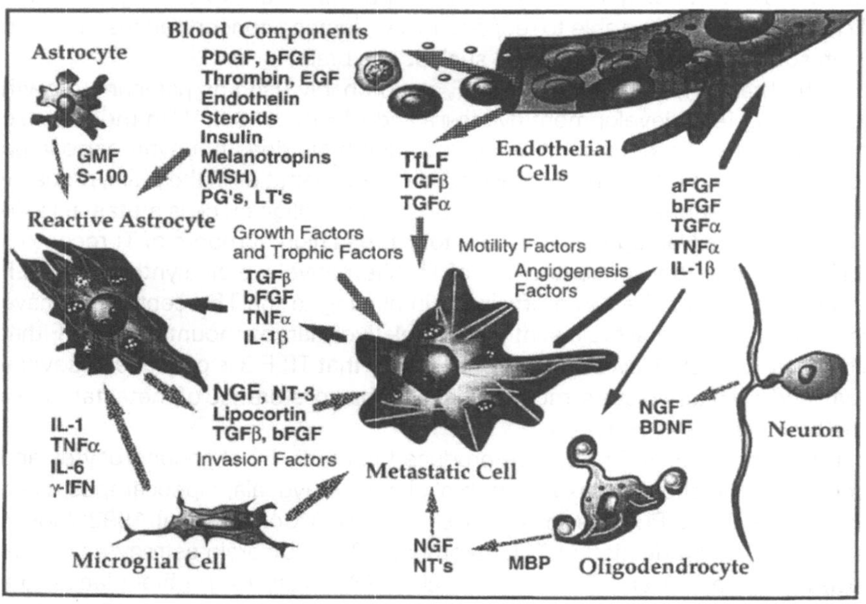

Fig. 3. Reciprocal interactions between brain-invading melanoma cells and normal cells in the brain microenvironment. Tumor cells release cytokines that can effect host cells, such as endothelial cells, parenchymal cells, glial cells, oligodendrocytes, astrocytes, and host tissue extracellular matrix. Reactive astrocytes can arise from stiumulation by blood-derived vasogenic factors, factors released by brain-invading melanoma cells, and factors released from other brain cells. In turn, the host cells release factors that stimulate or inhibit tumor cell motility and proliferation. Astrocytes, oligodendrocytes, and neurons can release neurotrophins in response to brain-invading melanoma cells. *IL*, interleukin; *TNF*, tumor necrosis factor; *IFN*, interferon; *PDGF*, platelet-derived growth factor; *bFGF*, basic fibroblast growth factor; *EGF*, epithelial growth factor; *MSH*, melanocyle-stimulating hormone; *TGF*, transforming growth factor; *NGF*, nerve growth factor; *NT*, Neurotrophin; *TfLF*, transferin-like factor; *aFGf*, acidic *FGF*; *BDNF*, brain-derived neurotrophic factor

reciprocal cytokine regulation of growth probably also extends to parenchymal cell types as well as to extracellular matrix (Fig. 3). The observation of metastatic growth explosion at certain organ sites can be easily explained by the notion that reciprocal release of cytokines and other factors by tumor and host cells stimulate the motility, invasion, and growth of both tumor and host cells.

In the isolated environment of the brain, inhibitory cytokines probably moderate the growth-stimulatory effects of many other growth factors. Following brain injury, there is an increase in TGF-β production coincident with an inhibition of astrocyte proliferation. It is thought that TGF-β suppresses the mitotic effects of bFGF and EGF on astrocytes and microglial cells. Interestingly, inhibition of astrocyte cell division is accompanied by a transient increases in NGF production, whereas BDNF levels remain unchanged (LINDHOLM et al. 1990, 1992; HUNTER et al. 1993).

Tumor cells that colonize the brain can have unusual responses to various cytokines. For example, melanoma cells that colonize the meninges and ventricles are growth stimulated by TGF-β, while others that colonize the brain parenchyma are growth inhibited by TGF-β (FUJIMAKI et al. 1993). Other growth factors, such as EGF, IL-1β, TNF-α, acidic FGF (aFGF), and bFGF, can stimulate NGF synthesis by reactive astrocytes, a response that can be potentiated by treatment with dibutyry1-cAMP (ONO et al. 1991; YOSHIDA and GAGE 1991, 1992; YOSHIDA et al. 1992). Other cell types that increase the synthesis of NGF in response to bFGF treatment are the meningeal fibroblasts (YOSHIDA and GAGE 1991). An important inhibitory cytokine for melanoma cells is IL-6 (LU et al. 1992). IL-6 can be produced by astrocytes after TNF treatment (SAWADA et al. 1992; LU et al. 1992). BDNF and NGF production are regulated in hippocampal neurons and astrocytes by glutamate and γ-aminobutyrate (GABA) systems in response to neuronal activities and cytokines (ZAFRA et al. 1992). Another neurotrophin, ciliary neurotrophic factor (CTNF), can also be produced by astrocytes (SENDTNER et al. 1991).

10 Brain Environment and Tumor Metastasis

Cellular and molecular passage into the brain are strictly regulated by the BBB. Anatomically, the BBB is defined by specialized endothelial cells that are joined by an extensive network of tight junctions. The endothelial barrier is supported by a thick basement membrane and underlying astrocytes that control the traffic of ions, nutrients, and cells into the brain. Metastatic cells must breach this barrier to invade and colonize the brain parenchyma. As discussed above, invasion into brain requires that metastatic cells increase their expression of certain cell surface receptors, degradative enzymes, growth factors, and possibly cytokines, and they must respond to invasion-stimulating cytokines such as neurotrophins and paracrine growth factors.

To penetrate the BBB, brain-metastasizing melanoma cells express relatively high levels of basement membrane hydrolytic enzymes, such as type IV collagenases, cathepsins, plasminogen activators, and heparanase (Nicolson et al. 1994a). For example, we found that murine and human melanoma cells that possess high brain-colonization properties secreted the highest amounts of various basement membrane-degrading enzymes (Nicolson et al. 1994a). Although highly metastatic cells generally expressed higher amounts of degradative enzymes than nonmetastatic cells, as discussed above, some of these enzymes may be induced to even higher levels by the microenvironment (paracrine invasion factors, such as neurotrophins), or these enzymes can be provided by certain normal cells, such as microvessel endothelial cells. If the appropriate paracrine signals are received by malignant cells, they can be stimulated to increase their synthesis and release of BBB-degrading enzymes. For example, as discussed in previous sections, we found that brain-metastatic human and murine melanoma cells are sensitive to exogenous NGF (Herrmann et al. 1993; Marchetti et al. 1993), and treatment of brain metastatic cells with NGF increases their expression of MMP-2 type IV collagenase, gelatinase A, and heparanase (Herrmann et al. 1993; Marchetti et al. 1993; Nicolson et al. 1994a).

10.1 Cellular Responses to Brain Tissue Injury as a Paradigm for Brain Metastasis

The primary cellular response following brain injury is mediated by astroglial cells (Norenberg 1994). Astrocytes are the predominant cell type in the brain, and they outnumber neurons by a factor of 10:1. Astrocytes make up one third of the cerebral cortex mass; however, as a population of cells they are very heterogenous (Wilken et al. 1990). They are usually organized into a well-developed syncytium containing gap junctions that mediate homeostasis and intercellular communication (Kettenmann et al. 1983). Astrocytes are influenced by neuronal interactions and, as a consequence, regulate brain function, including the BBB, water and ion flow, basal metabolism, immune responses, neuronal cell migration, neurite outgrowth, and functional synapse formation (Norenberg 1994). The integrated organization of the astrocytic cellular compartment of the brain provides tremendous potential for large-scale recruitment of astrocytes in response to tumor cell invasion.

Once the brain is injured, the earliest pathological response involve astrocyte swelling, predominantly in the perivascular astrocytic endings (Kimelberg and Ransom 1986; Hirano et al. 1994). In experimental brain tumors, cerebral edema has been associated with significant alterations in vascular permeability (Lantos et al. 1984). If the BBB is compromised, astrocyte swelling may involve vasogenic edema. In this case, the astrocytes swell as they take up proteins and water and may become cytotoxic due to increases in potassium ion and glutamate uptake (Klatzo et al. 1980). This may also include the production of arachidonic acid metabolites (prostaglandins and leukotrienes) and diffusion of cytokines into the astrocytic cell compartment (Norenberg 1994; Fig. 3).

It is generally believed that astrocyte swelling is caused by increases in intracellular osmolarity followed by water influx. This may occur without loss of BBB integrity and may simply represent a redistribution of water from the neuronal cell compartment to the astrocytic cell compartment. This mild form of astrocyte swelling is generally not as severe as the astrocyte swelling that can result from vasogenic edema associated with trauma caused by tumor cell invasion. If astrocyte swelling becomes too severe it can cause astroglial cells to depolarize, leading to the loss of homeostatic ion gradients and membrane rupture resulting in cell death. These dynamic astrocyte changes in response to tumor cell invasion can lead to increased intracranial pressure and further complications. The massive tumor induced response by astrocytes is a possible reason for small metastases causing severe symptoms, such as paralysis, headache, seizures, and impaired cognition.

In addition to astrocyte swelling, the most profound cellular response that brain tissue elicits to invasive injury is the production of reactive astrocytes (fibrous astrocytes). This is a condition known as reactive astrocytosis or reactive gliosis. Histologically reactive astrocytes exhibit cytoplasmic hypertrophy in the form of dense, elongated cellular processes or a fibrous-appearing glial scar. These fibrous processes stain positively for glial fibrillary acidic protein and vimentin intermediate filaments (NORENBERG 1994). Unlike fibrosis in other scar tissues, gliosis consists predominantly of cellular processes or glial fibers and lacks collagen or equivalent fibrous extracellular matrix proteins. Reactive astrocytes often contain enlarged nuclei having multiple nucleoli (Fig. 3). This is accompanied by increases in the numbers of organelles such as mitochondria, Golgi apparatus, endoplasmic reticulum, lysosomes microtubules, and dense bodies. Membrane alterations include increases in hemidesmosomes and gap junctions (NORENBERG 1994). Increased expression of particular receptors includes $p185^{neu}$, $p145^{kit}$, and class II histocompatability antigens (FRANK et al. 1986; KRISTT et al. 1993).

The induction of reactive astrocyte formation involves a number of cellular products from different brain cells. These include the following: glial maturation factor, S100 protein from astrocytes; IL-1, TNF-α, IL-6, and γ-IFN from microglial cells; myelin basic protein from oligodendrocytes; and K^+, Adenosine Triphosphate (ATP), and bFGF from neurons (Fig. 3). Vasogenic edema leads to the influx of thrombin, PDGF, steroids, insulin, and various cytokines from the blood and lymphocytes as well as endothelin, ATP, and bFGF from endothelial cells (Fig. 3). The induction of reactive astrocytes, when associated with tumor cell invasion, is probably initiated by endogenous factors in the brain in addition to those provided by the invading tumor cells (Fig. 3). The reactive astrocytes for example in addition to NGF and NT-3, can synthesize S100 protein, lipocortin (a precursor to β-melanocyte-stimulating hormone, β-MSH), TGF-β, and bFGF, which can affect the reciprocal cytokine loops described previously (Fig. 3).

We have observed extensive reactive astrocytosis or gliosis in brain tissue associated with melanoma invasion. This occurs apparently only at the invasion front, illustrating the cellular response of the adjacent brain tissue

(MENTER et al. 1995a). As discussed above, the brain tissue next to invading melanoma tumors undergoes morphological changes and produces high levels of NGF in comparison to uninvolved brain tissue (MENTER et al. 1995b). Thus, brain-metastatic melanoma cells may induce the production of brain cytokines such as NGF that aid in the brain invasion and survival of melanoma cells.

10. 2 Melanotropic Peptides May Modulate Astroglial and Tumor Cell Behavior

Melanotropins are small, stress-released brain neuropeptides derived from the pituitary and hypothalamic neurons (JEGOU et al. 1993) that are capable of bidirectional passage across the BBB (BANKS and KASTIN 1992). The melanotropins areknown to control adrenal steriodogenesis (via synthesis of adenocorticotropic hormone, ACTH) and melanocyte stimulating hormone (MSH), and they have a wide range of additional mitogenic and trophic functions, including mediating nerve regeneration and functional recovery from CNS trauma, modulating electrophysiological activity of nerve cells, regulating neuromuscular synapse formation, promoting growth stimulation, and stimulating morphological differentiation of brain astrocytes (EBERLE et al. 1993).

The melanocortin receptors belong to a family of G protein-coupled receptors that rely on extracellular calcium for activity. The melanoma MSH receptor consists of 317 amino acids of a 34.8-kDA protein that has seven transmembrane domains (MOUNTJOY et al. 1992).

MSH stimulation of melanoma cells results in elevation of cAMP leading to melanogenesis, differentiation, and growth inhibition (SALOMON et al. 1993; BENNETT et al. 1994; SIEGRIST et al. 1994). For example, treatment of murine melanoma cells with α-MSH can either enhance or suppress metastasis formation depending on the melanoma cells that are analyzed (KAMEYAMA et al. 1990; BENNETT et al. 1994). Melanotropins can affect not only invading melanoma cells, but also the morphology and behavior of brain astroglial cells (Fig. 3). α-MSH acts within the brain to mediate a neurogenic anti-inflammatory response to a wide range of cytokines (HILTZ et al. 1992; CERIANI et al. 1994). If melanotropins alter the responsiveness of astrocytes to swelling or gliosis, this could significantly modulate their effectiveness in responding to invading melanoma cells. The pleiotrophic nature of melanotropin effects may have a important systemic effects following stress-induced release. A greater understanding of the signaling mechanisms used by MSH receptors and their role in melanoma metastasis to brain could provide useful new targets for melanoma diagnosis and therapy.

11 Conclusions

The brain is a unique microenvironment that lacks lymphatic drainage, maintains a highly regulated vascular transport barrier, and is enclosed by the skull. Only very unique tumor cells with certain properties have the capacity to home to, invade, and colonize this organ. They have to attach to brain microvessel endothelial cells, invade the BBB by expressing high concentrations of degradative enzymes, survive by responding to brain trophic factors, particularly neurotrophins, and proliferate by responding to paracrine growth factors. Although we now know much more about brain metastasis than ever before, there remains much to learn.

Certain aspects of brain metastasis require additional information and further examination:

1. The response by brain tissues to invading melanoma cells is far from understood and requires further research emphasis. In particular, we need a better understanding of what triggers reactive astrocytosis and the interactions between brain cells and invading tumor cells.
2. We need a better understanding of the brain microenvironment and the reciprocal cytokine signaling circuits that enable tumor cells to invade, survive, and colonize the brain. Since many of the cytokines involved in brain metastasis stimulate both brain cells and invading tumor cells, this will be a difficult problem to decipher. It is important to determine the paracrine growth factors that are essential in maintaining brain homeostasis as well as those that stimulate survival and growth of invading malignant cells.
3. The precise role of neurotrophins and melanotrophins in promoting melanoma cell invasion, triggering differentiation, and maintaining survival or initiating apoptosis will have to be elucidated to understand the role of these factors in melanoma brain colonization.
4. The trophic responses of brain-metastatic cells will have to be determined. Trophic support as a tumor cell survival mechanism will have to be carefully examined, because the brain environment may require unique survival responses. Although neurotrophins and melanotrophins are some of the best examples of trophic substances, certain growth factors not typically thought of as trophic factors may actually support cell survival of certain tumor cells in particular environments. Trophic factors, autocrine growth factors, paracrine growth factors, and other factors and the responses of brain cells to the invasion of foreign cells may determine whether metastatic cells can successfully invade, colonize, and grow in the CNS. Elucidating their mechanisms of action may some day result in the development of new therapeutic approaches to the treatment of brain metastases.

Acknowledgments. These studies were supported by grants from the U.S. National Institute of Health and the National Foundation for Cancer Research, Inc., to G.L. Nicolson.

References

Aaronson SA (1991) Growth factors and cancer. Science 254: 1146–1153

Albino AP, Davis BM, Nanus DM (1991) Induction of growth factor RNA expression in human malignant melanoma: markers of transformation. Cancer Res 51: 4815–4820

Alvarez JA, Baird A, Tatum A, Daucher J, Chorsky R, Gonzalez AM, Stopa EG (1992) Localization of basic fibroblast growth factor and vascular endothelial growth factor in human glial neoplasms. Mod Pathol 5: 303–307

Avruch J, Zhang X, Kyriakis JM (1994) Raf meets Ras: completing the signal transduction pathway. Trends Biochem Sci 19: 279–283

Banks WA, Kastin AJ (1992) Bidirectional passage of peptides across the blood brain barrier. Prog Brain Res 91: 139–148

Barbacid M (1993) Nerve growth factor: a tale of two receptors. Oncogene 8: 2033–2042

Barker PA, Shooter EM (1994) Disruption of NGF binding to the low affinity neurotrophin receptor p75LNTR reduces NGF binding to TrkA on PC12 cells. Neuron 13: 203–215

Barrett GL, Bartlett PF (1994) The p75 nerve growth factor receptor mediates survival or death depending on stage of sensory neuron development. Proc Natl Acad Sci USA 91: 6501–6505

Batistatou A, Volonte C, Greene LA (1992) Nerve growth factor employs multiple pathways to induce primary response genes in PC12 cells. Mol Biol Cell 3: 363–371

Bennett DC, Holmes A, Devlin L, Hart IR (1994) Experimental metastasis and differentiation of murine melanoma cells: actions and interactions of factors affecting different intracellular signaling pathways. Clin Exp Metastasis 12: 385–397

Berg MM, Sternberg DW, Hempstead BL, Chao MV (1991) The low-affinity p75 nerve growth factor (NGF) receptor mediates NGF-induced tyrosine phosphorylation. Proc Natl Acd Sci USA 88: 7106–7110

Beutler B, van Huffel C (1994) Unraveling function in the TNF ligand and receptor families. Science 264: 667–668

Birren SJ, Lo L, Anderson DJ (1993) Sympathetic neuroblasts undergo a developmental switch in trophic dependence. Development 119: 597–610

Borrello MG, Pelicci G, Arighi E, De Philippis L, Greco A, Bongarzone I, Rizzetti M, Pelicci PG, Pierotti MA (1994) The oncogenic versions of the Ret and Trk tyrosine kinases bind Shc and Grb2 adaptor proteins. Oncogene 9: 1661–1668

Bradshaw RA, Blundell TL, Lapatto R, McDonald NQ, Murray RJ (1993) Nerve growth factor revisited. Trends Biochem Sci 18: 48–52

Bachman VL, Davies AM (1993) Different neurotrophins are expressed and act in a developmental sequence to promote the survival of embryonic sensory neurons. Development 118: 989–1001

Buxser S, Puma P, Johnson GL (1985) Properties of the nerve growth factor receptor. Relationship between receptor structure and affinity. J Biol Chem 260: 1917–1926

Cavanaugh PG, Nicolson GL (1989) Purification and some properties of lung-derived growth factor that differentially stimulates the growth of tumor cells metastatic to the lung. Cancer Res 89: 3928–3933

Cavanaugh PG, Nicolson GL (1991) Lung-derived growth for lung-metastasizing tumor cells: identification as a transferrin. J Cell Biochem 47: 261–267

Ceriani G, Macaluso A, Catania A, Lipton JM (1994) Central neurogenic anti-inflammatory action of alpha-MSH: modulation of peripheral inflammation induced by cytokines and other mediators of inflammation. Neuroendcrinology 59: 138–143

Chao MV (1992) Neurotrophin receptors: a window into neuronal differentiation. Neuron 9: 583–593

Connor JR, Menzies SL, StMartin SM, Mufson EJ (1990) Cellular distribution of transferrin, ferritin, and iron in normal and aged human brains. J Neurosci Res 27: 595–611

Constam DB, J.P, Maltipiero UV, ten Dijke P, Schachner M, Fontana A (1992) Differential expression of transforming growth factor-beta 1, -beta 2, and -beta 3 by glioblastoma cells, astrocytes, and microglia. J Immunol 148: 1404–1410

Crowley C, Spencer SD, Nishimura MC, Chen KS, Pitts-Meek S, Armanini, MP, Ling LH, MacMahon SB, Shelton DL, Levinson AD, Phillips HS (1994) Mice lacking nerve growth factor display perinatal loss of sensory and sympathetic neurons yet develop basal forebrain cholinergic neurons. Cell 76: 1001–1011

DiCicco BE, Friedman WJ, Black IB (1993) NT-3 stimulates sympathetic neuroblast proliferation by promoting precursor survival. Neuron 11: 1101–1111

Dobrowsky RT, Werner MH, Castellino AM, Chao MV, Hannun YA (1994) Activation of the sphyngomyelin cycle through the low-affinity neurotrophin receptor. Science 265: 1596–1599
Eberle AN, Siegrist W, Bagutti C, Chluba-De Tapia J, Solca F, Wikberg JE, Chhajlani V (1993) Receptors for melanocyte stimulating hormone on melanoma cells. Ann NY Acad Sci 680: 320–341
Ernfors P, Lee KF, Kucera J, Jaenisch R (1994a) Lack of neurotrophin-3 leads to deficiencies in the peripheral nervous system and loss of limb proprioceptive afferents. Cell 77: 503–512
Ernfors P, Lee KF, Jaenisch R (1994b) Mice lacking brain-derived neurotrophic factor develop with sensory deficits. Nature 368: 147–150
Fabry Z, Fitzsimmons KM, Herlein JA (1993) Production of the cytokines interleukin 1 and 6 by murine brain microvessel endothelium and smooth muscle pericytes. J Neuroimmunol 4: 23–34
Feinstein DL, Larhammar D (1990) Identification of a conserved protein motif in a group of growth factor receptors. FEBS Lett 272: 7–11
Fischer HG, Nitzgen B, Germannt T (1993) Differentiation driven by granulocyte-macrophage colony-stimulating factor endows microglia with interferon-gamma-independent antigen presentation function. J Neuroimmunol 42: 87–95
Frank E, Pulver M, DeTribolet N (1986) Expression of class II major histocompatability antigens on reactive astrocytes and endothelial cells within gliosis surrounding metastases and abscesses. J Neuroimmunol 12: 29–36
Fressinaud C, Laeng P, Labourdette G, Durand J, Vallat JM (1993) The proliferation of mature oligodendrocytes in vitro is stimulated by basic fibroblast growth factor and inhibited by oligodendrocyte-type 2 astrocyte precursors. Dev Biol 158: 317–329
Fujimaki T, Fan D, Staroselsky AH, Gohji K, Bucana CD, Fidler IJ (1993) Critical factors regulating site-specific brain metastasis of murine melanomas. Int J Oncol 3: 789–799
Hantzopoulos PA, Chitra S, Glass DJ, Goldfarb MP, Yancopoulos GD (1994) The low affinity NGF receptor, p75, can collaborate with each of the trks to potentiate functional responses to the neurotrophins. Neuron 13: 187–201
Hempstead BL, Schleifer LS, Chao MV (1989) Expression of functional nerve growth factor receptors after gene transfer. Science 243: 373–375
Hempstead BL, Patil N, Thiel B, Chao MV (1990) Deletion of cytoplasmic sequences of the nerve growth factor receptor leads to loss of high affinity ligand binding. J Biol Chem 265: 9595–9598
Hempstead BL, Martin ZD, Kaplan DR, Parada LF, Chao MV (1991) High-affinity NGF binding requires coexpression of the *trk* proto-oncogene and the low-affinity NGF receptor. Nature 350: 678–683
Herlyn M, Thurin J, Balaban G, Bennicelli JL, Herlyn D, Elder DE, Bondi E, Guerry D, Nowell P, Clark WH, Koprowski H (1985) Characteristics of cultured human melanocytes isolated from different stages of tumor progression. Cancer Res 45: 5670–5676
Herlyn M, Kath R, Williams N, Valyi-Nagy I, Rodeck U (1989) Growth regulatory factors for normal, premalignant and malignant human cells. Adv Cancer Res 54: 213–234
Herrmann JL, Menter DG, Hamada J, Nakajima M, Nicolson GL (1993) Mediation of NGF-stimulated extracellular matrix invasion by the human melanoma low-affinity p75 neurotrophin receptor: melanoma p75 functions independently of *trkA*. Mol Biol Cell 4: 1205–1216
Hiltz ME, Catania A, Lipton JM (1992) Alpha-MSH peptides inhibit acute inflammation induced in mice by rIL-1 beta, rIL-6, rTNF-alpha and endogenous pyrogen but not that caused by LTB-4, PAF and rIL-8. Cytokine 4: 320–328
Hirano A, Kawanami T, Llena JF (1994) Electron microscopy of the blood brain barrier. Microsc Res Tech 27: 543–556
Hunter KE, Sporn MB, Davies AM (1993) Transforming growth factor-betas inhibit mitogen-stimulated proliferation of astrocytes. Glia 7: 203–211
Inoue T, Cavanaugh PG, Steck PA, Brünner N, Nicolson GL (1993) Differences in transferrin response and numbers of transferrin receptors in rat and human mammary carcinoma lines of different metastatic potentials. J Cell Physiol 156: 212–217
Jegou S, Blasquez C, Delbende C, Bunel DT, Vaudry H (1993) Regulation of a-melanocyte-stimulating hormone release from hypothalmic neurons. Ann NY Acad Sci 680: 260–278
Jia LB, Cavanaugh PG, Nicolson GL (1994) Paracrine growth factors for metastatic breast cancer cells: cloning of three new transferrin-like growth factor cDNAs that may be involved in the growth stimulation of breast cancer cells at secondary sites. Proc Am Assoc Cancer Res 35: 44
Johnson D, Lanahan A, Buck CR, Chow M (1986) Expression and structure of the human NGF receptor. Cell 47: 545–554
Jones KR, Farinas I, Backus C, Reichardt LF (1994) Targeted disruption of the BDNF gene perturbs brain and sensory neuron development but not motor neuron development. Cell 76: 989–999

Kalcheim C, Carmeli C, Rosenthal A (1992) Neurotrophin 3 is a mitogen for cultured neural crest cells. Proc Natl Acad Sci USA 89: 1661–1665

Kameyama K, Vieira WD, Tsukamoto K, Law LW, Hearing VJ (1990) Differentiation and the tumorigenic and metastatic phenotype of murine melanoma cells. Int J Cancer 45: 1151–1158

Kannan Y, Usami K, Okada M, Shimizu S, Matsuda H (1992) Nerve growth factor suppresses apoptosis of murine neutrophils. Biochem Biophys Res Commun 186: 1050–1056

Kettenmann H, Orkand RK, Schachner M (1983) Coupling among identified cells in mammalian nervous system cultures. J Neurosci 3: 506–516

Kimelberg HK, Ransom BR (1986) Physiological aspects of astrocyte swelling. In: Fedoroff S, Verandakis A (eds) Astrocytes. Academic, Orlando, pp 129–166

Klatzo I, Chui E, Fujiwara K, Spatz M (1980) Resolution of vasogenic brain edema (VBE). Adv Neurol 28: 359–373

Klein R, Smeyne RJ, Wurst W, Long LK, Auerbach BA, Joyner AL, Barbacid M (1993) Targeted disruption of the *trkB* neurotrophin receptor gene results in nervous system lesions and neonatal death. Cell 75: 113–122

Klein R, Silos-Santiago I, Smeyne RJ, Lira SA, Brambilla R, Bryant S, Zhang L, Spider WD, Barbacid M (1994) Distruption of the neurotrophin-3 receptor gene *trkC* eliminates Ia muscle afferents and results in abnormal movements. Nature 368: 249–251

Knipper M, Beck A, Rylett J, Breer H (1993) Neurotrophin induced cAMP and IP3 responses in PC12 cells. Different pathways. FEBS Lett 324: 147–52

Kristt DA, Reedy E, Yarden Y (1993) Receptor tyrosine kinase expression in astrocytic lesions: similar features in gliosis and glioma. Neurosurgery 33: 106–115

Lange-Carter CA, Johnson GL (1994) Ras-dependent growth factor regulation of MEK kinase in PC12 cells. Science 265: 1458–1461

Lantos PL, Luthert PJ, Deane BR (1984) Vascular permeability and cerebral oedema in experimental brain tumors. In: Inaba Y, Klatzo I, Spatz M (eds) Brain edema. Springer, Berlin Heidelberg New York, pp 40–47

Lee KF, Li E, Huber LJ, Landis SC, Sharpe AH, Chao MV, Jaenisch R (1992) Targeted mutation of the gene encoding the low affinity NGF receptor p75 leads to deficits in the peripheral sensory nervous system. Cell 69: 737–749

Lee KF, Bachman K, Landis S, Jaenisch R (1994) Dependence on p75 for innervation of some sympathetic targets. Science 263: 1447–1449

Leevers SJ, Paterson HF, Marshall CJ (1994) Requirement for Ras in Raf activation is overcome by targeting Raf to the plasma membrane. Nature 369: 411–414

Lindholm D, Hengerer B, Zafra F, Theonin H (1990) Transforming growth factor-beta 1 stimulates expression of nerve growth factor in the rat CNS. Neuroreport 1: 9-12

Lindholm D, Castren E, Kiefer R, Zafra F, Thoenen H (1992) transforming growth factor-beta 1 in the rat brain: increase after injury and inhibition of astrocyte poliferation. J Cell Biol 117: 395–400

Lu C, Vickers MF, Kerbel R (1992) Interleukin 6: a fibroblast-derived growth inhibitor of human melanoma cells from early but not advanced stages of tumor progression. Proc Natl Acad Sci USA 89: 9215–9219

Maher PA (1988) Nerve growth factor induces protein-tyrosine phosphorylation. Proc Natl Acad Sci USA 85: 6788–6791

Marchetti D, Menter D, Jin L, Nakajima M, Nicolson GL (1993) Nerve growth factor effects on human and mouse melanoma cell invasion and heparanase production. Int J Cancer 55: 692–699

Marchetti D, McCutcheon IE, Ross ML, Nicolson GL (1995) Inverse expression of neurotrophins and neurotrophin receptors at the invasion front of human melanoma brain metastases. Int J Oncol 7: 87–94

Meakin SO, Shooter EM (1991) Molecular investigations on the high-affinity nerve growth factor receptor. Neuron 6: 153–163

Meakin SO, Shooter EM (1992) The nerve growth factor family of receptors. Trends Neurosci 15: 323–331

Menter DG, Herrmann JL, Nicolson GL (1995a) The role of trophic factors and autocrine/paracrine growth factors in brain metastasis. Clin Exp Metastasis 13: 67–88

Menter DG, Herrmann JL, Marchetti D, Nicolson GL (1995b) Involvement of neurotrophins and growth factors in brain metastasis formation. Invasion Metastasis 14: 372–384

Menter DG, Herrmann JL, Nicolson GL (1994) The metastatic melanoma neurotrophin receptor (p75) is a cell survival (menocytosis) receptor. Clin Exp Metastasis 12: 82a

Merrill JE (1992) Tumor necrosis factor alpha, interleukin 1 and related cytokines in brain development: normal and pathological. Dev Neurosci 14: 1–10

Mescher AL, Muniam SI (1988) Transferrin and the growth-promoting effect of nerves. Int Rev Cytol 110: 1–26

Miyasaka T, Chao MV, Sherline P, Saltiel AR (1990) Nerve growth factor stimulates a protein kinase in PC-12 cells that phosphorylates microtubule-associated protein-2. J Biol Chem 265: 4730–4735

Morris CM, Candy JM, Bloxham CA, Edwardson JA (1992) Immunocytochemical localisation of transferrin in the human brain. Acta Anat (Basel) 143: 14–18

Morse HG, Gonzalez R, Moore GE, Robinson WA (1992) Preferential chromosome 11q and or 17q aberrationsin short-term cultures of the metastatic melanoma resections from the brain. Cancer Genet Cytogenet 64: 118–126

Mountjoy KG, Robbins LS, Mortrund MT, Cone RD (1992) The cloning of a family of genes that encode the melanocortin receptors. Science 257: 1248–1251

Nicolson GL (1987) Tumor cell instability, diversification and progression to the metastatic phenotype: from oncogene to oncofetal expression. Cancer Res 47: 1473–1487

Nicolson GL (1988) Cancer metastasis: Tumor cell and host organ properties important in colonization of specific secondary sites. Biochim Biophys Acta 948: 175–224

Nicolson GL (1993a) Paracrine and autocrine growth mechanisms in tumor metastasis to specific sites with particular emphasis on brain and lung metastasis. Cancer Metastasis Rev 12: 325–343

Nicolson GL (1993b) Cancer progression and growth: Relationship of paracrine and autocrine growth mechanisms to organ preference of metastasis. Exp Cell Res 204: 171–180

Nicolson GL (1993c) Paracrine/autocrine growth mechanisms in tumor metastasis. Oncol Res 4: 389–399

Nicolson GL, Inoue T, Van Pelt C, Cavanaugh PG (1990) Differential expression of a M_r 90,000 cell surface transferrin-related glycoprotein on murine B16 metastatic melanoma sublines selected for enhanced brain or ovary colonization. Cancer Res 50: 515–520

Nicolson GL, Nakajima M, Herrmann JL, Menter DG, Cavanaugh PG, Park JS, Marchetti D (1994a) Malignant melanoma metastasis to brain: role of degradative enzymes and responses to paracrine growth factors. J Neurooncol 18: 139–149

Nicolson GL, Menter D, Herrmann J, Cavanaugh P, Jia L-B, Hamada J, Yun Z, Marchetti D (1994b) Tumor metastasis to brain: role of endothelial cells, neutrophins and paracrine growth factors. Crit Rev Oncogenesis 15: 451–471

Norenberg MD (1994) Astrocyte responses to CNS injury. J Neuropathol Exp Neurol 53: 213–220

Nowell PC (1976) The clonal evolution of tumor cell. Science 194: 23–28

Obermeier A, Lammers R, Wiesmuller KH, Jung G, Schlessinger J, Ullrich A (1993) Identification of Trk binding sites for SHC and phosphatidylinositol 3'-kinase and formation of a multimeric signaling complex. J Biol Chem 268: 22963–22966

Obermeier A, Bradshaw RA, Seedorf K, Choidas A, Schlessinger J, Ullrich A (1993) Neuronal differentiation signals are controlled by nerve growth factor receptor/Trk binding sites for SHC and PLC-gamma. EMBO J 13: 1585–1590

Oh YJ, Markelonis GJ, Oh TH (1993) Effects of interleukin-1 beta and tumor necrosis factor-alpha on the expression of glial fibrillary acidic protein and transferrin in cultured astrocytes. Glia 8: 77–86

Ohmichi M, Decker SJ, Pang L, Saltiel AR (1991) Phospholipase C-gamma 1 directly associates with the p70 trk oncogene product through its src homology domains J Biol Chem 266: 14858–14861

Ohmichi M, Decker SJ, Saltiel AR (1992a) Nerve growth factor stimulates the tyrosine phosphorylation of a 38-kDa protein that specifically associates with the src homology domain of phospholipase C-gamma 1. J Biol Chem 267: 21601–21606

Ohmichi M, Decker SJ, Saltiel AR (1992b) Activation of phosphatidylinositol-3 kinase by nerve growth factor involves indirect coupling of the trk proto-oncogene with src homology-2 domains. Neuron 9: 769–777

Ohmichi M, Pang L, Decker SJ, Saltiel AR (1992c) Nerve growth factor stimulates the activities of the raf-1 and the mitogen-activated protein kinases via the trk protooncogene. J Biol Chem 267: 14604–14610

Ono T, Sato H, Kashimoto T, Okumoto T, Miyamoto K (1991) Stimulation of biosynthesis of nerve growth factor by acidic fibroblast growth factor in cultured mouse astrocytes. Neurosci Lett 126: 18–20

Orita T, Akimura T, Nishizaki T, Kamiryo T, Ikeyama Y, Aoki H, Ito H (1990) Transferrin receptors in injured brain. Acta Neuropathol (Berl) 79: 686–688

Peacocke M, Yaar M, Mansur CP, Chao MV, Gilchrest BA (1988) Induction of nerve growth factor receptors pn cultured human melanocytes. Proc Natl Acad Sci USA 85: 5282–5286

Rabizadeh S, Oh J, Zhong LT, Yang J, Bitler CM, Butcher LL, Bredesen DE (1993) Induction of apoptosis by the low-affinity NGF receptor. Science 261: 345–348

Raff CM (1992) Social controls on cell survival and cell death. Nature 356: 397–400
Raff MC, Barres BA, Burne JF, Coles HS, Ishizaki Y, Jacobson MD (1993) Programmed cell death and the control of cell survival: lessons from the nervous system. Science 262: 695–700
Rodeck U, Herlyn M (1991) Growth factors in melanoma. Cancer Met Rev 10: 89–101
Ross AH, Grob P, Bothwell M, Elder DE, Ernst CS, Marano N, Ghrist BF, Slemp CC, Herlyn M, Atkinson B, Koprowski H (1984) Characterization of nerve growth factor receptor in neural crest tumors using monoclonal antibodies. Proc Natl Acad Sci USA 81: 6681–6685
Rozakis-Adcock M, McGlade J, Mbamalu G, Pelicci G, Daly R, Li W, Batzer A, Thomas S, Brugge J, Pelicci PG, Schlessinger J, Pawson T (1992) Association of the Shc and Grb2/Sem5 SH2-containing proteins is implicated in activation of the Ras pathway by tyrosine kinases. Nature 360: 689–692
Salomon Y, Zohar M, Dejordy JO, Eshel Y, Shafir I, Leiba H, Garty NB, Schmidt-Sole J, Azrad A, Shai E, Degani H (1993) Signaling mechanisms controlled by melanocortins in melanoma, lacrimal and brain astroglial cells. In: Vaudry H, Eberle AN (eds) The melanotropic peptides. Ann NY Acad Sci 680: 364–379
Sawada M, Suzumura A, Marunochi T (1992) TNF alpha induces IL-6 production by astrocytes but not by microglia. Brain Res 583: 296–299
Schnell L, Schneider R, Kolbeck R, Barde YA, Schwab ME (1994) Neurotrophin-3 enhances sprouting of corticospinal tract during development and after adult spinal cord lesion. Nature 367: 170–173
Sendtner M, Arakawa Y, Stockli KA, Kreutzberg GW, Thoenen H (1991) Effect of ciliary neurotrophic factor (CNTF) on motoneuron survival. J Cell Sci Suppl 15: 103–109
Siegrist W, Stutz S, Eberle AN (1994) Homologous and heterologous regulation of a-melanocyte stimulating hormone receptors in human and mouse melanoma cell lines. Cancer Res 54: 2604–2610
Smeyne RJ, Klein R, Schnapp A, Long LK, Bryant S, Lewin A, Lira SA, Barbacid M (1994) Severe sensory and sympathetic neuropathies in mice carrying a disrupted Trk/NGF receptor gene. Nature 368: 246–249
Smith CA, Farrah T, Goodwin RG (1994) The TNF receptor superfamily of cellular and viral proteins: activation, costimulation, and death. Cell 76: 959–962
Snider WD (1994) Functions of the neurotrophins during nervous system development: what knock-outs are teaching us. Cell 77: 627–638
Steck PA, Nicolson GL (1993) Metastasis to the central nervous system. In: Levine A, Schmidek H (eds) Molecular genetics of nervous system tumors. Wiley, New York, pp 371–379
Stehlin JS, Hills WJ, Rufino C (1967) Disseminated melanoma: biologic behavior and treatment. Arch Surg 94: 495–501
Stephens RM, Loeb DM, Copeland TD, Pawson T, Greene LA, Kaplan DR (1994) Trk receptors use redundant signal transduction pathways involving SHC and PLC-gamma 1 to mediate NGF responses. Neuron 12: 691–705
Stokoe D, Macdonald SG, Cadwallader K, Symons M, Hancock JF (1994) Activation of Raf as a result of recruitment to the plasma membrane. Science 264: 1463–1467
Sugarbaker EV (1981) Patterns of metastases in human malignancies. Cancer Biol Rev 2: 235–278
Taylor LK, Swanson KD, Kerigan J, Mobley W, Landreth GE (1994) Isolation and characterization of a nerve growth factor-regulated Fos kinase from PC12 cells. J Biol Chem 269: 308–318
Verdi JM, Birren SJ, Ibanez CF, Persson H, Kaplan DR, Benedetti M, Chao MV, Anderson DJ (1994) p75LNGFR regulates Trk signal transduction and NGF-induced neuronal differentiation in MAH cells. Neuron 12: 733–745
Volonte C, Greene LA (1990) Induction of ornithine decarboxylase by nerve growth factor in PC12 cells: dissection by purine analogues. J Biol Chem 265: 11050–11055
Volonte C, Greene LA (1992) Nerve growth factor-activated protein kinase N. Characterization and rapid near homogeneity purification by nucleotide affinity-exchange chromatography. J Biol Chem 267: 21663–21670
Volente C, Ross AH, Greene LA (1993a) Association of a purine-analogue-sensitive protein kinase activity with p75 nerve growth factor receptors. Mol Biol Cell 4: 71–78
Volente C, Angelastro JM, Greene LA (1993b) Association of protein kinases ERK1 and ERK2 with p75 nerve growth factor receptors. J Biol Chem 268: 21410–21415
von Bartheld CS, Kinoshita Y, Prevette D, Yin QW, Oppenheim RW, Bothwell M (1994) Positive and negative effects of neurotrophins on the isthmo-optic nucleus in chick embryos. Neuron 12: 639–654
Weskamp G, Reichardt LF (1991) Evidence that biological activity NGF is mediated through a novel subclass of high affinity receptors. Neuron 6: 649–663
Wilken GP, Marriot DR, Chlolewinski AJ (1990) Astrocyte heterogeneity. Trends Neurosci 13: 43–46
Wolff RA, Dobrowsky RT, Bielawaska A, Obeid LM, Hannun YA (1994) Role of ceramide-activated protein phosphatase in ceramide-mediated signal transduction. J Biol Chem 269: 19605–19609

Wright DC, Delaney TF (1989) Treatment of metastatic cancer to the brain. In: DeVita VT, Hellman S, Rosenburg SA (eds) Cancer: principals and practice of oncology. Lippincott, New York, pp 2245–2261

Yaar M, Eller MS, DiBenedetto P, Reenstra WR, Zhai S, McQuaid T, Archambault M, Gilchrest BA (1994) The trk family of receptors mediates nerve growth factor and neurotrophin-3 effects in melanocytes. J Clin Invest 94: 1550–1562

Yaar M, Gilchrest BA (1991) Human melanocyte growth and differentiation: a decade of new data. J Invest Dermatol 97: 611–617

Yoshida K, Gage FH (1991) Fibroblast growth factors stimulate nerve growth factor synthesis and secretion by astrocytes. Brain Res 538: 118–126

Yoshida K, Gage FH (1992) Cooperative regulation of nerve growth factor synthesis and secretion in fibroblasts and astrocytes by fibroblast growth factor and other cytokines. Brain Res 569: 14–25

Yoshida K, Kakihana M, Chen LS, Ong M, Baird A, Gage FH (1992) Cytokine regulation of nerve growth factor-mediated cholinergic neurotrophic activity synthesized by astrocytes and fibroblasts. J Neurochem 59: 919–931

Zafra F, Lindholm D, Castren J, Thoenen H (1992) Regulation of brain-derived neurotrophic factor and nerve growth factor mRNA in primary cultures of hippocampal neurons and astrocytes. J Neurosci 12: 4793–4799

Epithelial Differentiation and the Control of Metastasis in Carcinomas

W. Birchmeier, J. Behrens, K.M. Weidner, J. Hülsken, and C. Birchmeier

1 Structural and Molecular Characteristics of Epithelia 118
2 Molecular Control of Epithelial-Mesenchymal Transitions 120
3 Epithelial-Mesenchymal Transitions Induced by Tyrosine Kinase Receptors 122
4 Changes in Cadherin-Mediated Cell Adhesion in Invasive and Metastatic Tumors 125
5 Tyrosine Kinase Receptors and Their Ligands in the Control of Growth, Motility, and Metastasis of Carcinomas 126
6 Future Prospects 128
References 128

Tumors of epithelial origin (carcinomas) are of major medical importance (DeVita et al. 1993). Multiple genetic changes in dominant and recessive oncogenes that contribute to carcinoma formation and progression have been identified. The most frequent mutations found in carcinomas include activating mutations in the *ras* family of genes, mutations that interfere with the function of p53, and amplification or overexpression of genes encoding tyrosine kinase receptors (Barbacid 1987; Aronson 1991; Hollstein et al. 1991; Vogelstein and Kinzler 1993; Hinds and Weinberg 1994). These mutations affect both the growth and the genetic stability of affected cells. A further critical step in the development of malignant carcinomas is their ability to invade the underlying tissue and to metastasize to distant sites. Early genetic approaches for the analysis of oncogenes have focused mainly on the induction of transformation, i.e., unregulated growth in fibroblasts, which has led in the past decade to a wealth of data on the pathways controlling normal and aberrant growth. In comparison, there has been a lag in understanding the molecular causes leading to the acquisition of the metastatic potential of carcinoma cells.

Max-Delbrück-Centrum für Molekulare Medizin, Robert-Rössle-Str. 10, 13125 Berlin, Germany

1 Structural and Molecular Characteristics of Epithelia

Epithelia typically form continuous sheets of tightly adhering cells that are usually cuboidal in shape. In culture, epithelial cells grow in aggregates and show little motility. Characteristic are specialized organelles, tight junctions, adherens junctions, and desmosomes, which are responsible for the tight intercellular contacts between the individual cells. A consequence of the laterally located junctional complexes is the polarity characteristic for epithelial cells. Distinct proteins are expressed on the basolateral or apical surfaces, and free diffusion of membrane proteins to all surfaces of the cells is inhibited (RODRIGUEZ-BOULAN and NELSON 1989; SCHWARZ et al. 1990; BIRCHMEIER and BEHRENS 1994). Polar epithelial cells have evolved special mechanisms that allow the transport of membrane proteins to either the apical and basolateral surface (SIMONS and WANDINGER-NESS 1990).

Adherens junctions are specialized structures containing the transmembrane cell adhesion molecule E-cadherin, which recognizes and binds E-cadherin present on the neighboring cells in a Ca^{2+}-dependent manner. (The other epithelial junctions are reviewed in CITI 1993; GARROD 1994). E-cadherin is a 120-kDa transmembrane glycoprotein, of which an extracellular 80-kDa tryptic fragment can be released in the presence of Ca^{2+}. The cDNA of E-cadherin encodes a signal peptide and a presequence at the amino-terminus, a large extracellular sequence with four repeated domains important in Ca^{2+}-binding, a single transmembrane sequence, and a short cytoplasmic domain that is important for catenin binding (TAKEICHI 1991; KEMLER 1992). E-cadherin is the prototype of the family of Ca^{2+}-dependent cell adhesion molecules; close relatives include N-cadherin (expressed in neurones) and P-cadherin (expressed in placenta and epithelia). Further members are M-cadherin (in muscles), OB-cadherin (in osteoblasts), LI-cadherin (in liver and intestine), the desmosomal proteins desmoglein and desmocollin, and others (TAKEICHI 1991; DONALIES et al. 1991; BUXTON and MAGEE 1992; KEMLER 1992; SANO et al. 1993; OKAZAKI et al. 1994; BERNDORFF et al. 1994).

The cytoplasmic portion of E-cadherin interacts with α-, β-, and γ-catenins (plakoglobin; KEMLER 1992; TSUKITA et al. 1993; TROYANOVSKY et al. 1994; HÜLSKEN et al. 1994). In addition, other cytoskeletal proteins (vinculin, α-actinin, radixin, ezrin, moesin, fodrin) are located on the cytoplasmatic side of the junctional complex that anchor the actin-containing filaments (TSUKITA et al. 1992, 1993). The integrity of this junctional complex is critical for the maintenance of the functional characteristics of epithelia. Its disruption leads to the dissociation of epithelial sheets, a change in cell morphology, and an increased motility of the cells. The cells thus acquire characteristics of mesenchymal cells (see also below). A controlled loss of epithelial character concomitant with dissociation and increased motility of the cells is a prerequisite for normal morphogenic processes in development and occurs, for example, during mesoderm formation (BEDDINGTON and SMITH 1993) or migration of myogenic precursors (BLADT et al. 1995). A similar but uncontrolled epithelial-mesenchymal conversion is observed during progression of human tumors (VALLES et al. 1991; BIRCHMEIER and BEHRENS 1994; see also

BOYER et al., this issue). Here the loss of epithelial differentiation and the acquisition of mesenchymal characteristics, such as the ability to move, is correlated with malignancy of the carcinoma cells.

Hemidesmosomes are located on the basal surface of epithelia, which are junctions contacting the basement membrane (A. SONNENBERG et al. 1991; GARROD 1993). The basement membrane separates the epithelial cell compartment from underlying mesenchymal cells and is formed by both cell types. Characteristic constituents of basement membranes include laminin, collagen IV, nidogen/entactin, and basement membrane proteoglycan. The mesenchymal reticular lamina which lies below the basement membrane contains collagen types I and II as well as fibronectin (TIMPL 1989). The epithelial receptors for basement membrane components are located on the basal surface of the plasma membrane. Expressed on the epithelial cells are, for example, the integrins $\alpha6/\beta1$ and $\alpha6/\beta4$ which are receptors for laminin, and the integrin $\alpha1/\beta1$ which is a receptor for collagen. The hemidesmosomes are enriched for the integrin receptor $\alpha6/\beta4$ (A. SONNENBERG et al. 1991). A variety of experimental evidence is accumulating that demonstrates a function of integrins not only in cell adhesion but also in the transduction of signals that can influence growth and motility of cells (JULIANO and VARNER 1993).

In addition, various receptor with tyrosine kinase activity exist, that are expressed predominantly on epithelial cells, such as c-Ros, c-Met, c-Neu, c-Ret, and the receptor for keratinocyte growth factor (KGF); (KOKAI et al. 1987; QUIRKE et al. 1989; MORI et al. 1989; PRESS et al. 1990; E. SONNENBERG et al. 1991, 1993; ORR-URTREGER et al. 1993; PACHNIS et al. 1993). Since most of these receptors were identified via their transforming activity in NIH3T3 cells, they can clearly give mitogenic signals (ARONSON 1991). However, experiments with epithelial cells in culture have shown that the receptors do not only control growth but can also influence the motility and shape of epithelia as well as the differentiation or morphogenesis (BIRCHMEIER and BIRCHMEIER 1993). Genetic analysis in the mouse also demonstrates the pivotal role of these paracrine signaling molecules in the normal development and physiology of epithelia (SCHUCHARDT et al. 1994; WERNER et al. 1993, 1994; PETERS et al. 1994; SCHMIDT et al. 1995; cf. BIRCHMEIER and BIRCHMEIER 1993, for a review). Moreover, mutations or deregulated expression of such receptors as well as the formation of autocrine loops are observed in carcinoma cells and contribute to the formation of tumors (ARONSON 1991).

In contrast to epithelial cells (which display strong intercellular adhesion and basolateral polarization, and which are largely nonmobile), mesenchymal cells are generally loosely associated to one other; they are nonpolarized and surrounded by extracellular matrix. The major adhesive interactions of mesenchymal cells occur with the cell substrate (HAY 1990). Integrin molecules on the cell surface mediate the adhesion to extracellular matrix proteins of the connective tissues, for instance, to fibronectin, tenascin, or collagen (see HYNES 1992, for a review). Mesenchymal cells also express specific ligands for various epithelial receptor tyrosine kinases, such as SF/HGF, neuregulin, and KGF (WERNER et al. 1992; SONNENBERG et al. 1993; MEYER and BIRCHMEIER 1994; MASON et al. 1994) and thereby play an important role in the control of epithelial growth, morphogenesis and differentiation (see BIRCHMEIER and BIRCHMEIER 1993).

2 Molecular Control of Epithelial-Mesenchymal Transitions

Epithelial-mesenchymal transitions can be observed during development, for example, in gastrulation, when the first mesenchymal cells are formed (BEDDINGTON and SMITH 1993; FAUST and MAGNUSON 1993). This demonstrates the intrinsic ability of epithelia to change their cellular characteristics. Epithelial-mesenchymal conversions in development occur in a temporarily and spatially restricted and controlled manner. The signals responsible for mesoderm induction have been studied extensively in *Xenopus* and recently also in the mouse (BEDDINGTON and SMITH 1993; HARLAND 1994; DE ROBERTIS 1995). Similar morphological changes are observed during tumor progression. Here the acqusition of mesenchymal characteristics, i.e., loss of differentiated appearance and intercellular adhesion together with increased motility, are observed in late stages of tumor progression and are correlated with poor prognosis (see BIRCHMEIER and BEHRENS 1994, for a review).

Transitions of epithelial to mesenchyme-like cells can be induced in vitro by interfering with the functional integrity of the adherens junctions. One of the first experiments to demonstrate this was performed by the addition of antibodies that interfere with the function of E-cadherin (IMHOF et al. 1983; BEHRENS et al. 1989). MDCK epithelial cells treated in this manner change their shape; they become dissociated and resemble fibroblasts in morphology. When cocultured with heart tissue, the polar MDCK cells form a single layer of epithelial sheets that surrounds the tissue; in the presence of anti-E-cadherin antibodies, the cells become motile and invade the heart tissue. The presence of functional adherens junctions is therefore necessary for the epithelial morphology; its disruption leads to an epithelial-mesenchymal transition paralleled by an increase in motility and invasiveness of the cells (BEHRENS et al. 1989; FRIXEN et al. 1991; VLEMINCKX et al. 1991; CHEN and ÖBRINK 1991; NAVARRO et al. 1991). More recently the importance of E-cadherin in the maintenance of epithelial cell morphology was also demonstrated by the use of transdominant E-cadherin mutants expressed in epithelial (and embryonal) cells (KINTNER 1992; FUJIMORI and TAKEICHI 1993). Truncated variants of E-cadherin lacking large parts of the extracellular domain but with intact transmembrane and cytoplasmic sequences are unable to bind to E-cadherin expressed on neighboring cells. However, they can interact with the cytoplasmatically located catenins. When expressed in large amounts, they can deplete the cellular pool of catenins and interfere thus with normal E-cadherin-catenin interactions. Expressed on cells with epithelial morphology, transdominant E-cadherin induces a fibroblast morphology and an increase in cellular motility.

E-cadherin is important not only for the maintenance of morphology and adhesion in differentiated epithelia but also for the acquisition of epithelial characteristics early in embryogenesis. This was originally suggested by experiments with anti-E-cadherin antibodies (HYAFIL et al. 1981) and recently also demonstrated genetically (LARUE et al. 1994; RIETHMACHER et al. 1995). A targeted mutation was introduced into the E-cadherin gene of the mouse via homologous recombination and embryonic stem cell technology. This mutation removes sequences essential for Ca^{2+}-binding and thus for the adhesive function of the molecule (RIETHMACHER et al. 1995). Animals that carry this mutation in a heterozygous state appear normal and are fertile. However, homozygous mutant embryos do not develop normally. They can reach the morula stage and also compact properly, but the compacted state is not sustained. The individual morula cells in the embryo loose their morphological polarization; they become rounded and continue to divide. As a consequence, the embryos appear totally distorted at a time when wild-type or heterozygous mutant embryos form a well-organized blastocyst with a well-formed blastocoel. Since the mutant embryos never emerge from the zona pellucida, further development, particularly implantation into the uterus, cannot take place. It has previously been reported that removal of Ca^{2+} ions or treatment with anti-E-cadherin antibodies interferes with the compaction of the mouse morula, indicating that the initial compaction and polarization of the epithelial-like cells depends on E-cadherin (HYAFIL et al. 1981; VESTWEBER and KEMLER 1984). Nevertheless, embryos which lack a functional E-cadherin gene can undergo compaction. This appears to be due to maternally derived E-cadherin and not to functional compensation by other cell adhesion molecules. The maternal E-cadherin suffices thus for the initial compaction and for the formation of the first polarized, epithelial-like cells but not for further development beyond the morula stage.

The integrity of the adherens junction also requires functional catenin molecules. Mutant cells that lack α-catenin have been described; the cadherin mediated cell adhesion is not functional in such cells and can be restored by the expression of α-catenin cDNA driven by an appropriate promoter (HIRANO et al. 1992; ODA et al. 1993; SHIMOYAMA et al. 1992; MORTON et al. 1993). Similar experiments with a cell line mutant for the β-catenin gene have demonstrated its importance for the function of the adherens junction (OYAMA et al. 1994; KAWANISHI et al. 1995). By transient expression of cDNAs encoding variant α- and β-catenin cDNAs which lack sequences for different subdomains, the interactions of the molecules in the adherence complex have been elucidated. E-cadherin binds directly to β-catenin; the binding site in β-catenin consists of armadillo repeat sequences (named after the DROSOPHILA armadillo gene in which this sequence motive was first identified) located in the central part of the molecule. The N-terminus of β-catenin interacts with α-catenin, and via α-catenin the cytoskeletal proteins bind to the complex. γ-Catenin (plakoglobin) is similar in structure to β-catenin, and it interacts in a analogous manner with E-cadherin and α-catenin (NÄTHKE et al. 1994; HÜLSKEN et al. 1994; RUBINFELD et al. 1995). Also the adenomatous polyposis coli (APC) tumor suppresser gene product can bind to β-catenin.

Mutations of APC are frequently observed in colon carcinomas (Fearon et al. 1990). APC and E-cadherin have similar binding-sites on β-catenin, i.e., they can compete for β-catenin binding (Rubinfeld et al. 1993; Su et al. 1993; Hülsken et al. 1994). Tyrosine phosphorylation of β-catenin also appears to be critical for the function of adherens junctions: expression of a temperature-sensitive v-*src* converts epithelial cells to a mesenchymal cell shape (Behrens et al. 1993; see also Hamaguchi et al. 1993). The signaling molecule Wnt upregulates the expression of E-cadherin and plakoglobin (Bradley et al. 1993), stabilizes the association between β-catenin and E-cadherin (Hinck et al. 1994), and induces epithelial transformation of PC12 cells (Shackleford et al. 1993). This indicates that β-catenin is a member of the Wnt signaling pathway, as is true for armadillo in Drosophila.

3 Epithelial-Mesenchymal Transitions Induced by Tyrosine Kinase Receptors

In addition, various soluble factors that function as ligands for tyrosine kinase receptors can induce transient epithelial-mesenchymal conversion and increased motility in cell culture, such as SF/HGF, acidic fibroblast growth factor, and epidermal growth factor (Barrandon and Green 1987; Weidner et al. 1990; Jouanneau et al. 1991; Hoschuetzky et al. 1994). The best characterized of these factors is the ligand for the c-Met tyrosine kinase, which has been named scatter factor (SF) because of its activity on epithelial cells in culture (Stoker et al. 1987; Weidner et al. 1990). In addition, SF can induce growth of hepatocytes (hepatocyte growth factor, HGF) and other cells (Miyazawa et al. 1989; Nakamura et al. 1989; Weidner et al. 1991). A third activity of SF/HGF is the morphogenic action that was first discovered as the ability to induce epithelial cells to form tubules in vitro (Montesano et al. 1991 a,b; see also below). SF/HGF has a unique structure since it resembles closely proteases such as plasminogen (40% sequence identity) but not other ligands for tyrosine kinase receptors (Miyazawa et al. 1989; Nakamura et al. 1989; Weidner et al. 1991). The protein is produced as an inactive precursor molecule (90 kDa) that is cleaved outside of the producing cells into a heavy (60 kDa) and a light (30 kDa) chain (Hartmann et al. 1992; Lokker et al. 1992; Naldini et al. 1992). The heavy chain contains a N-terminal hairpin loop and four kringle domains; the light chain shows extensive homologies to serine proteases. However, two of the three amino acids which form the catalytic triad of serine proteases are altered in SF/HGF, and therefore the factor has no catalytic activity.

As with many other tyrosine kinase receptors, c-Met was initially identified because of its transforming activity when mutated. The c-*met* oncogene was derived from a *N*-methyl-*N*-nitro-*N*-nitrosoguanidine treated osteosarcoma cell line that was used in a transfection/tumorigenicity assay (Cooper et al. 1984).The gene transferred from the osteosarcoma cells was the product of a rearrangement that fused TPR (translocated promoter region) on chromosome 1 to c-*met* on chromosome 7 (PARK et al. 1986). The oncogenic variant of *c-met* encodes a cytoplasmatically located tyrosine kinase. In contrast, the proto-oncogene product is a transmembrane glycoprotein of 190 kDa that is cleaved postranslationally into an α- and a β-chain (GONZATTI-HACES et al. 1988; GIORDANO et al. 1989). A major breakthrough in the understanding of the *c-met* receptor was the identification of its ligand, which is scatter factor/hepatocyte growth factor (BOTTARO et al. 1991; NALDINI et al. 1991). All the known biological activities of SF/HGF are mediated by the c-Met receptor (WEIDNER et al. 1993) and require activation of the *ras* pathway (HARTMANN et al. 1994). In addition, other downstream targets play an important role in c-Met specific signal transduction ((PONZETTO et al. 1994; WEIDNER et al. 1995; FIXMANN et al. 1995).

We have recently analyzed the normal physiological role of the SF/HGF and the *c-met* gene by the introduction of targeted mutations in the mouse, via homologous recombination and embryonic stem cell technology (SCHMIDT et al. 1995; UEHARA et al. 1995; BLADT et al. 1995). Whereas animals with a heterozygous mutation in SF/HGF or *c-met* are normal and fertile, a homozygous mutation is not compatible with normal development. SF/HGF -/- or *c-met* -/-embryos die between days 13 and 16.5 (E13-E16.5) of development. The mutant embryos appear externally normal; however, on E12.5 their livers are considerably reduced in size. Histological examination shows damage to the embryonic liver on E14.5 that varies in severity and is not observed on E12.5. The essential function of SF/HGF and *c-met* in the development of the liver is also supported by the analysis of embryonic stem cells that carry two mutant alleles in *c-met*. Such cells cannot contribute to the liver but participate in the development of a variety of other organs and cell types. In contrast, ES cells with only one mutant SF/HGF or *c-met* allele participate in the development of the hepatic lineage (BLADT et al. 1995). In addition, placental development is impaired in the SF/HGF -/- or in *c-met* -/- embryos. Particularly the trophoblast cells are affected, and consequently a disorganization of the labyrinth layer is observed. In this layer exchange of nutrients and oxygen between the maternal and fetal blood occurs. Thus two epithelial cell types in the embryo, hepatocytes and trophoblast cells of the placenta, are affected by the SF/HGF and *c-met* mutations. A third phenotype observed in the mutant embryos is an absence of skeletal muscle, which develops in limbs, diaphragm, and tip of the tongue (BLADT et al. 1995). In wild-type animals these muscles develop from precursor cells which migrate from the somites. The precursor cells do not migrate in the mutant animals, and as a consequence muscle formation at these sites does not occur (BLADT et al. 1995). The c-Met receptor is widely expressed on epithelial cells, both during development and in the adult, whereas SF/HGF is usually produced by mesenchymal cells in the vicinity (SONNENBERG et al. 1993). SF/HGF and *c-met* therefore form a paracrine signalling system, a concept originally suggested by STOKER et al. (1987), which is essential for normal development.

Among the activities of SF/HGF that were described in cell culture, the morphogenic activity is unique and no other factor with similar properties has been described (MONTESANO et al. 1991a,b). MDCK cells grown in a three-dimensional collagen matrix for several days form hollow cysts. When SF/HGF is added, individual cells dissociate and move away from the cysts. Consequently the cells reassociate and form continuous tubules. These tubules have a lumen surrounded by well-polarized epithelial cells with a smooth basal surface in contact with the collagen matrix, and a apical surface rich in microvilli that faces the lumen. The structures formed in vitro thus resemble the tubular epithelia present in many organs (MONTESANO et al. 1991a). SF/HGF plays a similar role in the development of the mammary gland since it can induce the branching and growth of the tubular epithelia in organ cultures (YANG et al. 1995). Recent experimental evidence indicates, however, that not only the formation of tubular epithelia can be induced by SF/HGF but also other morphogenic programs, such as formation of cryptlike and alveolar structures from colon and lung epithelial cells (BRINKMANN et al. 1995). Thus, SF/HGF seems to be able to activate the intrinsic morphogenic potential of various epithelial cell types, but does not instruct the cells to form one particular structure.

In vitro data indicate that SF/HGF induces not only morphogenesis or motility of epithelial cell but also invasiveness into collagen matrices (WEIDNER et al. 1990). Furthermore, it has recently been shown that SF/HGF is expressed in breast carcinomas, and that the level of expression is correlated with the state of tumor progression (YAMASHITA et al. 1994). How can the same factor be responsible for epithelial morphogenesis and invasiveness of epithelial cells? Interestingly, the different morphogenic programs can be induced only by SF/HGF in cells with intact epithelial characteristics, including expression of E-cadherin and functional adherens junctions (BRINKMANN et al. 1995). On the other hand, SF/HGF was shown to induce metastasis to the lymph nodes and lung when expressed by transfection of cDNA in dedifferentiated breast carcinoma cells (MEINERS et al., in preparation; see also ROSEN et al. 1994). Thus the differentiation state of the epithelial cells seems to determine the response towards SF/HGF: morphogenesis requires in intact epithelial program, whereas metastatic behavior is observed primarily in cells that have already undergone epithelial-mesenchymal transitions.

Transitions of mesenchymal cells to epithelia (i.e., the reserve direction), are accomplished by several means. Ectopic expression of E-cadherin in fibroblasts results in cells which become adhesive to each other and show polar characteristics (NAGAFUCHI et al. 1987; RINGWALD et al. 1987; GALLIN et al. 1987). Similarily, poorly differentiated, fibroblastlike carcinoma cells can be converted to epitheloid cells by the expression of E-cadherin and then loose their invasive potential (VLEMINCKX et al. 1991; FRIXEN et al. 1991). During normal development, kidney mesenchyme is induced to form new epithelia by signals derived from the ureter bud; the newly induced cell then expresses Int-4 (STARK et al. 1994) and the transcription factors N-Myc and Pax-2 (MUGRAUER et al. 1988; DRESSLER et al. 1993). Recent experimental evidence indicates that the closely related factor, Int-1, is a signal that can convert kidney mesenchyma to epithelium (HERZLINGER et al., personal communication). Also, a retrovirus that transduces the transcription factor Pax-2 is sufficient to convert kidney mesenchyma to epithelium (DRESSLER et al. 1993). Expression of c-Met in NIH3T3 fibroblasts induced epithelial characteristics (TSARFATY et al. 1994). This may indicate that SF/HGF can actually interconvert epithelial and mesenchymal cells (see also below). Additionally, the adenovirus Ela gene causes diverse tumor cells of mesenchymal origin to adapt epithelial morphology and induces expression of epithelial-specific genes (FRISCH 1994).

4 Changes in Cadherin-Mediated Cell Adhesion in Invasive and Metastatic Tumors

In human carcinomas the loss of epithelial characteristics is an important prognostic markers and is correlated with a poor outcome of the disease. The loss of epithelial differentiation is particularly prominent at invasive fronts of the tumors, where the carcinoma cells break into the surrounding mesenchymal tissues. The invading cells can loose their epithelial appearance; they become spindle-shaped and fibroblastlike, and in the electron microscope a reduced number of desmosomes are often observed (WEINSTEIN et al. 1976; GABBERT et al. 1985). Desmosomes (spot desmosomes) are the type of intercellular junctions that is most easily detected in intact tissues. Together with adherens junctions (belt desmosomes) they constitute the main adhering junctions of epithelial cells.

The prominent change in morphology of malignant carcinoma cells is associated with a loosening of intercellular adhesion. This is a consequence of a functional disturbance of cell-cell contacts. The malignant cells apparently downmodulate intercellular adhesiveness. That loss of intercellular adhesion is a crucial step in carcinomas progression was first proposed 50 years ago (COMAN 1944; MCCUTCHEON et al. 1948). Since the importance of intact E-cadherin mediated cell adhesion in the maintenance of the epithelial phenotype and the prevention of invasion and metastasis was first recognized (BEHRENS et al. 1989), various types of primary human carcinomas have been examined for defects in one of the constituents of intercellular junctions. A correlation between dysfunction of E-cadherin-mediated cellular adhesion, by downregulation of E-cadherin and α-catenin expression or mutation in the E-cadherin gene, and invasive potential of many types of carcinomas has emerged from these studies. We discuss here selected examples, with the goal of reviewing particularly the various mechanisms of alternations in tumors that have been observed.

Diffuse-type gastric carcinomas contain poorly adhesive cells that often do not express E-cadherin protein or mRNA. When the mRNA from these tumors was analyzed by reverse polymerase chain reaction, a high frequency of altered E-cadherin transcripts were identified. One-half of all the tumors tested express shortened transcripts that lack exon 9 or 10 (BECKER et al. 1994); these exons encode part of the first Ca^{2+} binding site of E-cadherin and thus sequences known to be essential for adhesive function. Analysis of the genomic E-cadherin DNA demonstrated that mutations are present in the corresponding splice junctions, and that the second, intact copy of the gene is frequently lost (Becker et al. 1994). Mutations in one copy and mutation or loss of the second intact allele are typical for genes that play a causative role in the development or progression of human tumors, i.e. for tumor supressor genes. Mutations of E-cadherin in ovarian and lobular breast carcinomas have also been detected (Risinger et al. 1994; Kanai et al. 1994).

A frequently observed alteration in malignant carcinoma cells of various tissues of origin is downregulation or even complete absence of E-cadherin expression. In head and neck cancers, for example, downregulation of E-cadherin expression is correlated with loss of epithelial differentiation of the tumors and with the formation of lymph node metastasis (Schipper et al. 1991). In breast carcinomas a correlation between downregulation of E-cadherin expression and malignancy is observed in lobular but not in ductal carcinomas (Moll et al. 1993, Gamallo et al. 1993). In colorectal carcinomas downregulation of E-cadherin has similarly been observed to be correlated with late stages of tumor progression (Dorudi et al. 1993; Kinsella et al. 1993). In prostate carcinomas dowregulation of E-cadherin is correlated with poor prognosis (Umbas et al. 1994; Otto et al. 1994; see Birchmeier and Behrens 1994, for an extended review). The new data reported since our previous review on this subject clearly indicate a relationship between altered cadherin and catenin-mediated cell adhesion and metastasis as well as patients survival. In renal cell carcinomas Kaplan-Meier analysis shows better prognosis in the group with preserved E-cadherin expression than the group without expression (Katagiri et al. 1995). Similarly, decreased E-cadherin immunoreactivity is correlates with poor survival in patients with bladder (Bringuier et al. 1993; Otto et al. 1994) and prostata carcinomas (Umbas et al. 1994). The level of expression of E-cadherin mRNA in colorectal cancer is also correlated with clinical outcome (Dorudi et al. 1995; see also Mattijssen et al. 1993).

5 Tyrosine Kinase Receptors and Their Ligands in the Control of Growth, Motility, and Metastasis of Carcinomas

A variety of tyrosine kinase receptors are expressed predominantly or exclusively on epithelia and can control growth, differentiation, and motility of epithelial cells (Birchmeier and Birchmeier 1993). Since abnormal growth and motility are essential in tumor progression, deregulated activity of tyrosine kinase receptors can profoundly influence carcinogenesis. Indeed, c-*neu*/HER2, c-*met* and c-*ret* are found to be mutated or overexpressed in particular types of human carcinomas (Slamon et al. 1987,1989; Santoro et al. 1990; Liu et al. 1992; Di Renzo et al. 1992; Kuniyasu et al. 1992; Rege-Cambrin et al. 1992; Hofstra et al. 1994; Jucker et al. 1994; Boix et al. 1994). The mutations observed can often relieve the receptors from the presence of the specific ligand required for signaling under normal conditions and therefore constitutively activate the receptors. In addition, expression of high levels of the intact ErbB2 receptor has been reported to suffice for ligand-independent activation and the induction of transformation in fibroblast (Di Fiore and Hartwell 1987; Hudziak et al. 1987), and high levels of autophosphorylated c-Met have been observed in carcinoma cells that overexpress the intact receptor (Giordano et al. 1989; Ponzetto et al. 1991). However, the presence of the specific ligands, whether produced in an autocrine or paracrine fashion, is also a variable to be evaluated and of potential importance for tumor progression.

Mainly the mitogenic effect of tyrosine kinase receptors has been extensively studied, and growth and transformation are well-known responses to the signals mediated by the receptors in fibroblasts. Recently the biological responses of epithelial (carcinoma) cells have gained more attention, since receptor tyrosine kinases can also control motility and invasive behavior. This is particularly well documented for c-Met and its ligand, SF/HGF (Stoker et al. 1987; Weidner et al. 1990). In addition, acidic fibroblast growth factor has been found to increase motility of a bladder carcinoma cell line (Jouanneau et al. 1991), and by the use of chimeric receptors, the active Ros, Trk, or Erb2 tyrosine kinases have been demonstrated to dissociate epithelia and to increase their motility (Sachs et al., in preparation). Such signaling specificities of tyrosine kinase receptors in epithelial or carcinoma cells may be more common than previously assumed. Since motility of carcinoma cells is essential for their ability to form metastasis, receptor tyrosine kinases may therefore contribute significantly to the development of malignancies.

In recent years, various investigators have examined the involvement of tyrosine receptors in human cancers. We concentrate here on the discussion of new data on *c-met* and its ligand SF/HGF in the progression of human carcinomas. In has been shown in hepatocellular carcinomas that overexpression of *c-met* correlates with poor differentiation of cancer cells Suzuki et al. 1994; see also Selden et al. 1994). Gastric carcinomas showed amplification of *c-met*, particularly in scirrous type of stomach cancer (Kuniyasu et al. 1992). Overexpression of *c-met* and SF/HGF was detected in pancreatic carcinomas (Ebert et al. 1994; see also Di Renzo et al. 1995). In breast cancer, high expression of SF/HGF is correlated with shorter survival of the patients, and SF/HGF was found to be the most important independent prognostic factor for survival prediction (Yamashita et al. 1994). Overexpression of *c-met* and SF/HGF are reported for thyroid and bladder carcinomas (Di Renzo et al. 1991; Joseph et al. 1995). Aberrant expression of SF/HGF and *c-met* are also found in various tumors of mesnchymal origin such as osteosarcomas, sarcomas of patients with Li-Fraumeni syndrome, Kaposi sarcoma, melanoma, and leukemia (Naidu et al. 1994; Jucker et al. 1994; Natali et al. 1993; Tsarfaty et al. 1994; Rong et al. 1995, Ferracini et al. 1995). A direct association of tyrosine kinase receptors with development of metastasis has as yet been examined in only a few cases. Expression of *c-erbB2* in mice under the control of the mouse mammary tumors resulted in focal malignant mammary tumors which frequently form metastasis in the lung (Guy et al. 1992). Creation of an autocrine loop by transfection of SF/HGF into bladder carcinoma cells that express endogenous *c-met* leads to the generation of highly malignant and invasive cells (Bellusci et al. 1994). Similarly, SF/HGF modulates the metastatic behavior of mouse mammary tumors (Rosen et al. 1994). We have recently expressed the SF/HGF cDNA in MDA MB 435 human mammary carcinoma cells followed by transplantation of the cells into the mammary fat pad of nude mice. Expression of SF/HGF strongly stimulates the ability to metastasize into lymph nodes and lungs (Meiners et al., in preparation).

6 Future Prospects

The molecular aspects of junctional deficiencies and the loss of differentiation of epithelial cells in carcinomas has thus turned out to be quite complex; many junctional proteins (E-cadherin, catenins) but also proteins of various signal cascades (*src* or tyrosine kinase receptors) can be involved. The critical proteins are affected by downregulation of expression, by mutation or by biochemical modification. Down-regulation of E-cadherin expression and overexpression of *c-met* are often correlated with loss of tumor differentiation and progression to lymph node metastasis. The prognostic value of the measurement of E-cadherin, catenin, SF/HGF, and *c-met* expression is certainly important for future clinical research. Also, several components of the epithelial differentiation program appear to be deregulated in concert in carcinomas, which indicates that epithelial master genes are affected during tumorigenesis. Developmental studies aimed at the elucidation of mechanisms that regulate the epithelial phenotype will obviously also contribute to our understanding of malignant aberrations of epithelial cells during carcinogenesis.

An interesting interrelationship between epithelial-mesenchymal transitions and metastasis has recently been uncovered. A novel gene was found by subtractive hybridization which codes for a cytoskleleton-associated, fibroblast-specific protein (FSP1, Strutz et al. 1995). When the FSP1 cDNA is overexpressed in tubular epithelial cells, it induces epithelial-mesenchymal transitions. FSP1 was found to be identical to the metastasis-specific gene *mts*-1 previously identified by Lukanidin's group (Ebralidze et al. 1989; see also the article in this issue) which is a member of the S100 family of Ca^{2+} binding proteins. This is thus corroborative evidence that epithelial-mesenchymal transitions contribute to metastasis.

References

Aaronson SA (1991) Growth factors and cancer. Science 254: 1146–1153
Barbacid M (1987) Ras genes. Annu Rev Biochem 56: 779–827
Barrandon Y Green H (1987) Cell migration is essential for sustained growth of keratinocyte colonies: the roles of transforming growth factor-alpha and epidermal growth factor. Cell 50: 1131–1137

Becker KF, Atkinson MJ, Reich U, Becker I, Nekarda H, Siewert JR, Hofler H (1994) E-cadherin gene mutations provide clues to diffuse type gastric carcinomas. Cancer Res 54: 3845–3852

Beddington RS, Smith JC (1993) Control of vertebrate gastrulation: inducing signals and responding genes. Curr Opin Genet Dev 3: 655–661

Behrens J, Mareel MM, Van Roy FM, Birchmeier W (1989) Dissecting tumor cell invasion: epithelial cells acquire invasive properties after the loss of uvomorulin-mediated cell-cell adhesion. J Cell Biol 108: 2435–2447

Behrens J, Vakaet L, Friis R, Winterhager E, Van Roy F, Mareel MM, Birchmeier W (1993) Loss of epithelial differentiation and gain of invasiveness correlates with tyrosine phosphorylation of the E-cadherin/beta-catenin complex in cells transformed with a temperature-sensitive v-SRC gene. J Cell Biol 120: 757–766

Bellusci S, Moens G, Gaudino G, Comoglio P, Nakamura T, Thiery JP, Jouanneau J (1994) Creation of an hepatocyte growth factor/scatter factor autocrine loop incarcinoma cells induces invasive properties associated with increased tumorigenicity. Oncogene 9: 1091–1099

Berndorff D, Gessner R, Kreft B, Schnoy N, Lajous Petter AM, Loch N, Reutter W, Hortsch M, Tauber R (1994) Liver-intestine cadherin: molecular cloning and characterization of a novel Ca(2+)-dependent cell adhesion molecule expressed in liver and intestine. J Cell Biol 125: 1353–1369

Birchmeier C, Birchmeier W (1993) Molecular aspects of mesenchymal-epithelial interactions. Annu Rev Cell Biol 9: 511–540

Birchmeier W, Behrens J (1994) Cadherin expression in carcinomas: role in the formation of cell junctions and the prevention of invasiveness. Biochim Biophys Acta 1198: 11–26

Bladt F, Riethmacher D, Isenmann S, Aguzzi A, Birchmeier C (1995) Essential role for the c-met receptor in the migration of myogenic precursor cells into the limb bud. Nature (in press)

Boix L, Rosa JL, Ventura F, Castells A, Bruix J, Rodes J, Bartrons R (1994) c-met mRNA overexpression in human hepatocellular carcinoma. Hepatology 19: 88–91

Bottaro DP, Rubin JS, Faletto DL, Chan AM, Kmiecik, TE, Vande Woude GF, Aaronson SA (1991) Identification of the hepatocyte growth factor receptor as the c-met proto-oncogene product. Science 251: 802–804

Bradley RS, Cowin P, Brown AM (1993) Expression of Wnt-1 in PC12 cells results in modulation of plakoglobin and E-cadherin and increased cellular adhesion. J Cell Biol 123: 1857–1865

Bringuier PP, Umbas R, Schaafsma HE, Karthaus HF, Debruyne FM, Schalken JA (1993) Decreased E-cadherin immunoreactivity correlates with poor survival in patients with bladder-tumors. Cancer Res 53: 3241–3245

Brinkmann V, Foroutan H, Sachs M, Weidner KM, Birchmeier W (1995) Hepatocyte growth factor/scatter factor induces a variety of tissue-specific morphogenic programs in epithelial cells. J Cell Biol 131: 1573–1586

Buxton RS, Magee AI (1992) Structure and interactions of desmosomal and other cadherins. Semin Cell Biol 3: 157–167

Chen W, Öbrink B (1991) Cell-cell contacts mediated by E-cadherin (Uvomorulin) restrict inasive behavior of L-cells. J Cell Biol 114(2): 319–327

Citi S (1994) Molecular mechanisms of epithelial cell junctions: from development to disease. Landes, Austin

Coman DR (1944) Decreased mutual adhesiveness, a property of cells from squamous cell carcinomas. Cancer Res 1: 625–629

Cooper CS, Park M, Blair DG, Tainsky MA, Huebner K, Croce CM, Vande Woude GF (1984) Molecular cloning of a new transforming gene from a chemically transformed human cell line. Nature 311: 29–33

De Robertis EM (1995) Developmental biology – dismantling the organizer. Nature 374: 407–408

De Vita VT, Hellman S, Rosenberg SA (1993) Cancer: principles and practice of oncology. Lippincott, Philadelphia

Di Fiore PM, Hartwell GR (1987) Median mandibular lateral periodontal cyst. Oral Surg Oral Med Oral Pathol 63: 545–550

Di Renzo MF, Narsimhan RP, Olivero M, Bretti S, Giordano S, Medico E, Gaglia P, Zara P, Comoglio PM (1991) Expression of the Met/HGF receptor in normal and neoplastic human tissues. Oncogene 6: 1997–2003

Di Renzo MF, Olivero M, Ferro S, Prat M, Bongarzone I, Pilotti S, Belfiore A, Costantino A, Vigneri R, Pierotti MA et al. (1992) Overexpression of the c-MET/HGF receptor gene in human thyroid carcinomas. Oncogene 7: 2549–2553

Di Renzo MF, Poulsom R, Olivero M, Comoglio PM, Lemoine NR (1995) Expression of the Met/hepatocyte growth factor receptor in human pancreatic cancer. Cancer Res. 55: 1129–1138

Donalies M, Cramer M, Ringwald M, Starzinski Powitz A (1991) Expression of M-cadherin, a member of the cadherin multigene family, correlates with differentiation of skeletal muscle cells. Proc Natl Acad Sci USA 88: 8024–8028

Dorudi S, Sheffield JP, Poulsom R, Northover JM, Hart IR (1993) E-cadherin expression in colorectal cancer. An immunocytochemical and in situ hybridization study. Am J Pathol 142: 981–986

Dorudi S, Hanby AM, Poulsom R, Northover J, Hart IR (1995) Level of expression of E-cadherin mRNA in colorectal cancer correlates with clinical outcome. Br J Cancer 71: 614–616

Dressler F, Whalen JA, Reinhardt BN, Steere AC (1993) Western blotting in the serodiagnosis of Lyme disease (comments) J Infect Dis 167: 392–400

Ebert M, Yokoyama M, Friess H, Buchler MW, Korc M (1994) Coexpression of the c-met protooncogene and hepatocyte growth factor in human pancreatic cancer. Cancer Res 54: 5775–5778

Ebralidze A, Tulchinsky E, Grigorian A, Afanasyeva A, Senin V, Revazova E, Lukanidin E (1989) Isolation and characterization of a gene specifically expressed in different metastatic cells and whose deduced gene product has a high degree of homology to a Ca^{2+}-binding protein family. Genes Dev 3: 1086–1093

Faust C, Magnuson T (1993) Genetic control of gastrulation in the mouse. Curr Opin Genet Dev 3: 491–498

Fearon ER, Cho KR, Nigro JM, Kern SE, Simons JW, Ruppert JM, Hamilton SR, Preisinger AC, Thomas G, Kinzler KW et al. (1990) Identification of a chromosome 18q gene that is altered in colorectal cancers. Science 247: 49–56

Ferracini R, Di Renzo MF, Scotlandi K, Baldini N, Olivero M, Lollini P, Cremona O, Campanacci M, Comoglio PM (1995) The Met/HGF receptor is over-expressed in human osteosarcomas and is activated by either a paracrine or an autocrine circuit. Oncogene 10: 739–749

Fixman ED, Naujokas MA, Rodrigues GA, Moran MF, Park M (1995) Efficient cell transformation by the Tpr-Met oncoprotein is dependent upon tyrosine 489 in the carboxy-terminus. Oncogene 10: 237–249

Frisch SM (1994) E1a induces the expression of epithelial characteristics. J Cell Biol 127: 1085–1096

Frixen UH, Behrens J, Sachs M, Eberle G, Voss B, Warda A, Lochner D, Birchmeier W (1991) E-cadherin-mediated cell-cell adhesion prevents invasiveness of human carcinoma cells. J Cell Biol 113: 173–185

Fujimori T, Takeichi M (1993) Disruption of epithelial cell-cell adhesion by exogenous expression of a mutated nonfunctional N-cadherin. Mol Biol Cell 4: 37–47

Gabbert H, Wagner R, Moll R, Gerharz CD (1985) Tumor dedifferentiation: an important step in tumor invasion. Clin Exp Metastasis 3: 257–279

Gallin WJ, Sorkin BC, Edelman GM, Cunningham BA (1987) Sequence analysis of a cDNA clone encoding the liver cell adhesion molecule, L-CAM. Proc Natl Acad Sci USA 84: 2808–2812

Gamallo C, Palacios J, Suarez A, Pizarro A, Navarro P, Quintanilla M, Cano A (1993) Correlation of E-cadherin expression with differentiation grade and histological type in breast carcinoma. Am J Pathol 142: 987–993

Garrod DR (1993) Desmosomes and hemidesmosomes. Curr Opin Cell Biol 5: 30–40

Giordano S, Di Renzo MF, Narsimhan RP, Cooper CS, Rosa C, Comoglio PM (1989) Biosynthesis of the protein encoded by the c-met protooncogene. Oncogene 4: 1383–1388

Gonzatti-Haces M, Seth A, Park M, Copeland T, Oroszlan S, Vande Woude GF (1988) Characterization of the TPR-MET oncogene p65 and the MET protooncogene p140 protein-tyrosine kinases. Proc Natl Acad Sci USA 85: 21–25

Guy CT, Cardiff RD, Muller WJ (1992) Induction of mammary tumors by expression of polyomavirus middle T oncogene: a transgenic mouse model for metastatic disease. Mol Cell Biol 12: 954–961

Hamaguchi M, Matsuyoshi N, Ohnishi Y, Gotoh B, Takeichi M, Nagai Y (1993) p60v-src causes tyrosine phosphorylation and inactivation of the N-cadherin-catenin cell adhesion system. EMBO J 12: 307–314

Harland RM (1994) Neural induction in Xenopus. Curr Opin Genet Dev 4: 543–549

Hartmann G, Naldini L, Weidner KM, Sachs M, Vigna E, Comoglio PM, Birchmeier W (1992) A functional domain in the heavy chain of scatter factor/hepatocyte growth factor binds the c-Met receptor and induces cell dissociation but not mitogenesis. Proc Natl Acad Sci USA 89: 11574–11578

Hartmann G, Weidner KM, Schwarz H, Birchmeier W (1994) The motility signal of scatter factor/hepatocyte growth factor mediated through the receptor tyrosine kinase met requires intracellular action of ras. J Biol Chem 269: 21936–21939

Hay ED (1990) Role of cell-matrix contacts in cell migration and epithelial-mesenchymal transformation. Cell Differ Dev 32: 367–375
Hinck L, Nelson WJ, Papkoff J (1994) Wnt-1 modulates cell-cell adhesion in mammalian cells by stabilizing beta-catenin binding to the cell adhesion protein cadherin. J Cell Biol 124: 729–741
Hinds PW, Weinberg RA (1994) Tumor suppressor genes. Curr Opin Genet Dev 4: 135–141
Hirano S, Kimoto N, Shimoyama Y, Hirohashi S, Takeichi M (1992) Identification of a neural alpha-catenin as a key regulator of cadherin function and multicellular organization. Cell 70: 293 –301
Hofstra RM, Landsvater RM, Ceccherini I, Stulp RP, Stelwagen T, Luo Y, Pasini-B, Hoppener JW, van Amstel HK, Romeo G et al. (1994) A mutation in the RET proto-oncogene associated with multiple endocrineneoplasia type 2B and sporadic medullary thyroid carcinoma (comments). Nature 367: 375–376
Hollstein M, Sidransky D, Vogelstein B, Harris CC (1991) p^{53} mutations in human cancers. Science 253: 49–53
Hoschuetzky H, Aberle H, Kemler R (1994) Beta-catenin mediates the interaction of the cadherin-catenin complex with epidermal growth factor receptor. J Cell Biol 127: 1375–1380
Hudziak RM, Schlessinger J, Ullrich A (1987) Increased expression of the putative growth factor receptor p185HER2 causes transformation and tumorigenesis of NIH 3T3 cells. Proc Natl Acad Sci USA 84: 7159–7163
Hülsken J, Birchmeier W, Behrens J (1994) E-cadherin and APC compete for the interaction with beta-catenin and the cytoskeleton. J Cell Biol 127: 2061–2069
Hyafil F, Babinet C, Jacob F (1981) Cell-cell interaction in early embryogenesis: a molecular approach to the role of calcium. Cell 21: 927–934
Hynes RO (1992) Integrins: versatility, modulation, and signaling in cell adhesion. Cell 69: 11–25
Imhof BA, Vollmers HP, Goodman SL, Birchmeier W (1983) Cell-cell interaction and polarity of epithelial cells. Specific perturbation using a monoclonal antibody. Cell 35: 667–675
Joseph A, Weiss GH, Jin L, Fuchs A, Chowdhury S, O'Shaugnessy P, Goldberg ID, Rosen EM (1995) Expression of scatter factor in human bladder carcinoma. J Natl Cancer Inst 87: 372–377
Jouanneau J, Gavrilovic J, Caruelle D, Jaye M, Moens G, Caruelle JP, Thiery JP (1991) Secreated or nonsecreted forms of acidic fibroblast growth factor produced by transfected epithelial cells influence cell morphology, motility, and invasive potential. Proc Natl Acad Sci USA 88: 2893–2897
Jucker M, Gunther A, Gradl G, Fonatsch C, Krueger G, Diehl V, Tesch H (1994) The Met/hepatocyte growth factor receptor (HGFR) gene is overexpressed in some cases of human leukemia and lymphoma. Leuk Res 18: 7–16
Juliano RL, Varner JA (1993) Adhesion molecules in cancer: the role of integrins. Curr Opin Cell Biol 5: 812–818
Kanai Y, Oda T, Tsuda H, Ochiai A, Hirohashi S (1994) Point mutation of the E-cadherin gene in invasive lobular carcinoma of the breast. Jpn J Cancer Res 85: 1035–1039
Katagiri A, Watanabe R, Tomita Y (1995) E-cadherin expression in renal cell cancer and its significance in metastasis and survival. Br J Cancer 71: 376–379
Kawanishi J, Kato J, Sasaki K, Fujii S, Watanabe N, Niitsu Y (1995) Loss of E-cadherin-dependent cell-cell adhesion due to mutation of the β-catenin gene in a human cancer cell line, HSC-39. Mol Cell Biol 15: 1175–1181
Kemler R (1992) Classical cadherins. Semin Cell Biol 3: 149–155
Kemler R (1993) From cadherins to catenins: cytoplasmic protein interactions and regulation of cell adhesion. Trends Genet 9: 317–321
Kinsella AR, Green B, Lepts GC, Hill CL, Bowie G, Taylor BA (1993) The role of the cell-cell adhesion molecule E-cadherin in large bowel tumour cell invasion and metastasis. Br J Cancer 67: 904–909
Kintner C (1992) Regulation of embryonic cell adhesion by the cadherin cytoplasmic domain. Cell 69: 225–236
Kokai Y, Cohen JA, Drebin JA, Greene MI (1987) Stage- and tissue-specific expression of the neu oncogene in rat development. Proc Natl Acad Sci USA 84: 8498–8501
Kuniyasu H, Yasui W, Kitadai Y, Yokozaki H, Ito H, Tahara E (1992) Frequent amplification of the c-met gene in scirrhous type stomach cancer. Biochem Biophys Res Commun 189: 227–232
Kuniyasu H, Yasui W, Yokozaki H, Kitadai Y, Tahara E (1993) Aberrant expression of c-met mRNA in human gastric carcinomas. Int J Cancer 55: 72–75
Larue L, Ohsugi M, Hirchenhain J, Kemler R (1994) E-cadherin null mutant embryos fail to form a trophectoderm epithelium. Proc Natl Acad Sci USA 91: 8263–8267
Liu C, Park M, Tsao MS (1992) Overexpression of c-met protooncogene but not epidermal growth factor receptor or c-erbB-2 in primary human colorectal carcinomas. Oncogene 7: 181–185

Lokker NA, Mark MR, Luis EA, Bennett GL, Robbins KA, Baker JB, Godowski PJ (1992) Structure-function analysis of hepatocyte growth factor: identification of variants that lack mitogenic activity yet retain high affinity receptor binding. EMBO J 11: 2503–2510

Mason IJ, Fuller Pace F, Smith R, Dickson C (1994) FGF-7 (keratinocyte growth factor) expression during mouse development suggests roles in myogenesis, forebrain regionalisation and epithelial-mesenchymal interactions. Mech Dev 45: 15–30

Mattijssen V, Peters HM, Schalkwijk L, Manni JJ, Van't Hof-Grooteenboer B, De Mulder PHM, Ruiter DJ (1993) E-cadherin expression in head and neck squamous-cell carcinoma is associated with clinical outcome. Int J Cancer 55: 580–585

Mayer B, Johnson JP, Leitl F, Jauch KW, Heiss MM, Schildberg FW, Birchmeier W, Funke I (1993) E-cadherin expression in primary and metastatic gastric cancer: down-regulation correlates with cellular dedifferentiation and glandular disintegration. Cancer Res 53: 1690–1695

McCutcheon M, Coman DR, Moore F (1948) Studies on invasiveness of cancer. Cancer 1: 460–467

Meyer D, Birchmeier C (1994) Distinct isoforms of neuregulin are expressed in mesenchymal and neuronal cells during mouse development. Proc Natl Acad Sci USA 91: 1064–1068

Miyazawa K, Tsubouchi H, Naka D, Takahashi K, Okigaki M, Arakaki N, Nakayama H, Hirono S, Sakiyama O et al. (1989) Molecular cloning and sequence analysis of cDNA for human hepatocyte growth factor. Biochem Biophys Res Commun 163: 967–973

Moll R, Mitze M, Frixen UH, Birchmeier W (1993) Differential loss of E-cadherin expression in infiltrating ductal and lobular breast carcinomas. Am J Pathol 143: 1731–1742

Montesano R, Matsumoto K, Nakamura T, Orci L (1991a) Identification of a fibroblast-derived epithelial morphogen as hepatocyte growth factor. Cell 67: 901–908

Montesano R, Schaller G, Orci L (1991b) Induction of epithelial tubular morphogenesis in vitro by fibroblast-derived soluble factors. Cell 66: 697–711

Mori S, Akiyama T, Yamada Y, Morishita Y, Sugawara I, Toyoshima K, Yamamoto T (1989) C-erbB-2 gene product, a membrane protein commonly expressed of human fetal epithelial cells. Lab Invest 61: 93–97

Morton RA, Ewing CM, Nagafuchi A, Tsukita S, Isaacs WB (1993) Reduction of E-cadherin levels and deletion of the alpha-catenin gene in human prostate cancer cells. Cancer Res 53: 3585–3590

Mugrauer G, Alt FW, Ekblom P (1988) N-myc proto-oncogene expression during organogenesis in the developing mouse as revealed by in situ hybridization. J Cell Biol 107: 1325–1335

Nagafuchi A, Takeichi M (1988) Cell binding function of E-cadherin is regulated by the cytoplasmic domain. EMBO J 7: 3679–3684

Nagafuchi A, Shirayoshi K, Okataki K, Yasuda, Takeichi M (1987) Transformation of cell adhesion properties by exogenously introduced E-cadherin cDNA. Nature 329: 341–343

Naidu YM, Rosen EM, Zitnick R, Goldberg I, Park M, Naujokas M, Polverini PJ, Nickoloff BJ (1994) Role of scatter factor in the pathogenesis of AIDS-related Kaposi sarcoma. Proc Natl Acad Sci USA 91: 5281–5285

Nakamura T, Nishizawa T, Hagiya M, Seki T, Shimonishi M, Sugimura A, Tashiro K, Shimizu S (1989) Molecular cloning and expression of human hepatocyte growth factor. Nature 342: 440–443

Naldini L, Weidnes KM, Vigna E, Gaudino E, Bardelli A, Ponketto C, Narsimhan RP, Hartmann F, Zarnegar R, Michaloponlos FK, Birchmeier W, Lomoglio PM (1991) Scatter Factor and hepatocyte growth Factor are in distinguishable ligents for the MET receptor. EMBO J 10: 2867–2878

Naldini L, Tamagnone L, Vigna E, Sachs M, Hartmann G, Birchmeier W, Daikuhara Y, Tsubouchi H, Blasi F, Comoglio PM (1992) Extracellular proteolytic cleavage by urokinase is required for activation of hepatocyte growth factor/scatter factor. EMBO J 11: 4825–4833

Natali PG, Nicotra MR, Di Renzo MF, Prat M, Bigotti A, Cavaliere R, Comoglio PM (1993) Expression of the c-Met/HGF receptor in human melanocytic neoplasms: demonstration of the relationship to malignant melanoma tumour progression. Br J Cancer 68: 746–750

Näthke IS, Hinck L, Swedlow JR, Papkoff J, Nelson WJ (1994) Defining interactions and distributions of cadherin and catenin complexes in polarized epithelial cells. J Cell Biol 125: 1341–1352

Navarro P, Gomez M, Pizarro A, Gamallo C, Quintanilla M, Cano A (1991) A role for the E-cadherin cell-cell adhesion molecule during tumor progression of mouse epidermal carcinogenesis. J Cell Biol 115: 517–533

Oda T, Kanai Y, Shimoyama Y, Nagafuchi A, Tsukita S, Hirohashi S (1993) Cloning of the human alpha-catenin cDNA and its aberrant mRNA in a human cancer cell line. Biochem Biophys Res Commun 193: 897–904

Okazaki M, Takeshita S, Kawai S, Kikuno R, Tsujimura A, Kudo A, Amann E (1994) Molecular cloning and characterization of OB-cadherin, a new member of cadherin family expressed in osteoblasts. J Biol Chem 269: 12092–12098

Orr Urtreger A, Bedford MT, Burakova T, Arman E, Zimmer Y, Yayon A, Givol D, Lonai P (1993) Developmental localization of the splicing alternatives of fibroblast growth factor receptor-2 (FGFR2). Dev Biol 158: 475–486

Otto T, Birchmeier W, Schmidt U, Hinke A, Schipper J, Rubben H, Raz A (1994) Inverse relation of E-cadherin and autocrine motility factor receptor expression as a prognostic factor in patients with bladder carcinomas. Cancer Res 54: 3120–3123

Oyama T, Kanai Y, Ochiai A, Akimoto S, Oda T, Yanagihara K, Nagafuchi A, Tsukita S, Shibamoto S, Ito F et al. (1994) A truncated beta-catenin disrupts the interaction between E-cadherin and alpha-catenin: a cause of loss of intercellular adhesiveness in human cancercell lines. Cancer Res 54: 6282–6287

Pachnis V, Mankoo B, Costantini F (1993) Expression of the c-retproto-oncogene during mouse embryogenesis. Development 119: 1005–1017

Park M, Dean M, Cooper CS, Schmidt M, O'Brien SJ, Blair DG, Vande Woude GF (1986) Mechanism of met oncogene activation. Cell 45: 895–904

Peters K, Werner S, Liao X, Wert S, Whitsett J, Williams L (1994) Targeted expression of a dominant negative FGF receptor blocks branching morphogenesis and epithelial differentiation of the mouse lung. EMBO J 13: 3296–3301

Ponzetto C, Giordano S, Peverali F, Della Valle G, Abate ML, Vaula G, Comoglio PM (1991) c-met is amplified but not mutated in a cell line with an activated met tyrosine kinase. Oncogene 6: 553–559

Ponzetto C, Bardelli A, Zhen Z, Maina F, Dallazonca P, Giordano S, Graziani A, Panayotou G, Comoglio PM (1994) A multifunctional docking site mediates signaling and transformation by the hepatocyte growth factor scatter receptor family. Cell 77: 261–271

Press MF, Cordon Cardo C, Slamon DJ (1990) Expression of the HER-2/neu proto-oncogene in normal human adult and fetal tissues. Oncogene 5: 953–962

Quirke P, Pickles A, Tuzi NL, Mohamdee O, Gullick WJ (1989) Pattern of expression of c-erbB-2 oncoprotein in human fetuses. Br J Cancer 60: 64–69

Rege-Cambrin G, Scaravaglio P, Carozzi F, Giordano S, Ponzetto C, Comoglio PM, Saglio G (1992) Karyotypic analysis of gastric carcinoma cell lines carrying an amplified c-met oncogene. Cancer Genet Cytogenet 64: 170–173

Riethmacher D, Brinkmann V, Birchmeier C (1995) A targeted mutation in the mouse E-cadherin gene results in defective preimplantation development. Proc Natl Acad Sci USA 92: 855–859

Ringwald M, Schuh R, Vestweber D, Eistetter H, Lottspeich F, Engel J, Dolz R, Jahnig F, Epplen J, Mayer S et al. (1987) The structure of cell adhesion molecule uvomorulin. Insights into the molecular mechanism of Ca^{2+}-dependent cell adhesion. EMBO J 6: 3647–3653

Risinger JI, Berchuck A, Kohler MF, Boyd J (1994) Mutations of the E-cadherin gene in human gynecologic cancers. Nature Genet 7: 98–102

Rodriguez-Boulan E, Nelson WI (1989) Morphogenesis of the polarized epithelial cell phenotype Science 245: 718–725

Rong S, Donehower LA, Hansen MF, Strong L, Tainsky M, Jeffers M, Resau JH, Hudson E, Tsarfaty I, Vande Woude GF (1995) Met protooncogene product is overexpressed in tumors of P53-deficient mice and tumors of Li-Fraumeni patients. Cancer Res 55: 1963–1970

Rosen EM, Knesel J, Goldberg ID, Jin L, Bhargava M, Joseph A, Zitnik R, Wines J, Kelley M, Rockwell S (1994) Scatter factor modulates the metastatic phenotype of the EMT6 mouse mammary tumor. Int J Cancer 57: 706–714

Rubinfeld B, Souza B, Albert I, Muller O, Chamberlain SH, Masiarz FR, Munemitsu S, Polakis P (1993) Association of the APC gene product with beta-catenin. Science 262: 1731–1734

Rubinfeld B, Souza B, Albert I, Munemitsu S, Polakis P (1995) The APC protein and E-cadherin form similar but independent. Bio Chem 270: 5549–5555

Sano K, Tanihara H, Heimark RL, Obata S, Davidson M, St John T, Taketani S, Suzuki S (1993) Protocadherins: a large family of cadherin-related molecules in centralnervous system. EMBO J 12: 2249–2256

Santoro M, Rosati R, Grieco M, Berlingieri MT, D'Amato GL, de Franciscis V, Fusco A (1990) The ret proto-oncogene is consistently expressed in human pheochromocytomas and thyroid medullary carcinomas. Oncogene 5: 1595–1598

Schipper JH, Frixen UH, Behrens J, Unger A, Jahnke K, Birchmeier W (1991) E-cadherin expression in squamous cell carcinomas of head and neck: inverse correlation with tumor dedifferentiation and lymph node metastasis. Cancer Res 51: 6328–6337

Schmidt C, Bladt F, Goedecke S, Brinkmann V, Zschiesche W, Sharpe M, Gherardi E, Birchmeier C (1995) Scatter factor/hepatocyte growth factor is essential for liver development. Nature 373: 699–702

Schuchardt A, D'Agati V, Larsson Blomberg L, Costantini F, Pachnis V (1994) Defects in the kidney and enteric nervous system of mice lacking thetyrosine kinase receptor Ret comments. Nature 367: 380–383

Schwarz MA, Owaribe K, Kartenbeck J, Franke WW (1990) Desmosomes and hemidesmosomes: constitutive molecular components. Annu Rev Cell Biol 6: 461–491

Selden C, Farnaud S, Ding SF, Habib N, Foster C, Hodgson HJ (1994) Expression of hepatocyte growth factor mRNA, and c-met mRNA (hepatocyte growth factor receptor) in human liver tumours. Hepatol J 21: 227–234

Shackleford GM, Willert K, Wang J, Varmus HE (1993) The Wnt-1 proto-oncogene induces changes in morphology, gene expression and growth factor responsiveness in PC12 cells. Neuron 11: 865–875

Shimoyama Y, Nagafuchi A, Fujita S, Gotoh M, Takeichi M, Tsukita S, Hirohashi S (1992) Cadherin dysfunction in a human cancer cell line: possible involvement of loss of alpha-catenin expression in reduced cell-cell adhesiveness. Cancer Res 52: 5770–5774

Simons K, Wandinger-Ness A (1990) Polarized sorting in epithelia. Cell 62: 207–210

Slamon DJ, Clark GM, Wong SG, Levin WJ, Ullrich A, McGuire WL (1987) Human breast cancer: correlation of relapse and survival with amplification of the HER-2/neu oncogene. Science 235: 177–182

Slamon DJ, Godolphin W, Jones LA, Holt JA, Wong SG, Keith DE, Levin WJ, Stuart SG, Udove J, Ullrich A et al. (1989) Studies of the HER-2/neu proto-oncogene in human breast and ovarian cancer. Science 244: 707–712

Sonnenberg A, Calafat J, Janssen H, Daams H, van der Raaij Helmer LM, Falcioni R, Kennel SJ, Aplin JD, Baker J, Loizidou M et al. (1991) Integrin alpha 6/beta 4 complex is located in hemidesmosomes, suggesting a major role in epidermal cell-basement membrane adhesion. J Cell Biol 113: 907–917

Sonnenberg E, Godecke A, Walter B, Bladt F, Birchmeier C (1991) Transient and locally restricted expression of the rosl protooncogene during mouse development. EMBO J 10: 3693–3702

Sonnenberg E, Meyer D, Weidner KM, Birchmeier C (1993) Scatter factor/hepatocyte growth factor and its receptor, the c-met tyrosine kinase, can mediate a signal exchange between mesenchyme and epithelia during mouse development. J Cell Biol 123: 223–235

Stark K, Vainio S, Vassileva G, McMahon AP (1994) Epithelial transformation of metanephric mesenchyme in the developing kidney regulated by Wnt-4. Nature 372: 679–683

Stoker M, Gherardi E, Perryman M, Gray J (1987) Scatter factor is a fibroplast-derived modulator of epithelial cellmobility. Nature 327: 239–242

Strutz F, Okada H, Lo CW, Danoff T, Carone RL, Tomaszewski JE, Neilson G (1995) Identification and characterization of a fibroblast marker: FSPI. J Cell Biol 130: 393–405

Su L, Vogelstein B, Kinzler K (1993) Association of the APC tumor supressor protein with catenins. Science 262: 1734–1737

Suzuki K, Hayashi N, Yamada Y, Yoshihara H, Miyamoto Y, Ito Y, Ito T, Katayama K, Sasaki Y, Ito A et al. (1994) Expression of the c-met protooncogene in human hepatocellular carcinoma. Hepatology 20: 1231–1236

Takeichi M (1991) Cadherin cell adhesion receptors as a morphogenetic regulator. Science 251: 1451–1455

Timpl R (1989) Structure and biological activity of basement membrane proteins. Eur J Biochem 180: 487–502

Troyanovsky SM, Troyanovsky RB, Eshkind LG, Leube RE, Franke WW (1994) Identification of amino acid sequence motifs in desmocollin, a desmosomal glycoprotein, that are required for plakoglobulin binding and plaque formation. Proc Natl Acad Sci USA 91: 10790–10794

Tsarfaty I, Rong S, Resau JH, Rulong S, da Silva PP, Vande Woude GF (1994) The Met proto-oncogene mesenchymal to epithelial cell conversion. Science 263: 98–101

Tsukita S, Nagafuchi A, Yonemura S (1992) Molecular linkage between cadherins and actin filaments in cell-cell adherens junctions. Curr Opin Cell Biol 4: 834–839

Tsukita S, Itoh M, Nagafuchi A, Yonemura S (1993) Submembranous junctional plaque proteins include potential tumor suppressor molecules. J Cell Biol 123: 1049–1053

Uehara Y, Minowa O, Mori C, Shiota K, Kuno J, Noda T, Kitamura N (1995) Placental defect and embryonic lethality in mice lacking hepatocyte growth factor/scatter factor. Nature 373: 702–705

Umbas R, Isaacs WB, Bringuier PP, Schaafsma HE, Karthaus HF, Oosterhof GO, Debruyne FM, Schalken JA (1994) Decreased E-cadherin expression is associated with poor prognosis in patients with prostate cancer. Cancer Res 54: 3929–3933

Valles AM, Boyer B, Thiery JP (1991) Adhesion systems in embryonic epithelial to mesenchymal transformation and in cancer invasion and metastasis. In: Goldenberg ID (ed) Cell motility factors. Birkhäuser, Basel, pp 17–34

Vestweber D, Kemler R (1984) Rabbit antiserum against a purified surface glycoprotein decompacts mouse preimplantation embryos and reacts with specific adult tissues. Exp Cell Res 152: 169–178
Vleminckx K, Vakaet L, Mareel M Jr , Fiers W, Van Roy F (1991) Genetic manipulation of E-cadherin expression by epithelial tumor cells reveals an invasion suppressor role. Cell 66: 107–119
Vogelstein B, Kinzler KW (1993) The multistep nature of cancer. Trends Genet 9: 138–141
Weidner KM, Behrens J, Vandekerckhove J, Birchmeier W (1990) Scatter factor: molecular characteristics and effect on the invasiveness of epithelial cells. J Cell Biol 111: 2097–2108
Weidner KM, Arakaki N, Hartmann G, Vandekerckhove J, Weingart S, Rieder H, Fonatsch C, Tsubouchi H, Hishida T, Daikuhara Y et al. (1991) Evidence for the identity of human scatter factor and human hepatocyte growth factor. Proc Natl Acad Sci USA 88: 7001–7005
Weidner KM, Sachs M, Birchmeier W (1993) The Met receptor tyrosine kinase transduces motility, proliferation and morphogenic signals of scatter factor/hepatocyte growth factor in epithelial cells. J Cell Biol 121: 145–154
Weidner KM, Sachs M, Riethmaches D, Birchmeier W (1995) Mutation of juxta membrane tyrosine residue 1001 suppresses loss-of-function mutations of the met receptor in epithelial cells. Proc Natl Acad Sci NY 92: 2597–2601
Weinstein RS, Merck FB, Alroy J (1976) The structure and function of intercellular junctions in cancer. Adv Cancer Res 23: 23–89
Werner S, Peters KG, Longaker MT, Fuller Pace F, Banda MJ, Williams LT (1992) Large induction of keratinocyte growth factor expression in the dermis during wound healing. Proc Natl Acad Sci USA 89: 6896–6900
Werner S, Weinberg W, Liao X, Peters K, Blessing M, Yuspa S, Weiner R, Williams L (1993) Targetted expression of a dominant negative FGF receptor mutant in the epidermis of transgenic mice reveals a role of FGF in keratinocyte organization and differentiation. EMBO J 12: 2635–2643
Werner S, Smola H, Liao X, Longaker MT, Krieg T, Hofschneider PH, Williams LT (1994) The function of KGF in morphogenesis of epithelium and reepitheliazation of wounds. Science 266: 819–822
Yamashita J, Ogawa M, Yamashita S, Nomura K, Kuramoto M, Saishoji T, Shin S (1994) Immunoreactive hepatocyte growth factor is a strong and independent predictor of recurrence and survival in human breast cancer. Cancer Res 54: 1630–1633
Yang J, Spitzer E, Meyer D, Sachs M, Birchmeier C, Birchmeier W (1995) Sequential requirement of scatter factor/hepatocyte growth factor and neu differentiation factor/heregulin in the morphogenesis and differentiation of the mammary gland. J Cell Biol 131: 1–12

Regulation of Autocrine Motility Factor Receptor Expression in Tumor Cell Locomotion and Metastasis

S. SILLETTI[3] and A. RAZ[1,2]

1 Introduction 137

2 Specificity of the Autocrine Motility Factor 139
2.1 The Nature of Autocrine Motility Factor 139
2.2 Autocrine Motility Factor Response 141
2.3 Autocrine Motility Factor-Mediated Intracellular Signaling 141

3 Autocrine Motility Factor Receptor 143
3.1 Chromosomal Localization of the Gene Encoding Autocrine Motility Factor Receptor 148
3.2 Surface Expression of Autocrine Motility Factor Receptor 148
3.3 Autocrine Motility Factor Receptor Is Endocytosed and Localized to Tubulovesicles 150
3.4 Internalized Autocrine Motility Factor Receptor Is Recycled to the Leading Edge 152

4 Receptor Activation and Cellular Locomotion 154
4.1 Autocrine Motility Factor Initiates Motility by Reorganization of Cellular Focal Adhesions . 155
4.2 Regulation of Autocrine Motility Factor Receptor 157
4.3 Prognostic Value of the Autocrine Motility Factor System 161

5 Conclusion 163

References 163

1 Introduction

Active locomotion by invading tumor cells is thought to be a prerequisite step in the establishment of secondary neoplasms. Successful metastasis requires the invasion of surrounding normal tissue and crossing of vascular and/or lymphatic boundaries (NICOLSON1988; FIDLER 1990) and it has been suggested that motility of individual cells or groups of cells at the leading edge of a tumor protrusion might be responsible for such invasive movement (STRAULI and WEISS 1977). Analysis of previously characterized high- and low-metastatic variant melanoma subpopulations has demonstrated that low-metastatic cells are largely immobile, while their high-metastatic counterparts exhibit profoundly greater locomotory activity (RAZ and GEIGER 1982; VOLK et al. 1984; GEIGER et al. 1985; ZVIBEL and RAZ 1985; RAZ and BEN-ZE'EV 1987). Similar findings were obtained using the Lewis lung carcinoma

[1] Tumor Progression and Metastasis, Karmanos Cancer Institute, Detroit, MI 48201, USA
[2] Departments of Pathology and Radiation Oncology, Wayne State University School of Medicine, Detroit, MI 48202, USA
[3] Current Address: Departments of Immunology and Vascular Biology, Scripps Clinic and Research Foundation, LA Jolla, CA 92037-1092, USA

(Young et al. 1985) and a rat mammary adenocarcinoma model (Badenoch-Jones and Ramshaw 1984), while more recent work in the Dunning R-3327 rat prostatic adenocarcinoma model has further corroborated the relationship between motility and metastatic potential (Mohler et al. 1987, 1988; Partin et al. 1989).

The repeated observation of nonrandom patterns of metastatic dissemination suggests that there are specific parameters governing cell migration (Nicolson 1988; Raz and Ben-Ze'ev 1987); therefore, in an effort to understand the regulation and role of cell locomotion in invasion and metastasis, recent studies have focused on identifying potential motility-inducing and chemotactic factors in the tumor environment. Host serum proteins and extracellular matrix breakdown products were found to exert a chemotactic effect on various tumor cells (Lam et al. 1981; McCarthy et al. 1985; Nabeshima et al. 1986), and haptotactic attraction has been demonstrated for a number of components of the basement membrane and extracellular matrix as well (McCarthy et al. 1983, 1986; Taraboletti et al. 1987).

A group of secreted cytokines has been described as cell motility factors which specifically induce cellular locomotion. Rat ascites hepatoma AH109A cells secrete a chemotactic polypeptide thought to facilitate malignant invasion (Yoshida et al. 1970). A similar mode of migratory behavior was observed in both fetal and tumor-derived fibroblasts and both cell types were subsequently found to secrete a 70-kDa "migration-stimulating factor" which stimulates entry into collagen matrices (Schor et al. 1988; Grey et al. 1989). A fibroblast-derived and paracrine-acting group of "scatter factors" ranging from 32 to 92 kDa has been described which causes dispersion of epithelial cell aggregates (Stoker et al. 1987; Gherardi et al. 1989; Rosen et al. 1990; Weidner et al. 1990) and whose human homologue has been shown to be indentical with human hepatocyte growth factor (Weidner et al. 1991). A tumor-specific autocrine factor of less than 30 kDa was characterized from the Dunning R-3327 rat prostatic adenocarcinoma AT-2.1 subline which was shown to induce its motility-stimulation through the cyclic adenosine monophosphate (cAMP) pathway (Evans et al. 1991). A 53-kDa autocrine chemotactic factor produced by the 13762NF mammary adenocarcinoma cell line was found to have different secretion and response characteristics in the respective high- and low-metastatic variant subpopulations, prompting the suggestion that the production of motility-inducing cytokines may represent a phenotypic aspect modulating the difference between high- and low- metastatic clones, at least in the 13762NF system (Atnip et al. 1987). A 125-kDa motility-stimulating protein was recently identified from the conditioned medium of the A2058 human melanoma; it induces migration of the producer cells via a pertussis toxin-sensitive G protein and was termed autotaxin (Stracke et al. 1992). Previously, a motility factor from the A2058 conditioned medium had been described which evoked a similar, but distinct response in these cells. Tumor cell autocrine motility factor (AMF) is a cytokine or related group of 55-kDa cytokines named for their induction of both random and directed cell migration in self-producing cells (Liotta et al. 1986). These factors may represent a family of cytokines whose regulated expression induces proper motility responses in normal tissue during such processes as wound healing (scatter factor/hepatocyte growth factor) and embryogenesis (migration-stimulating factor) and whose aberrantly regulated autocrine expression (AMF) may confer or enhance invasive and metastatic capabilities on neoplastic cells.

2 Specificity of the Autocrine Motility Factor

The secretion of AMF is confined to transformed cells, but some untransformed cells are capable of responding to AMF as well. LIOTTA et al. (1986) determined that although untransformed parental NIH-3T3 fibroblasts did not secrete appreciable AMF (AMF/NIH-3T3) since their conditioned media was incapable of stimulating either the producer cells or the transformed lines, they were able to respond with an equivalent motile response when stimulated by AMF from three different *ras*-transformed NIH-3T3 clones, all of which were metastatic in animal tests. The role of AMF in normal tissue therefore could be that of a paracrine regulator of cellular motility, whereas transformed cells, which both express the receptor and are able to produce and secrete AMF, can bypass the normal requirement for external input and regulate their own motility in an autocrine manner.

Whereas AMF has been shown to stimulate the motility of A2058 human melanoma, HeLa human cervical carcinoma, MCF-7 human breast carcinoma, HT-1080 human fibrosarcoma, UV-2237-IP3 murine fibrosarcoma, J82 human bladder carcinoma, NIH-3T3, and Balb/c-3T3-A31 untransformed murine fibroblasts as well as the metastatic angiosarcoma variant of the BALB/c 3T3-A31 fibroblast and various subpopulations of the K-1735 and B16 murine melanoma cell lines, AMF does not stimulate T lymphocyte or neutrophil motility (LIOTTA et al. 1986; NABI et al. 1990, 1992; WATANABE et al. 1991a; SILLETTI et al. 1993, 1994; S. SILLETTI, unpublished observations). This data suggests that the activity of the 55-kDa AMF is at least partially tumor cell specific, and secretion has been only demonstrated for immortalized or transformed cells thus far.

2.1 The Nature of Autocrine Motility Factor

AMF has been purified to homogeneity from the A2058 human melanoma cell line (LIOTTA et al. 1986), the HT-1080 human fibrosarcoma (WATANABE et al. 1991a), and the B16-F1 murine melanoma cell lines (SILLETTI et al. 1991), and these three AMF appear to represent identical or homologous molecules. Recently, acidic and basic forms of AMF have been described from protein-independent fibrosarcoma cells which appear to have identical biochemical characteristics as the previously defined AMF and one another, except for the differential purification affinity (WATANABE et al. 1994). AMF is a protein which migrates with an apparent molecular mass of 55 kDa under nonreducing and 64 KDa under reducing polyacrylamide gel electrophoresis (PAGE), indicating the presence of a single-chain polypeptide with one or more disulfide bonds. The active conformation of AMF is stabilized by these disulfide bonds as reduction with dithiothreitol (DTT), a sulfhydryl-reducing agent, resulted in almost total elimination of activity. That AMF is a protein is supported by its susceptibility to proteolytic enzymes and heat inactivation (100°C) and its concomitant resistance to DNase, RNase, and prolonged exposure to moderate temperature (60°C, 60 min) (LIOTTA et al. 1986).

Isoelectric focusing of the B16-F1 murine melanoma AMF resolved the protein into two distinct species with isoelectric points (pI) of 6.35 and 6.4 (SILLETTI et al. 1991), whereas the HT-1080 human fibrosarcoma AMF separated into four species, two minor forms with pI values identical to the murine AMF and two major bands with pI values of 6.1 and 6.2 (WATANABE et al. 1991a). The reason for the appearance of multiple species with different pI values is unknown; however, analysis of both AMF to detect covalently bound carbohydrate moieties which could account for the multiple isoforms yielded negative results. Neuraminidase treatment to remove terminal sialic acid residues resulted in an unaltered migration on gel electrophoresis and an iodinated-wheatgerm agglutinin overlay procedure failed to detect *N*-acetyl glucosamine moieties, indicating that the AMF are probably not glycosylated and that some other alteration such as phosphorylation, modification of the AMF polypeptide core, alternative mRNA splicing, or the presence of more than one homologous gene for AMF might be responsible for the multiple species observed. Indeed, acidic and basic forms of AMF have recently been reported which exhibit identical pI values approximately of 6.5, confirming that the different biochemical properties of AMF subspecies may not be resolved sufficiently by standard molecular procedures (WATANABE et al. 1994).

Recent characterization of the AMF from the *ras*-transfected NIH-3T3 fibroblasts examined previously by LIOTTA et al. (1986) has demonstrated that the AMF/NIH-3T3 is most likely a distinct motility-inducing factor from the 55-kDa AMF previously described (SEIKI et al. 1991). Unlike the AMF from the A2058, HT-1080, and B16-F1 cell lines, AMF/NIH-3T3 elutes on gel filtration in a fraction corresponding to a molecular mass of 150–200 kDa and its action is not inhibited by pretreatment of the recipient cells with pertussis toxin. Similar to the previously described AMF, AMF/NIH-3T3 is also presumed to be a protein as it was susceptible to protease and heat inactivation (100°C), but whereas AMF from the A2058, HT-1080, and B16-F1 cell lines retains almost all activity upon exposure to moderate heat even after prolonged incubation (60°C, 60 min), AMF/NIH-3T3 was unstable at this temperature and all activity was lost within 1h. In addition, simultaneous stimulation of N-*ras*-transfected cells by both the A2058 AMF and the AMF/NIH-3T3 resulted in an additive effect, indicating that these two factors probably operate through different pathways. Due to complications in the purification procedure, this AMF could not be purified to homogeneity and its identity remains unclear; however, it represents another model of the numerous autocrine pathways which can be exploited by tumor cells in the metastatic cascade.

2.2 Autocrine Motility Factor Response

Stimulation of producer cells with AMF at concentrations as low as 15 pg/ml causes them to consistently clear an area two to three fold larger than controls when plated on a gold particle-coated substrate, as described by ALBRECHT-BUEHLER (1977) (WATANABE et al. 1991a; SILLETTI et al. 1991). This nondirectional motility induction (chemokinesis) was described previously in addition to the chemotactic effect observed in a modified Boyden chamber assay (LIOTTA et al. 1986). An AMF-like substance with similar biochemical characteristics as AMF was reported to lack haptotactic attraction, contrasting with fibronectin (MCCARTHY et al. 1986), laminin (MCCARTHY et al. 1985), and thrombospondin (TARABOLETTI et al. 1987).

It is not known at present which of these activities, chemokinesis or chemotaxis, is most important in the intact system or whether it is a combination of these processes which is exploited in vivo. It has been noted, however, that a wounded monolayer of NIH-3T3 or BALB/c 3T3-A31 murine fibroblasts could be induced to heal more rapidly in the presence of AMF (SILLETTI and RAZ 1993). This increased rate of healing appeared to result not only from an increase in the motility of cells at the wound edge, but also from more rapid cellular proliferation as well. Examination of the mitogenic activity of AMF on untransformed BALB/c 3T3-A31 fibroblasts as well as on the producer HT-1080 cells by analysis of [^{3}H] thymidine incorporation and cell number showed that AMF exerted a dose-dependent effect on the proliferation of both cell lines which was completely inhibited by preincubation with pertussis toxin at a concentration which abolishes migration stimulation by AMF as well. A similar biphasic dose–response curve was noted for both cell lines, i.e., growth stimulation by AMF increases to a maximum value at 0.1 ng/ml and then declines at higher concentrations, similar to growth curves reported for other mitogenic cytokines such as hepatocyte growth factor (MATSUMOTO et al. 1991). The concentration of AMF required for maximal migratory effect, suggesting a concentration-dependent mechanism of action whereby a single molecule may have different effects based on the local concentration encountered by the recipient cell. It is interesting to note that the growth-stimulatory effect of AMF is independent of the presence of serum factors, indicating that AMF acts as a competence growth factor, at least in this system.

2.3 Autocrine Motility Factor-Mediated Intracellular Signaling

Comparison of the motility profiles of the A2058 human melanoma and the human neutrophil showed that, whereas the nuetrophil responded strongly to the formyl peptide fMet-Leu-Phe (fMLP), a standard leukocyte chemoattractant, it showed no response to AMF. Similarly, the A2058 cells showed no response to the fMLP leukoattractant and exhibited a marked stimulation in response to AMF (LIOTTA et al. 1986). This indicates that these two motility-stimulating factors act through different receptors and that the AMF response is at least partially tumor cell specific.

Methylation of phospholipids had been implicated in the signaling pathway of phagocyte chemotaxis by earlier studies (GARCIA-CASTRO et al. 1983), and analysis of the AMF system has shown that concentrations of AMF which are capable of inducing a motile response also increased the incorporation of [^{14}C]-labeled methyl groups from methionine to phosphatidylcholine (LIOTTA et al. 1986). Both AMF-stimulated motility of A2058 cells and leukocyte chemotaxis are inhibited by 3-deazadenosine (cAdo), which inhibits the transmethylation production of methylated phosphoethanolamines, and the addition of another methyltransferase inhibitor, homocysteine thiolactone, decreased the concentration of cAdo required for inhibition but was inactive by itself, similar to the effect described for neutrophils (BAREIS et al. 1982; LIOTTA et al. 1986). Since inhibitors of fMLP-stimulated chemotaxis also inhibit AMF-stimulated motility, this suggests the presence of distinct receptors whose signals converge to a common pathway at some point downstream of ligand-induced receptor activation.

The binding of fMLP to the chemotactic receptor on leukocytes activates a G protein which stimulates phospholipase C (PLC)-mediated cleavage of phosphatidylinositol-4, 5-bisphosphate (PIP_2) into the second messengers inositol-1,4,5-trisphosphate (IP_3) and 1,2-diacylglycerol (DAG) (SMITH et al. 1986). *Bordella* pertussis toxin (PT) inactivates certain G proteins by adenosine diphosphate (ADP)-ribosylating a cysteine residue on the α-subunit of the heterotrimer, thereby uncoupling the activated receptor from its effector molecule (GILMAN et al. 1987; ROTROSEN et al. 1988), and it has been shown that more than one of these distinct phosphoinositide (pl)-specific G protein pathways, termed G_p and distinguished by their specific sensitivity to PT, may exist in the same cell, selectively coupling different receptors to PI hydrolysis (ASHKENAZI et al. 1989). Chemotaxis stimulated by fMLP and its associated receptor-mediated responses in neutrophils and leukocytes are specifically sensitive to pertussis toxin (BRANDT et al. 1985; SHEFCYK et al. 1985; VERGHESE et al. 1985), and these responses are reciprocally unaffected by cholera toxin and other inhibitors of the cAMP pathway (SHEFCYK et al. 1985). The motility-stimulating signal transduction pathway of AMF in melanoma and fibrosarcoma cells is specifically sensitive to pertussis toxin as well (STRACKE et al. 1987; SILLETTI and RAZ 1993), implicating a G_p-like protein in the AMF motility-stimulating pathway. Neither cholera toxin nor agents which interact directly with adenylate cyclase had any effect on AMF stimulation of motility.

At a sufficient concentration to produce a motility response, AMF stimulated the incorporation of inositol into cellular lipids and inositol phosphates, particularly inositol trisphosphates (SMITH et al. 1986; KOHN et al. 1990). This AMF-stimulated production of inositol phosphates was dose dependent, correlated with induction of motility, and was partially inhibited by pretreatment of the cells with PT (KOHN et al. 1990). PT was shown to specifically ADP-ribosylate a 40-kDa membrane-associated protein in this system, consistent with the presence of a G protein signal mediator similar to the inositol phosphate-associated G_p (GILMAN 1987; ASHKENAZI et al. 1989). Inositol trisphosphate was seen to increase over a 2-h period after stimulation with AMF, and this phenomenon may be representative of the time lag involved in adherence of the cells to the substratum and initiation of locomotion upon cellular adaptation to the levels of AMF in the environment (KOHN et al. 1990).

Whereas the chemotactic response of neutrophils to fMLP can be down-regulated by chronic exposure to low levels of this leukoattractant, actin polymerization and phosphatidylinositol-3,4,5-trisphosphate (PIP_3) production persist in response to stimulation, suggesting that PIP_3 may function to regulate actin polymerization in neutrophil chemotaxis (EBERLE et al. 1990). Motile fibroblasts exhibit actin networks which align perpendicular to the direction of movement at the leading edge and project longitudinally in lamellopodia being actively extended (SMALL et al. 1978; DE BIASIO et al. 1988). Stimulation of cells with AMF causes the extension of pseudopodia which exhibit prominent axial actin filament bundles, and isolated pseudopodia were highly enriched for laminin and integrin receptors after stimulation, containing over 20 times the laminin and fibronectin receptors of plasma membranes of unstimulated cells (GUIRGUIS et al. 1987). A lipoxygenase metabolite of arachidonic acid, 12-(*S*)- hydroxyeicosatetraenoic acid [12-(*S*)-HETE] has been shown to alter several cellular mechanisms including integrin $\alpha_{IIb}\beta_3$ receptor expression and extracellular matrix adhesion (GROSSI et al. 1989; CHOPRA et al. 1991), indicating that 12-(*S*)-HETE might play a role in such processes as adhesion and motility. Exogenous 12-(*S*)-HETE stimulated the motility of high-metastatic melanoma cells in a manner analogous to AMF without affecting the migration of the low-metastatic variant, which is also incapable of an AMF-induced locomotory response (TIMAR et al. 1993; SILLETTI et al. 1994). This effect was correlated with enhanced expression of the AMF receptor, and AMF treatment stimulated production of endogenous 12-(*S*)-HETE as well as upregulation of the 12-lipoxygenase enzyme in the migration-responsive cells exclusively (TIMAR et al. 1993; SILLETTI et al. 1994). Although pertussis toxin inhibits AMF action entirely, 12-(*S*)-HETE's effects are unaffected by pretreatment with PT, and this eicosanoid appears to act downstream of the PT-sensitive G protein in the AMF signaling pathway. However, the motility effects of both AMF and 12-(*S*)-HETE are abolished by inhibitors of protein kinase C (TIMAR et al. 1993; KANBE et al. 1994), and it has been shown recently that both arachidonic acid metabolism and protein kinase C activity are related to AMF's locomotory and not its proliferative effects (SILLETTI et al. 1995).

3 Autocrine Motility Factor Receptor

The receptor for AMF has been identified as a 78 kDa cell surface glycoprotein designated gp78 (glycoprotein 78). A role for gp78 in the metastatic process was first suggested by its increased O-linked glycosylation in B16-F1 cells grown in a spherical configuration on a nonadhesive substrate which exhibited enhanced lung-colonizing ability in mice (RAZ and BEN ZE'EV 1983; NABI and RAZ 1987, 1988). This growth pattern parallels tumor cell interactions in the tumor mass or as aggregates in the circulation prior to invasion through the endothelium (FIDLER et al. 1988; POSTE 1982; NICOLSON 1988), and the ability to grow in such a manner is distinct to the neoplastic phenotype (RAZ et al. 1987). The causal involvement of gp78 in metastasis was further demonstrated by the observation that treatment of B16-F1 cells grown in monolayer with F_{ab} fragments from a polyclonal anti-gp78 antibody prior to intravenous inoculation into the tail vein of syngeneic mice resulted in an increase in lung colonization similar to that seen with cells grown in a spherical configuration (NABI and RAZ 1987). In addition, cells treated with an mAb against gp78 displayed increased motility in vitro as evidenced by phagokinetic clearing of particles from gold-coated substrates as well as enhanced in vivo lung-colonizing capability compared to stimulation with control antibodies (NABI et al. 1990; WATANABE et al. 1991b). The induction of motility in vitro by either conditioned media from these B16-F1 cells or the anti-gp78 mAb was found to be of similar magnitude and each was specifically inhibited by PT (NABI et al. 1990), indicating use of a common signaling pathway.

Putative identification of gp78 as the AMF receptor came from the observation that binding of anti-gp78 mAb to its antigen was inhibited (ten fold) by preincubation of membranes with AMF-containing conditioned media (NABI et al. 1990). Since heat-inactivated AMF was incapable of blocking recognition by anti-gp78, it was concluded that AMF and anti-gp78 bind to the same epitope on the receptor. To confirm the identity of gp78 as the AMF receptor, gp78 and AMF were purified to homogeneity from B16-F1 cells by immunoaffinity chromatography of membrane extracts and molecular sieve chromatography of conditioned media, respectively, and binding analyses were performed. Iodinated AMF binds quantitatively to nitrocellulose-immobilized gp78, and soluble, free gp78 inhibited both AMF and anti-gp78 mAb-stimulated in vitro motility without any effect on the basal locomotory rate, presumably by competing for binding of the ligands with the cell surface receptors and thus diluting the stimulatory activity (SILLETTI et al. 1991).

The human fibrosarcoma gp78 cDNA has recently been cloned; it contains an open reading frame which encodes a 323-amino acid polypeptide with all the characteristics of an integral membrane glycoprotein (WATANABE et al. 1991a). A hydrophobic stretch of 25 amino acids is located between amino acids 111 and 137, consistent with a single-pass transmembrane helix. The mature protein is predicted to begin at Ala-18 following the putative leader sequence whose peptide cleavage signal site obeys the (–3,–1) rule (VON HEIJNE 1986). The N-terminal domain contains one potential N-linked and several potential O-linked glycosylation sites, supporting the previous finding that gp78 is glycosylated with both N- and O-linked oligosaccharides (NABI and RAZ 1987) and indicating that the NH_2 terminus is extracellularly exposed. The predicted amino acid sequence of the intracellular domain contains two regions which may play a role in gp78 signal transduction, the sequence Ser-Gly-Lys (residues 194–196) fits the Ser/Thr-X-Lys/Arg consensus for a phosphorylation site (potentially protein kinase C) (WOODGETT et al. 1986), while residues 157–162 conform to the consensus for a nucleotide-binding region (Gly-X-Gly-X-X-Gly) similar to that seen in several serine/threonine kinases (HANKS et al. 1988). Thus, activation of gp78 upon binding its ligand may result from receptor autophosphorylation or from a GTP-binding "coupling" protein which associates with the activated receptor on the cytoplasmic face of the plasma membrane, as described previously for rhodopsin, β-adrenergic and muscarinic cholinergic receptors, and the mating receptor of yeast (HERSKOWITZ and MARSH 1987).

Sequence analysis of the AMF receptor revealed that amino acids 247–260 (N-KELLVRYLEQRRGK-C) of the full-length reading frame match very closely the consensus sequence described for a G protein-activator motif. This sequence has approximately 60% homology with the G protein activator domain of the insulin-like growth factor II/mannose 6-phosphate receptor (OKAMOTO et al. 1990; OKAMOTO and NISHIMOTO 1991), which was first identified among G protein-activator sequences. Upon analysis of the functional features of this sequence, it appears that this sequence is optimized for activation of G_i, especially G_{i2} (IKEZU et al. 1992). This speculation is consistent with the fact that AMF stimulation is sensitive to *Bordella* pertussis toxin, which ADP-ribosylates and inactivates the G_α subunit of G_{i2}. Paradoxically, the fMLP neutrophil chemotaxis receptor is coupled to G_{i2} as well, suggesting that, since AMF is incapable of stimulating neutrophil chemotaxis and fMLP is incapable of stimulating tumor cell migration, these pathways may be mediated by different receptors coupled to the similar signal transduction pathways.

A computer-assisted search of several sequence data bases using the nucleotide sequence of the entire gp78 cDNA revealed a significant homology to only one other known gene, the human tumor suppressor p53 cDNA. The two sequences exhibit 50.1% identity at the DNA level, while optimal alignment of the deduced amino acid sequences of the two cDNA revealed 27.2% identity over 296 amino acids; this homology increased to 44.5% when conservative amino acid substitutions were considered (WATANABE et al. 1991a). The p53 gene product is a nuclear serine-phosphoprotein which acts as a tumor suppressor in its normal form, but can act as an oncogene when mutated. It is thought to be active in growth regulation and has been shown to form protein–protein complexes, mainly with viral antigens (WEINBERG 1991). Evolutionarily, p53 is restricted to vertebrates, and protein sequence conservation between species is concentrated in five domains (I–V) which are thought to be critical for function (reviewed in SOUSSI et al. 1990). Although the AMF receptor and p53 are probably functionally unrelated, based on the fact that gp78 is a transmembrane glycoprotein and it lacks the five distinct functional domains conserved in p53 molecules from different species, there are notable similarities which imply that p53 and gp78 may have diverged from a common ancestral gene. Both molecules exist as phosphoproteins and each contains one N-linked glycosylation site as well as a protein-binding domain. Comparison of the DNA and protein sequences implies that no selective pressures have caused retention of domains critical for function, yet the randomly distributed homology between gp78 and p53 is similar to or higher than that observed in the regions separating conserved domains in p53 from different species.

Examination of the gp78 protein sequence with the DOTPLOT program (University of Wisconsin Genetics Computer Group Software; Fig. 1a) demonstrated

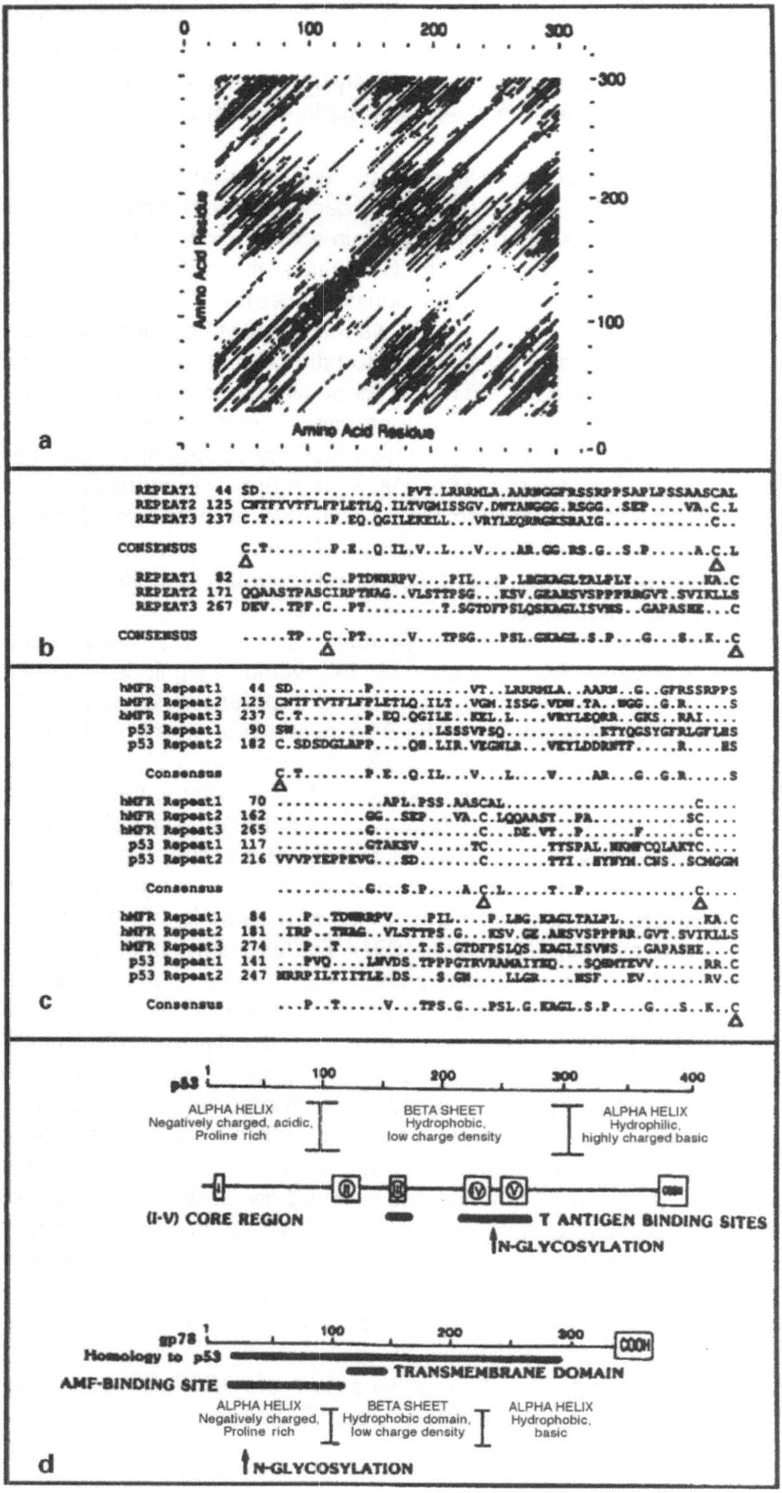

a
Amino Acid Residue
Amino Acid Residue
0
100
200
300
b
REPEAT1
REPEAT2
REPEAT3
CONSENSUS
c
hMFR Repeat1
hMFR Repeat2
hMFR Repeat3
p53 Repeat1
p53 Repeat2
Consensus
d
p53
ALPHA HELIX
Negatively charged, acidic,
Proline rich
BETA SHEET
Hydrophobic,
low charge density
ALPHA HELIX
Hydrophilic,
highly charged basic
(I-V) CORE REGION
T ANTIGEN BINDING SITES
N-GLYCOSYLATION
gp78
Homology to p53
COOH
TRANSMEMBRANE DOMAIN
AMF-BINDING SITE
ALPHA HELIX
Negatively charged,
Proline rich
BETA SHEET
Hydrophobic domain,
low charge density
ALPHA HELIX
Hydrophobic,
basic
N-GLYCOSYLATION

the presence of internally repeated sequences as described for the tumor necrosis factor (TNF) receptor (SCHALL et al. 1990), in the case of gp78 consisting of three domains linked by conserved cysteine residues (Fig. 1b). Comparison of the p53 protein alignment over these regions showed that this molecule exhibits a pattern of conserved cysteine residues as well, the sequence of which can be superimposed over that of the gp78 internal repeats (Fig. 1C). This pattern of internally repeated cysteine-rich domains is reminiscent of that described for steroid receptor (O'MALLEY 1990) and immunoglobulin superfamilies (MARCHALONIS and SCHLUTTER 1989), as well as numerous families of transmembrane proteins and cell surface receptors (LOETSCHER et al. 1990; SCHALL et al. 1990). Although it is not apparent whether there is a significant functional utilization of these residues in gp78, this is unlikely as this pattern extends through the transmembrane domain in a manner unprecedented in other molecules. The functional use of these repeated cysteines in the p53 molecule is also unknown; however, the substitution of serine for cysteine (a predictable but nonfunctional substitution) at more than one location in the repeated structure argues against this possibility. The retention of such a pattern reinforces the potential divergence of these two proteins from a common ancestral gene, however. In addition, the human cDNA for these two genes cross-hybridize under highly stringent conditions, further supporting the evolutionarily divergent relationship of p53 and gp78 (S. SILLETTI, unpublished observations).

Primary and secondary structure comparison of these two proteins demonstrates a number of other related factors which indicate that these proteins may have evolved divergently from an ancestral gene duplication (Fig. 1d). Both of these proteins contain a predicted helical region which is negatively charged and proline rich in their amino-terminal 100 amino acids, followed by a region of low charge density which is highly hydrophobic and is predicted to form primarily β-sheet conformation. The carboxy-terminal 100+ amino acids of each protein returns to a hydrophilic α-helical configuration which is high in basic amino acid residues. In addition, both p53 and gp78 are phosphoproteins involved in protein–protein-binding interactions, and both molecules contain an N-glycosylation consensus sequence; this macrostructural similarity is particularly interesting as it has been noted previously that structure is better conserved through evolution than sequence (SOUSSI et al. 1990).

Fig. 1a–d. Sequence repeats and structural similarities of the gp78 and p53 proteins. **a** DOTPLOT analysis of internal sequence similarities within the gp78 protein. gp78 amino acids are plotted and numbered in the horizontal and vertical directions. (Adapted from SOUSSI et al. 1990). **b** Internal repeats of the gp78 protein. Amino acid positions are indicated at the *left* and consensus residues are shown below the alignment. *Dots* indicate spaces and conserved cysteines are marked with open *arrowheads*. (Adapted from WATANABE et al. 1991a). **c** Comparison of the gp78 internal repeat domains with the p53 protein sequence as above. *hMFR*, human motility factor receptor. **d** Structural similarities between the human gp78 and human p53 proteins. *AMF*, autocrine motility factor

3.1 Chromosomal Localization of the Gene Encoding Autocrine Motility Factor Receptor

Fluorescence in situ hybridization analysis of normal human diploid metaphase preparations determined that the gene which encodes gp78 is located on chromosome 16, in band q21 (SILLETTI et al. 1993). There is ample interest in genes on the long arm of chromosome 16 in relation to common epithelial tumors as well as leukemias. The first report of specific involvement of this region was based on cytological abnormalities in acute nonlymphocytic leukemia M4 (ARTHUR and BLOOMFIELD 1983), and lesions involving chromosome 16 have been identified frequently in near-diploid metaphases from primary breast cancers (DUTRILLAUX et al. 1990). Since few or no other chromosomal aberrations were present in these breast cancer samples, the 16q alteration appears to represent an early (possibly initiating) change or to mark a biologically distinct subset of breast cancers. Recent reports show loss of heterozygosity at several chromosome 16 loci in breast cancer (LARSSON et al. 1990), and it has been shown that alterations and allelic deletions of chromosome 16q are among the most common genetic abnormalities in primary and metastatic human prostate cancer (BERGERHEIM et al. 1991). Interestingly, the locus for E-cadherin is located at 16q22.1 (MANSOURI et al. 1988), although the significance of this degree of proximity to the gp78 gene locus is unknown. It is possible to speculate that chromosome 16 alterations might result in coordinated loss of E-cadherin expression coupled with reciprocally increased levels of gp78 in epithelial cancers, as described for bladder cancer below, thereby promoting tumor invasion.

3.2 Surface Expression of Autocrine Motility Factor Receptor

The gp78 protein is localized to the leading edge, trailing edge, and a region proximal to the nucleus on the surface of motile cells (NABI et al. 1990). This distinctive polarized cell surface distribution most closely resembles that observed for the fibronectin receptor in Chinese hamster ovary (CHO) cells and for galactosyl transferase in mesenchymal cells plated on laminin, with which the galactosyl transferase was shown to interact (BRETSCHER 1989; ECKSTEIN and SHUR 1989). This pattern of surface distribution may be characteristic of proteins involved directly in the locomotory process which require receptor recycling to the leading edge for the maintenance of forward movement.

Localization and redistribution of cell surface proteins by temperature and energy-dependent processes has been characterized previously and is described by two hypotheses: molecules exhibit restricted movement if they are attached to the cytoskeletal submembrane architecture, whereas molecules which are not attached in such a way are carried rearward by membrane flow resulting from the insertion of new membrane and surface components at the leading edge (ABERCROMBIE et al. 1970; RYAN et al. 1974; SCHREINER et al. 1976; ISHIHARA et al. 1988). Surface molecules whose redistribution is characteristic to both of these models have been examined previously (KUCIK et al. 1990; DE BRABANDER et al. 1991), and more recently it has been suggested that the differences in redistribution may instead be due to stronger attachment of surface glycoproteins to the cytoskeleton at the leading edge and subsequent rearward movement in coordination with the cytoskeletal rearrangements involved in migration (KUCIK et al. 1991).

Protein exocytosis occurs specifically at the leading edge of motile cells and at the cell periphery in an unpolarized manner in nonmotile cells. Newly synthesized hemagglutinin is first detected at the periphery of round HeLa cells which have been virally infected, whereas the new protein first appears in membrane protrusions in those cells which are irregularly shaped (MARCUS 1962). In giant HeLa cells, the nonendocytosing surface proteins are distributed uniformly over the cell membrane, whereas proteins which are endocytosed and recycled localize to cell protrusions (BRETSCHER 1983). It has been proposed that exocytosis of vesicles at the leading edge in motile cells would provide membrane for extension of pseudopodia and lamellopodia and generate a rearward flow of membrane and unattached surface constituents in these migratory cells (BRETSCHER 1984; SINGER and KUPFER 1986). The distinct polarized surface distribution of the AMF receptor is not merely the result of cell movement, however, since gp78 has been shown to preferentially localize to one side of round cells which were only allowed to attach to the substrate for 2h, ruling out the possibility of membrane flow-directed localization (NABI et al. 1990).

Different high- and low-metastatic variant sublines of the B16 and K-1735 melanomas were examined for gp78 surface expression and localization. The high-metastatic sublines exhibit markedly higher in vivo lung-colonizing ability and in vitro motility in comparison to their low-metastatic counterparts, and fluorescence-activated cell sorter (FACS) analysis of gp78 on the cell surface showed that the low-metastatic sublines express high levels of the receptor in contrast to the relatively low levels seen on the corresponding high-metastatic lines (WATANABE et al. 1991b). Surface distribution of gp78 was seen to differ in these sublines as well; indirect-labeling immunofluorescence microscopy demonstrated that the less motile, low-metastatic cells have multiple focal areas of gp78 clustering, whereas the high-metastatic and more motile cells have a single intensely staining region at the cell periphery. It appears that the various stained areas on the low-metastatic cells correspond to multiple counterproductive paths of membrane flow, resulting in numerous pseudo-leading edges and lower net forward migration of the cell (NABI et al. 1990, 1992). Similarly, the single intensely stained region in the high-metastatic cells is most likely the result of unidirectional membrane flow and preferential relocation of gp78 to the advancing edge of the cell, resulting in extension of a single leading edge and forward displacement of the cell. These patterns of surface localization hold true in the BALB/c 3T3-A31 angiosarcoma system as well, since sublines which are capable of AMF motility responses exhibit polarized surface gp78 and those which are refractory to such a locomotory stimulus maintain multiple clusters along the cell periphery (SILLETTI et al. 1995).

3.3 Autocrine Motility Factor Receptor Is Endocytosed and Localized to Tubulovesicles

While gp78 is expressed on the surface of motile cells in discrete regions, the bulk of the receptor is intracellular. As described for B16-F1 melanoma cells (NABI et al. 1990), gp78 displays a distinct distribution pattern on the surface of Balb/c-3T3-A31 (A31) fibroblasts with areas of staining localized to the advancing protrusions, the trailing edge, and/or a region proximal to the nucleus. This pattern of staining suggests shuffling of the AMF receptor concomitant with migration (NABI et al. 1990, 1992). A31 cells were used to examine the intracellular mechanism by which shuffling might occur since their spread morphology allows extensive cytoplasmic visualization. Cells were permeabilized and labeled by indirect immunofluorescence using the anti-gp78 mAb. Cells fixed in cold methanol or a cytoskeleton-stabilizing buffer (SCHLIWA et al. 1981) showed that gp78 localized to a network of elongated vesicular structures in both the periphery and center of the cell. Intense staining was observed around the nucleus and in the Golgi complex. Peripheral gp78 localized in linear arrays of vesicles and tubules which extended to the edge of the cell and often corresponded to advancing protrusions of the plasma membrane (NABI et al. 1992). In K-1735 melanoma cells, this internal pool of receptor molecules is mostly perinuclear in the high-metastatic cells and elongates toward the cell periphery after stimulation with 12-(*S*)-HETE; however, these structures remain diffuse in the low-metastatic, AMF motility-unresponsive variant, with perinuclear shuffling, if any, in cells treated with this AMF second messenger (SILLETTI et al. 1994).

The peripheral pattern of gp78-containing structures strongly resembled the pattern previously reported for extended tubular lysosomal networks radiating from the cell center in macrophages and other cells (SWANSON et al. 1987; HEUSER 1989), suggesting that gp78 might be localized to lysosomes. To address the possibility that gp78 is indeed a lysosomal protein, A31 cells were immunofluorescently double-labeled with the anti-gp78 mAb and antibodies to P2B/LAMP-1, a lysosomal-associated major protein. The AMF receptor exhibits only partial colocalization (10%–20%) with LAMP-1, and lysosomes labeled with both anti-gp78 and anti-LAMP-1 are distributed proximal to the nucleus while anti-gp78 alone predominantly stains tubular vesicles distributed near the periphery of the cell, indicating that while gp78 is found within lysosomes, it is associated with non-lysosomal compartments as well (NABI et al. 1992). Indeed, gp78 has recently been shown to be a marker for a new class of smooth vesicular as well as tubular membranous intracellular organelles which is distinct from endosomal and lysosomal populations as well as the endoplasmic reticulum and Golgi apparatus (BENLIMAME et al. 1995).

Endocytosis of surface receptors through coated or noncoated pits and their transport through successive compartments from endosomes to lysosomes has been previously characterized (PASTAN and WILLINGHAM 1981; STEINMEN et al. 1983), and subsequent reports have described the movement of protein-containing vesicles to the plasma membrane as well (LIPPINCOTT-SCHWARTZ and FAMBROUGH

1987; de BRABANDER et al. 1988; HEUSER 1989). Lysosomal association with microtubules has been shown in fibroblast cells (COLLOT et al. 1984), and it has been demonstrated that the tubular morphology of lysosomes in macrophages (SWANSON et al. 1987) and lysosomal translocation and aggregation are dependent on an intact microtubule system (MATTEONI and KREIS 1987). Immunofluorescent staining of A31 cells with antibodies to both gp78 and tubulin demonstrated colocalization of gp78 compartments with microtubules (NABI et al. 1992). Disruption of microtubules in the macrophage results in collapse of the lysosomal tubular morphology and relocation of lysosomes to the center of the cell (SWANSON et al. 1987). In A31 cells treated with the microtubule-disrupting agent colchicine, it was observed that gp78-containing vesicles disappeared from the cell periphery and aggregated around the nucleus, demonstrating that the distribution of gp78 tubulovesicles within the cell is also dependent on microtubules. In addition, the endocytosis inhibitor monensin produced an immunoflorescence staining pattern which was identical to that observed in the presence of colchicine, suggesting that there is a relationship between endocytosis of the AMF receptor and its transport along the microtubular network (NABI et al. 1992).

Recent characterization of gp78 trafficking within Madin-Darby canine kidney (MDCK) cells has shown that, following disruption of the actin cytoskeleton, gp78 tubules extend into peripheral cellular extensions, where they exhibit a highly elongated morphology and can be observed a coalign with microtubules (BENLIMAME et al. 1995). Disruption of the microtubule cytoskeleton by nocodazole treatment or direct modulation of microtubule stability by sequential nocodazole and taxol treatment results in loss of the extended linear morphology and peripheral orientation of gp78 tubules. These results clearly indicate that the linear extention and peripheral cellular orientation of gp78 tubulovesicles in microtubule dependent and identify the gp78 class of tubule as microtubule associated, at least in these systems. Futhermore, the actin and microtubule cytoskeletons may coordinately interact to regulate gp78 tubule morphology and extension to the cell periphery (BENLINAME et al. 1995).

Lysosomal vesicles move both towards and away from the nucleus in viable cells, and movement of vesicles between the cell periphery and nuclear region can be experimentally manipulated by altering the intracellular pH, as described for horseradish peroxidase-labeled fibroblasts (HEUSER 1989). Such vesicular movement in response to pH modulation is indicative of the ability to participate in peripheral/perinuclear translocation events. Acidification of the cytoplasm causes the rapid fission and movement of discrete gp78-containing vesicles from the nuclear region to the periphery, while realkalinization results in the fusion of these discrete vesicles to form elongated tubulovesicles which migrate toward and aggregate around the nucleus of the cell (NABI et al. 1992). Vesicles which contain gp78 have been denoted tubulovesicles based on their morphology, colocalization with microtubules, and relocation in response to intracellular pH modulation; the AMF receptor may be an integral component of these tubulovesicles.

Endolyn-78 is a lysosomal membrane glycoprotein which exhibits numerous similarities to the AMF receptor (CROZE et al. 1989) and may be related to a family of heavily glycosylated lysosomal-associated proteins which exhibits extensive cross-species homology but whose functions remain unknown (KORNFELD and MELLMAN 1989). The common theme of extensive sugar modification in these lysosomal proteins suggests a functional role for both N- and O-linked glycosylation in these proteins. The presence of N-linked carbohydrates, however, had no effect on the rate of transport of these lysosomal glycoproteins to their lysosomal destinations following de novo synthesis, whereas the absence of these moieties decreased the half-life of the molecules significantly (BARRIOCANAL et al. 1986). Both the AMF receptor and endolyn-78 contain O-linked oligosaccharides, and endolyn-78 was shown to be present in the membranes of endosomes, lysosomes, and multivesicular bodies as well as the plasma membrane, similar to gp78. It has been suggested that endolyn-78 may be a constituent of peripheral endosomal compartments, while other lysosomal-associated membrane proteins such as the family of LAMP may be restricted to mature lysosomes (CROZE et al. 1989; KORNFELD and MELLMAN 1989). Some lysosomal-associated glycoproteins may have distinct functions in locations other than lysosomes. An endogenous canine LAMP was recently shown to be directed to the basolateral surface of the cell prior to translocation to lysosomes in MDCK epithelial cells (NABI et al. 1991) and the avian LAMP-1, LEP100, can be detected at low levels on the plasma membrane and in endosomal compartments in chick fibroblasts (LIPPENCOTT-SCHWARTZ and FAMBROUGH 1986,1987), suggesting the possibility of a distinct role for these proteins at other locations within the cell. LEP100 expression on the plasma membrane as well as in endosomal compartments is increased after treatment of the cells with chloroquine, indicating that LEP100 may function in transport between lysosomes and the cell surface (LIPPENCOTT-SCHWARTZ and FAMBROUGH 1987). Similarly, alterations in oligosaccharide branching of the P2B protein, a mouse homologue of LAMP-1, is associated with increased expression of P2B on the plasma membrane and efficient metastasis (HEFFERNAN et al. 1989). This is similar to the increased O-linked glycosylation of gp78 on B16-F1 cells with enhanced metastatic potential which initially led to the recognition of a role for gp78 in metastasis (NABI and RAZ, 1987,1988). In addition, the CD63 antigen is a lysosomal-associated membrane glycoprotein which is identical to the melanoma surface antigen ME491, whose surface expression is related to neoplastic transformation (HOTTA et al. 1988; METZELAAR et al. 1991), indicating again that altered glycosylation and/or surface expression of proteins normally restricted to intracellular compartments may be associated with the acquisition of enhanced invasive and metastatic phenotypes by neoplastic cells.

3.4 Internalized Autocrine Motility Factor Receptor Is Recycled to the Leading Edge

The requirement for polarized translocation of membrane vesicles to generate a leading edge and thereby direct cell movement has been demonstrated previously (Singer and Kupper 1986), and earlier work showed that an intact microtubule system was required for this process to operate. Disruption of microtubules resulted in a loss of directionality of vesicular transport to the leading edge, resulting in uncoordinated extension of multiple pseudo-leading edges and unpolarized surface expression of a membrane protein as well as loss of motility (Rogalski et al. 1984), demonstrating the critical role that microtubules play in directing extension of the leading edge. The AMF receptor colocalizes with microtubules, and disruption of the microtubular network negates the stimulatory effect of AMF and interferes with translocation of gp78 to the periphery of the cell, indicating that gp78 is involved in the directed endocytosis pathway whereby membrane vesicles move along microtubules for preferential relocation to the leading edge. Based on the proposed models of cell surface protein distribution and the data on gp78 surface expression, several predictions can be made. Transport of the AMF receptor and exocytosis at the leading edge should lead to the appearance of gp78 on the surface of the advancing protrusions of the cell. Cellular locomotion should result in the rearward movement of surface gp78 relative to the cell prior to its endocytosis at some point during migration. The pool of unendocytosed gp78 should accumulate at the rear of the cell as positive displacement is affected. Finally, the increased rate of extension of the leading edge associated with higher motility will produce an increased rate of membrane flow and a concomitantly enhanced rate of gp78 endocytosis, resulting in a smaller surface pool of receptor.

To determine directly whether gp78 is indeed transported to the leading edge, BALB/c 3T3-A31 fibroblasts were incubated with anti-gp78 mAb in the cold and then warmed to 37°C to allow internalization of gp78, and the antibody–gp78 complex was examined at different time intervals. The receptor was diffusely distributed within the cell after 10 min, indicating that endocytosis is not restricted to specific surface domains, contrasting starkly with the localized surface distribution of gp78. Large perinuclear complexes form by 20 min post-internalization in a manner analogous to the multivesicular bodies described for the transferrin receptor (de Branbander et al. 1988). After 40 min the complexes are grouped nondirectionally around the cell center, and between 40 and 60 min the labeled complexes become polarized and are redistributed and translocated to the leading edge of the cell (Nabi et al. 1992). This is similar to the case described for newly synthesized vasicular stomatic virus (VSV), G protein, which is targeted to the leading edge of motile fibroblasts (Bergmann et al. 1983). Indeed, low-metastatic K-1735 melanoma cells which exhibit multiple AMF receptor surface domains and do not respond to AMF with enhanced motility exhibit recycling of gp78 to the respective surface clusters of origin, whereas high-metastatic K-1735 melanoma cells which are capable of an AMF migratory response show gp78 recycling to the leading edge during cellular displacement, as described for the AMF-responsive BALB/c 3T3-A31 fibroblasts (Silletti et al. 1994).

Endocytosed transferrin receptor was observed to have a similar staining pattern to gp78, and Hopkins et al. (1990) noted irregularly spaced varicosities along tubular cisternae which contained the fluorescent marker. These swellings were shown to be large multivesicular bodies which could be seen moving along a filamentous network described as a continuous endosomal reticulum. This movement of multivesicular bodies along a filamentous network supports the previous observation that gp78 colocalizes with microtubules and that endocytosis of gp78 results in large complex structures which are recycled to the leading edge (NABI et al. 1992). In a similar manner, tubulovesicular processes have been described which interlink Golgi cisternae and which stem from the trans-Golgi and translocate along microtubules (COOPER et al. 1990). Thus, certain processes of intracellular membrane traffic may be mediated by tubulovesicles.

4 Receptor Activation and Cellular Locomotion

The AMF receptor's distinctive surface distribution, internalization, and localization to intracellular tubulovesicles followed by selective transport to the leading edge of the cell suggests the presence of a specific AMF-stimulated pathway of membrane flow. Receptor-containing vesicles which are transported to the leading edge would provide membrane for forward displacement of the cell, and gp78 thus inserted at the leading edge could move rearward in coordination with cell migration to interact with additional AMF molecules and thereby propagate the signal. Endocytosis of gp78 need not occur only at the trailing edge of the cell, but could occur over the entire cell surface such that there is a net positive membrane flow to the leading edge and remaining unendocytosed gp78 accumulates at the rear of the cell, as seen in immunofluorescently stained motile cells (NABI et al. 1992).

Binding of the anti-gp78 mAb to the AMF receptor mimics the motility-stimulating effect of the natural ligand, and these binding interactions could operate by inducing the endocytosis of gp78, thereby increasing the rate of membrane flow and stimulating extension of the leading edge. This implies that on motile cells which extend a single leading edge, receptor stimulation would result in enhanced directional motility, whereas less motile cells which extend multiple, independently acting pseudo-leading edges would respond with an increase in endocytosis and possibly random motility without necessarily stimulating directional migration. This prediction was verified by comparing the observed response of high- and low-metastatic melanoma sublines to stimulation with the anti-gp78 mAb in both the in vitro nondirectional motility assay and in vivo metastatic lung colonization assay. Only the high-metastatic cells displayed a two - to 2.5-fold increases in nondirectional phagokinetic motility, and correspondingly only the high-metastatic sublines responded to the anti-gp78 stimulation with increased lung colonization (WATANABE et al. 1991b; SILLETTI et al. 1994). This illustrates that in responding to receptor stimulation with increased motility, only the high-metastatic cells were able to increase the directional migration necessary to extravasate into the parenchyma, and it may be, therefore, that only in these cells did stimulation result in receptor endocytosis and specific transport to a single advancing region.

Ligand-induced receptor phosphorylation is a well-characterized signal for receptor internalization and signal transduction in many systems (ULLRICH and SCHLESSINGER 1990). Since internalization and translocation of the receptor appears to be an integral part of AMF-mediated motility stimulation and since the predicted protein sequence of the AMF receptor intracellular domain contains consensus sequences for nucleotide binding and phosphorylation the phosphorylation state of gp78 was examined. Immunoprecipitation analysis of cells labeled with radioactive phosphate showed that the AMF receptor becomes phosphorylated within 4 min after stimulation with AMF, suggesting a relationship between AMF binding its receptor (WATANABE et al. 1991a, unpublished data).

AMF-induced motility stimulation is blocked by PT (STRACKE et al. 1987; NABI et al. 1990); therefore, endocytosis of AMF-stimulated gp78 must be blocked by PT if this step is a critical mediator of the motility signal. B16-F1 melanoma cells pretreated with PT for 18 h did not exhibit any change in surface expression of gp78; however, upon washing out of the PT with fresh medium, uninhibited G proteins are recruited to the plasma membrane and downregulation of gp78 occurs within 1 h (WATANABE et al. 1991b). Similarly, overnight incubation of B16-F1 cells with PT and the anti-gp78 mAb resulted in a significant increase in lung-colonizing ability as compared to the antibody alone. Thus it appears that the PT freezes the cells in a state of surface-bound receptor– ligand complex, poised for rapid internalization and motility stimulation as soon as the PT is removed and new uninhibited G proteins can be recruited. PT therefore blocks AMF induction of motility by preventing endocytosis of the AMF receptor, and it was previously reported that treatment of A2058 melanoma cells with PT disrupts AMF-induced pseudopodial protrusion (GIORGUIS et al. 1987). Stimulation of motility by AMF is therefore associated with receptor phosphorylation and internalization as well as pseudopodial protrusion at the leading edge, and the following data provides insights into the biochemical linking of these phenomena.

4.1 Autocrine Motility Factor Initiates Motility by Reorganization of Cellular Focal Adhesions

Focal adhesion proteins and their structural regulation during cellular migration has been examined in some detail. Highly metastatic tumor cells exhibit decreased adhesive properties and increased endogenous migration rates (VOLK et al. 1984) as well as altered organization of cell substrate and membrane-associated cytoskeletal contacts compared with their low- or nonmetastatic counterparts (RAZ and GEIGER1982). Futhermore, suppression of the expression of the cytoskeletal-coupling protein vinculin increased the motility of 3T3 fibroblasts (FERNANDEZ et al. 1993), and recombinant expression of vinculin complemented the adhesion-defective phenotype of a mutant mouse embryonal carcinoma cell line (SAMUELS et al. 1993), demonstrating the importance of this cytoskeletal-associated protein in cell adhesion and migration. Similarly, regulation of adhesion is closely linked with the expression and activity of the 125-kDa tyrosine kinase $pp125^{FAK}$ (focal adhesion kinase, FAK; (SCHALLER et al. 1992), which appears to function in cytoskeletal rearrangement activities (BURRODGE et al. 1988). This enzyme is localized at focal adhesion plaques and is associated with other proximate proteins including vinculin, actin, zyxin, and paxillin (BURRODGE et al. 1988; SCHALLER et al. 1992). Focal adhesion structures maintain a dynamic interplay with the cytoskeleton, and this is a critical determinant of cellular processes including intracellular trafficking and rearrangements associated with alterations in cellular morphology and plasticity. In addition, oncogenic transformation associated with cellular infection with the $pp^{60v\text{-}src}$ protein tyrosine kinase enhances phosphorylation of a number of proteins, including FAK, resulting in alterations in cellular morphology and growth regulation (PARSONS and WEBER 1989; SCHALLER et al. 1992). The cytoskeleton is a crucial intermediary of the changes a cell must enact in the process of locomotion (THERIOT and MITCHISON 1991; CUNNINGHAM 1992) and, as such, the expression and distribution of cytoskeletal proteins such as actin, vimentin, and α-tubulin are associated with altered cellular phenotypes in malignant progression (RAZ and GEIGER 1982), apparently mediating some of the processes of surface receptors such as gp78 (NABI et al. 1992).

Analysis of BALB/c 3T3-A31 angiosarcoma cells which displayed progressively malignant phenotypes and differential responses to AMF showed that AMF-induced migratory responses were associated with marked rearrangement of focal adhesion plaque proteins. Reorganization of vinculin after AMF stimulation was paralleled by morphological redistribution of tyrosine-phosphorylated proteins and the tyrosine kinase FAK in the migration-responsive cells;however, we observed neither a concomitant change in the FAK phosphorylation state nor in the general level of cellular tyrosine phosphorylation in response to treatment, suggesting that the induction of cellular migration by AMF is independent of tyrosine phosphorylation events at the focal contacts. After stimulation of locomotory-responsive cells with AMF, the focal adhesion proteins (as detected by indirect immunofluorescence of FAK and vinculin) as well as phosphotyrosine-containing proteins reorganized from a diffuse array of plaques along the basal cell surface into thin filamentous structures oriented perpendicular to the cell edges, often showing a preponderance of accumulation within the advancing edges of pseudopodia. As can be seen in Fig. 2, the AMF-producing HT-1080 fibrosarcoma cells display a marked rearrangement of phosphotyrosine-containing proteins within 2 min and maintain this pattern for up to 60 min.

Although the cytoskeletal rearrangements induced through Protein Kinase C in other systems may also be present in the AMF pathway(s), the altered distribution of FAK and the accompanying reorientation of tyrosine-phosphorylated proteins is independent of de novo tyrosine phosphorylation events, unlike the pathway reported for motility signal transduction through the *met* receptor by scatter factor/hepatocyte growth factor (HARTMANN et al. 1994). This is also in contrast to the transient tyrosine kinase activation induced by the hyaluronan

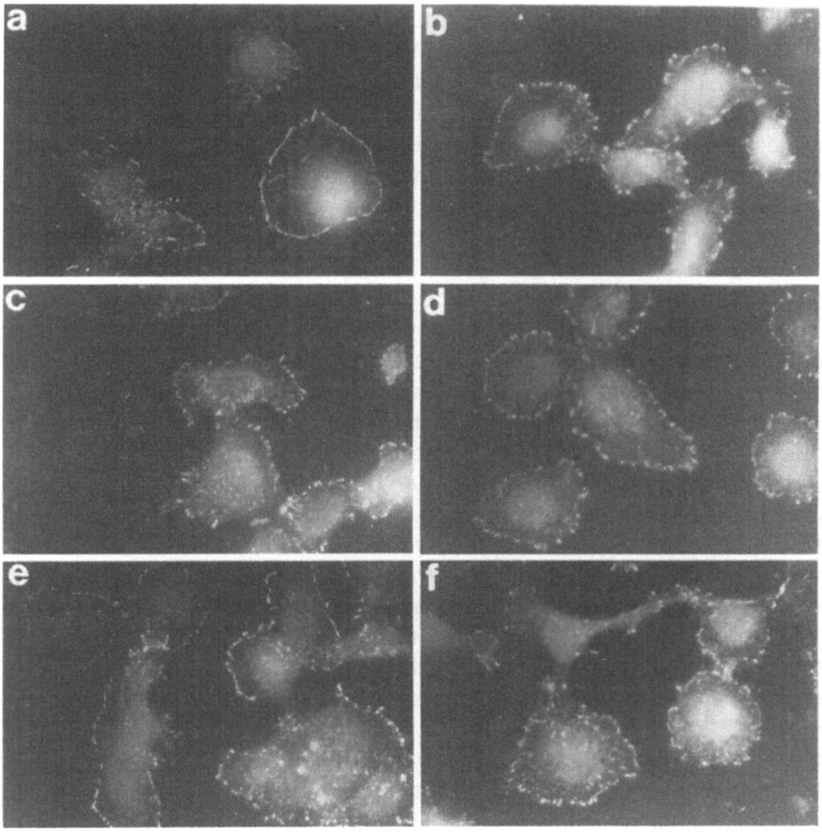

Fig. 2a–f. Reorganization of tyrosine-phosphorylated proteins in HT-1080 cells. The time course of phosphotyrosine reorganization was examined in the autocrine motility factor (AMF)-producing HT-1080 human fibrosarcoma cells after treatment for **a** 0 min, **b** 2 min, **c** 5 min, **d**10 min, **e** 30 min and, **f** 60 min with 50 pg HT-1080 AMF ml. Cells were plated on glass coverslips and grown for 1–3 days before fresh medium was replenished and the cells were stimulated with AMF for the indicated times. Culture medium was removed and the cells were fixed with paraformaldehyde prior to recognition of phosphotyrosine residues with the PY20 monoclonal antibody (mAb; Transduction Laboratories, Lexington, KY) and visualization of the primary antibody with rhodamine-conjugated goat anti-mouse immoglobulin G (IgG) secondary antibody (Zymed, San Francisco, CA). Immunofluorescence photomicrograph, x 600

receptor RHAMM, which has also been shown to promote focal adhesion turnover during stimulation of C-*Ha-ras*-transformed fibroblast migration (HALL et al. 1994).

4.2 Regulation of Autocrine Motility Factor Receptor

Metabolites of arachidonic acid have been implicated in several aspects of tumor biology, and the lipoxygenase metabolite 12-(*s*)-HETE has been shown to alter several cellular mechanisms including integrin $\alpha_{IIb}\beta_3$ receptor expression and extracellular matrix adhesion (GROSSI et al. 1989; CHOPRA et al. 1991), suggesting that 12-(*S*) -HETE might play a role in cellular motility as well. Activation of the cell surface receptor for AMF was shown to stimulate production of 12-(*S*)-HETE in highly metastatic melanoma cells exclusively (SILLETTI et al. 1994). Reciprocally, treatment with submicromolar concentrations of exogenous 12-(*S*)-HETE resulted in increased expression of total surface gp78, as well as more than doubling the proportion of gp78 high expressors, and caused gp78 translocation from an intracellular perinuclear pool to tubulovesicles which extended to the cell periphery in the high-metastatic cells exclusively (TIMAR et al. 1993; SILLETTI et al. 1994). This effect directly mimicked that of the natural gp78 ligand AMF. Furthermore, gp78 expression on the leading lamella and advancing pseudopodia of the high-metastatic melanoma cells was enhanced by stimulation with 12-(*S*)-HETE, and 12-(*S*)-HETE also induced phosphorylation of the AMF receptor in a manner analogous to AMF (TIMAR et al. 1993). This effect was directly correlated with the stimulation of motility and, as such, the low-metastatic cells which are incapable of an AMF-induced locomotory response did not respond to stimulation with 12-(*S*)-HETE (SILLETTI et al. 1994). This effect was specific for the 12-(*S*) enantiomer, as treatment with 12-(*R*)-HETE and other monohydroxy fatty acids had no effect. In addition, these effects were protein kinase C dependent, as described above for the AMF signaling cascade. This data indicates that 12-(*S*)-HETE does indeed play a role in AMF-mediated motility stimulation and suggests that this signaling pathway may be at least partially specific to highly metastatic cells. The mechanism by which 12-(*S*)-HETE mediates its motility effect may involve inducing cytoskeletal rearrangements which allow translocation of gp78 from intracellular stores to the leading edge where the receptor can interact with additional AMF to increase motility, analogous to the 12-(*S*)-HETE-induced overexpression of $\alpha_{IIb}\beta_3$ at focal contacts previously described by CHOPRA et al. (1991).

It has been demonstrated that metabolites and synthetic analogues of vitamin A (retinoids) can inhibit carcinogenesis and suppress the tumorigenicity of several tumor cell types by inducing terminal differentiation (LOTAN 1980; SPORN et al. 1984; SHERMAN 1986; LIPPMAN et al. 1987). In order to examine the possible effects of retinoids on the AMF pathway, four sublines of the A375 human melanoma which exhibit varying metastatic propensities were treated with β-*all-trans*-retinoic acid (RA) and examined for inhibited invasion through a reconstituted basement membrane and biochemical alterations including enzyme secretion and receptor expression levels. This study demonstrated the ability of RA to inhibit the invasiveness of human melanoma cells of different metastatic capabilities in a time- and dose-dependent manner, and it was shown that expression of the AMF receptor was decreased 13- to 50-fold compared with control cells in each of the four sublines after growth for 7 days in the presence of RA (HENDRIX et al. 1990). This data suggests that RA has either a direct or indirect effect on gp78 expression which may be directly related to decreased tumor cell invasiveness in this context. In all likelihood, the effect of RA is indirect, as recent characterization of the gp78 promoter failed to distinguish an RA-response element (HUANG et al. 1995). Furthermore, LOTAN et al. (1992) showed that decreased migratory responses of RA-treated melanoma cells to AMF and the AMF-mimicking anti-gp78 mAb correlated with decreased gp78 expression, suggesting that the two events are related and that previously reported observations of RA-mediated suppression of tumor cell invasion and metastasis may be related, at least in part, to suppression of cell motility resulting from decreased levels of the AMF receptor.

Cell density-dependent downregulation of gp78 appears to be a hallmark of lower malignant potential, as normal BALB/c 3T3-A31 fibroblasts as well as normal human bladder epithelial cells demonstrate complete downregulation of both the surface as well as total expression of AMF receptor when grown to confluence (SILLETTI and RAZ 1993; SILLETTI et al. 1993). This density-dependent regulation of receptor expression is not unique to the AMF system, as the receptor for acidic fibroblast growth factor is downregulated in NBT-II epithelial cells cultured at high density (BRENNER et al. 1989), and platelet-derived growth factor receptor expression is reduced in contact-inhibited 3T3 fibroblasts as well (GROTENDORST 1984). Cells which display progressively malignant and especially metastatic phenotypes appear to be capable of maintaining steady state expression of gp78 in high cell density conditions (SILLETTI et al. 1993,1995). The display of gp78 in a polarized fashion on the malignant cell surface coupled with the motile responses of these cells to AMF, irrespective of cell density, suggests a mechanism for their exploitation of this continued expression for invasion. Indeed, the fact that more malignant cells have lost the biphasic return to baseline function of their mitogenic response to AMF indicates that these cells are equipped to maintain any and all AMF responses regardless of the circumstance.

In order to determine whether the AMF migration responsiveness noted in MCF-7 breast carcinoma cells might facilitate locomotory responsiveness in density-arrested conditions such as those found within an organ in vivo, we analyzed the surface and total cellular expression of gp78 from sparse and confluent cultures of these breast cell lines. The expression of g78 on the surface of the migration-inducible MCF-7 breast carcinoma cells was capped at locations which are apparently associated with the leading and/or trailing edges of individual cells (Fig. 3A) in a manner reminiscent of that described previously for AMF migration-responsive cells (WATANABE et al. 1991b; SILLETTI et al. 1993,1994, 1995). Surface expression of gp78 is maintained in these malignant cells under high-density conditions as well (Fig. 3B), suggesting loss of the contact-mediated inhibition observed in normal and nonmalignant cells of other types (SILLETTI and RAZ 1993; SILLETTI et al. 1993). A decreased adhesivity of these cells was also observed, in that cells which were growing apart from large masses of cells tended to be individual (Fig. 3C) and not in smaller clumps of cells grouped together as is seen in the nonmalignant MCF-10A cells (Fig. 3E). There is also a tendency to form a less tightly packed monolayer in the MCF-7 cells (Fig. 3D), in contrast to the more differentiated appearance of the MCF-10A cells in monolayer culture (Fig. 3F). This decreased adhesivity was associated with a decreased staining of MCF-7 monolayer cultures when immunocytochemically

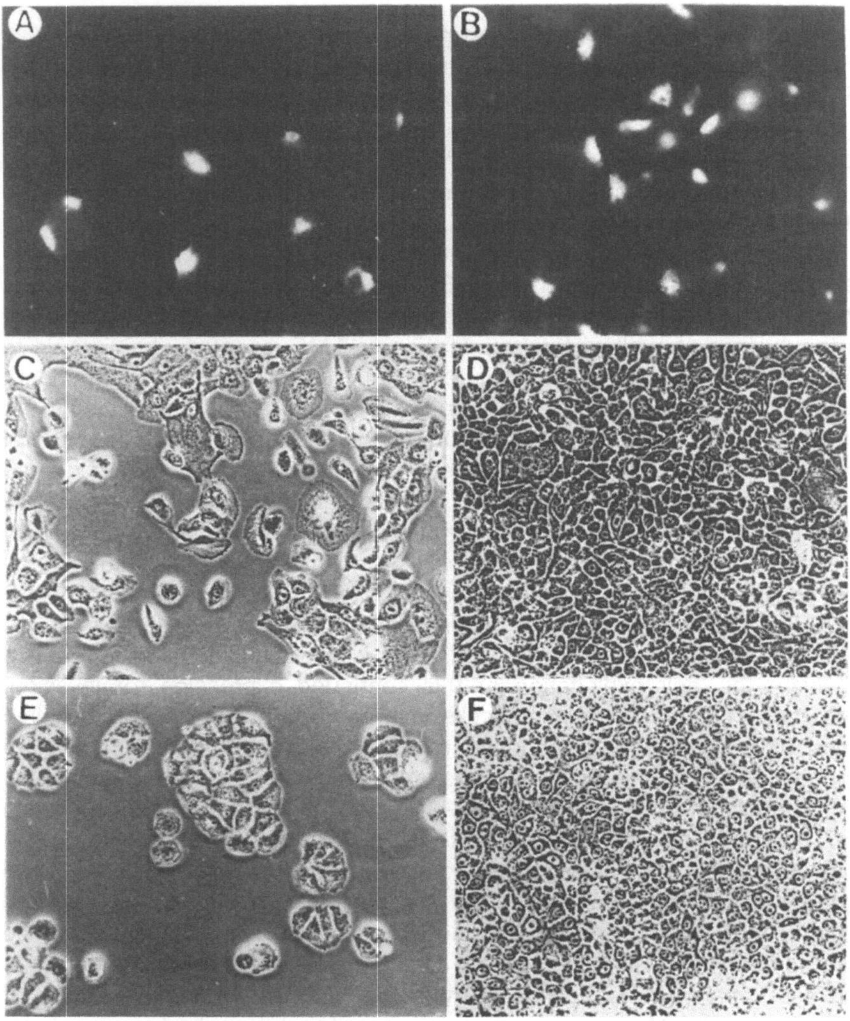

Fig. 3A–F. Cell-density regulation of gp78 and adhesive phenotype of MCF breast cells. Indirect immunofluorescence photomicrographs of surface gp78 expression (**A,B**) and phase-contrast morphology photomicrographs (**C–F**) of MCF-7 breast carcinoma (**A–D**) and MCF-10A breast epithelial cells (**E,F**) grown to spare (**A,C,E**) or contact-inhibited densities (**B,D,F**). Cells were plated on glass coverslips, grown for 1–3 days, and then fixed with paraformaldehyde prior to detection of gp78 by antibody hybridization. **A,B** x 600. **C–F** x 200

analyzed wiith an anti-E-cadherin mAb (data not shown). Although these MCF-7 breast carcinoma cells maintain some level of E-cadherin positivity, the functionality of this expression is questioned by the lack of intercellular adhesivity, suggesting that expression of this cell–cell homotypic aggregation molecule may not be representative of function, similar to the observation in prostate cancer cells whereby deletion of the cadherin–cytoskeleton coupling protein α-catenin produced identical changes in cellular phenotype in the absence of alteractions in E-cadherin itself by short-circuiting the coupling of E-cadherin homotypic binding at the level of the cytoskeletal interaction (MORTON et al. 1993). E-cadherin was detected at significant levels in the nonmalignant MCF-10A cells, however, supporting the association of this protein with the nonhomotypically adhesive phenotype of these cells.

MCF-7 cells display a pattern of gp78 surface expression similar to that described previously for high-metastatic cells (WATANABE et al. 1991b) and normal cells which respond to AMF motility stimulation (SILLETTI and RAZ 1993). This is not surprising since these malignant breast carcinoma cells responded to paracrine stimulation with tumor-derived AMF as well as the anti-gp78 mAb. MCF-10A cells are nonmotile in the absence of stimulation, being unable to invade Matrigel-reconstituted basement membrane and only minimally migration stimulated by treatment with the anti-gp78 mAb, while MCF-10A cells transfected with an activated Ha-ras oncogene (MCF-10AneoT cells) exhibited enhanced basal invasion of matrigel (Biomedical Products, Bedford, MA) and a four to fivefold increase in motility response to the anti-gp78 mAb (OCHIENG et al. 1991). Interestingly, the parental and transformed MCF-10 cells both displayed surface labeling of gp78 to discrete spots along the edge of the membrane, with one to three spots per cell. The absence of a discernible leading or trailing edge in the rounded MCF-10A cells suggests that the motility response exhibited by these cells may be nondirectional and incapable of facilitating invasion/metastasis, while the morphology of the Ha-ras-transfected MCF-10AneoT cells appears more elongate, with more prominent localization of AMF receptor molecules to cell edges involved in directional locomotory processes (OCHIENG et al. 1991).

Quantitative immunoblotting analysis of these two cell lines under sparse and confluent conditions (Fig. 4) showed that the expression of gp78 in total cellular lysates from sparse cultures was lower in the sparse MCF-7 cells, a situation which has been described previously in the K-1735 melanoma system (SILLETTI et al. 1994). When density-arrested cultures were examined, however, the expression of gp78 in the MCF-10A cells was virtually abolished, while that of the MCF-7 cells was actually enhanced, possibly due to direct upregulation as proposed for the metastatic BALB/c 3T3-A31 variant cells (SILLETTI et al. 1995) or as a side effect of the disruption of the expression of other density-arrested proteins, causing an increase in the ratio of gp78 to total cellular protein and not necessarily an increase in the gp78 content of each individual cell (comparison based on protein concentration to compensate for differences in cell size).

4.3 Prognostic Value of the Autocrine Motility Factor System

AMF and its receptor have been postulated to play a role in the most devastating aspect of cancer, i.e., tumor cell metastasis, based upon their motility-regulating effects (VAN ROY and MAREEL 1992). Moreover, the presence of AMF in the urine is a marker of transitional cell carcinoma of the bladder (GUIRGUIS et al. 1988).

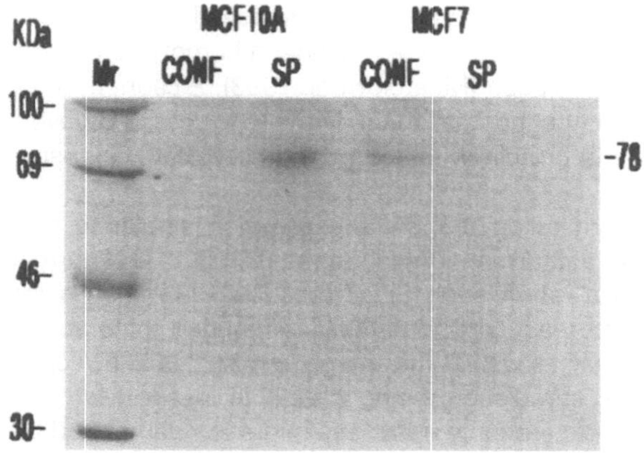

Fig. 4. Density-dependent regulation of gp78 in normal and neoplastic breast tissue. Quantitative immunoblotting of total cellular gp78 in whole-cell lysates extracted from sparse (*SP*) and confluent (*CONF*) MCF-10A normal and MCF-7 carcinoma breast cell cultures. Cells were harvested and lysed, and equal quantities of protein were separated by sodium dodecyl sulfate polyacrylamide gel electrophoresis (SDS-PAGE) before transfer to a nylon membrane and detection of gp78 by antibody hybridization and autoradiography. *Mr*, molecular weight protein standards shown in kilodaltons (kDa) at the *left*. The relative migration of gp78 is indicated at the *right*

Recently, it was shown that expression of the AMF receptor correlates with malignant potential in bladder cell lines (SILLETTI et al. 1993), as well as disease stage, recurrence, and incidence of metastasis in invasive and noninvasive human bladder cancer specimens (OTTO et al. 1994). Up to 90% of malignant bladder neoplasms are epithelial-derived transitional cell carcinomas, which exhibit a high rate of local recurrence and progress to advanced stages of disease and metastasize in 20% of the superficial and 60% of the invasive cases (GUIRGUIS et al. 1988; RAGHAVAN et al. 1990). That the AMF system may play a functional role in bladder cancer is supported by the finding that the invasion of invasive human bladder carcinoma cells can be suppressed by inhibiting protein kinase C, consistent with AMF-mediated intracellular signaling (SCHWARTZ et al. 1990). Upregulation of gp78 expression concomitant with disease progression in bladder cancer is associated with downregulation of the calcium-dependent homotypic cell adhesion molecule E-cadherin, a pattern similar to that observed in squamous cell carcinomas of the head and neck (SCHIPPER et al. 1991) and prostate cancer (UMBAS et al. 1992) and suggesting a shift from a sedentary to a motile (i.e., invasive) cellular phenotype. The expression of gp78 has been shown to be associated with colon cancer posttreatment recurrence (NAKAMORI et al. 1994). In addition, AMF-like activity has been reported recently in oral squamous cell carcinoma cells, which actively downregulate E-cadherin in response to such motile stimulation (ISHISAKI et al. 1994).

5 Conclusion

AMF is secreted by neoplastic cells, which in turn respond to external levels of this cytokine in a dose-dependent manner. The cellular response to AMF appears to be of a dual nature. AMF both stimulates motility and affects proliferation of competent recipient cells, although the ability to respond to one or both of these inputs rests with the phenotype of the target cell. Whereas stimulation of gp78 produces a mitogenic response which is dose dependent and limited to a narrow effective range, this effect is exploited in cells with increasingly malignant phenotypes.

Upon binding of AMF to its receptor, a 78-kDa cell surface glycoprotein, the AMF response is mediated by a pertussis toxin-sensitive G protein, inositol trisphosphate production, 12-lipoxygenase activity with subsequent production of 12-(*S*)-HETE, protein kinase C (PKC) activation, and receptor phosphorylation. The receptor is internalized and localized to tubulovesicles, whereby it is transported along the microtubular network to a region near the nucleus and subsequently recycled to the leading edge of the cell, resulting in pseudopodial extension and increased migration in responsive cells. Surface localization appears to be paramount in the ability of a cell to respond to AMF's locomotory impetus with polarized surface expression to the leading and trailing cell edges requisite for motility stimulation; however, this paradigm does not hold true for a cell's capacity for proliferative response to AMF. Expression of gp78 is regulated by cell–cell contact in normal cells, a regulatory scheme which is lost during malignant progression and which appears to make the AMF system a useful tool in predicting patient prognosis in the clinical setting. Further characterization of cellular responses to AMF and similar cytokines will increase our understanding of the fundamental processes which control motility in mammalian cells and may provide the basis for the development of more effective clinical anticancer therapies and specific modalities which inhibit tumor cell motility in vitro and invasion and metastasis in vivo.

Acknowledgments. This work was supported in part by NIH Grant CA-51714–01A2 and by the Paul Zuckerman Support Foundation for Cancer Research to A. Raz. The University of Wisconsin Genetics Computer Group Software Package (version 6.1) was employed for all genetic analyses as described (Devereux et al. 1984).

References

Abercrombie M, Heaysman JEM, Pegrum SM (1970) The locomotion of fibroblasts in culture. III. Movement of particles on the dorsal surface of the leading lamella. Exp Cell Res 62: 389–398

Albrecht-Buehler G (1977) The phagokinetic tracks of 3T3 cells. Cell 11: 395–404

Arthur DC, Bloomfield CD (1983) Partial deletion of the long arm of chromosome 16 and bone marrow eosinophilia in acute nonlymphocytic leukemia: a new association. Blood 61: 994–998

Ashkenazi A, Peralta EG, Winslow JW, Ramachandran J, Capon DJ (1989) Functionally distinct G proteins couple different receptors to PI hydrolysis in the same cell. Cell 56: 487–493
Atnip KD, Carter LM, Nicolson GL, Dabbous MK (1987) chemotactic response of rat mammary adenocarcinoma cell clones to tumor-derived cytokines. Biochem Biophys Res Commun 146: 996–1002
Badenoch-Jones P, Ramshaw IA (1984) Spontaneous capillary tube migration of metastatic rat mammary adenocarcinoma cells. Invasion Metastasis 4: 98–110
Bareis DL, Hirata F, Schiffmann E, Axelrod J (1982) Phospholipid metabolism, calcium flux, and the receptor mediated induction of chemotaxis in rabbit neutrophils. J Cell Biol 93: 690–697
Barriocanal JG, Bonifacio JS, Yuan L, Sandoval IV (1986) Biosynthesis glycosylation, movement through the Golgi system and transport to lysosomes by an N-linked carbohydrate-independent mechanism of three lysosomal integral membrane proteins. J Biol Chem 261: 16755–16763
Benlimame N, Simard D, Nabi IR (1995) Autocrine motility factor receptor is a marker for a distinct membranous tubular organelle. J Cell Biol 129: 459–471
Bergerheim USR, Kunimi K, Collins VP, Ekman P (1991) Deletion mapping of chromosomes 8, 10 and 16 in human prostatic carcinoma. Genes Chromosom Cancer 3: 215–220
Bergmann JE, Kupfer A, Singer SJ (1983) Membrane insertion at the leading edge of motile fibroblasts. Proc Natl Acad Sci USA 80: 1367–1371
Brandt SJ, Dougherty RW, Lapetina EG, Niedel JE (1985) Pertussis toxin inhibits chemotactic peptide-stimulated generation of inositol phosphates and lysosomal enzyme secretion in human leukemic (HL-60) Cells. Proc Natl Acad Sci USA 82: 3277–3280
Brenner DA, Koch KS, Leffert HL (1989) Transforming growth factor-α stimulates proto-oncogene c-jun expression and a mitogenic program in primary culture of adult rat hepatocytes. DNA 8: 279–285
Bretscher MS (1983) Distribution of receptors for transferrin and low density lipoprotein on the surface of giant HeLa cells. Proc Natl Acad Sci USA 80: 454–458
Bretscher MS (1984) Endocytosis: relation to capping and cell locomotion. Science 226: 681–686
Bretscher MS (1989) Endocytosis and recycling of the fibronectin receptor in CHO cells. EMBO J 8: 1341–1348
Burridge K, Fath K, Kelly T, Nuckolls G, Turner CE (1988) Focal adhesions: transmembrane junctions between the extracellular matrix and the cytoskeleton. Annu Rev Cell Biol 4: 487–525
Chopra H, Timar J, Chen YQ, Rong X, Grossi IR, Fitzgerald LA, Taylor JT, Honn KV (1991) The lipoxygenase metabolite 12(*S*)-HETE induces a cytoskeleton-dependent increase in surface expression of integrin αIIbβ3 on melanoma cells. Int J Cancer 49: 774–786
Collot M, Louvard D, Singer SJ (1984) Lysosomes are associated with microtubules and not with intermediate filaments in cultured fibroblasts. Proc Natl Acad Sci USA 81: 788–792
Cooper MS, Cornell-Bell AH, Chernjavsky A, Dani JW, Smith SJ (1990) Tubulovesicular processes emerge from the trans-Golgi cisternae, extend along microtubules, and interlink adjacent trans-Golgi elements into a reticulum. Cell 61: 135–145
Croze E, Ivanov IE, Kreibich G, Adesnik M, Sabatini DD, Rosenfeld MG (1989) Endolyn-78, a membrane glycoprotein present in morphologically diverse components of the endosomal and lysosomal compartments: implication for lysosome biogenesis. J Cell Biol 108: 1597–1613
Cunningham CC (1992) Actin structural proteins in cell motility. Cancer Metast Rev 11: 69–77
De Biasio RL, Wang LL, Fisher GW, Taylor DL (1988) The dynamic distribution of fluorescent analogues of actin and myosin at the leading edge of migrating Swiss 3T3 fibroblasts. J Cell Biol 107: 2631–2645
de Brabander M, Nuydens R, Geerts H, Hopkins CR (1988) Dynamic behavior of the transferrin receptor followed in living epidermoid carcinoma (A431) cells with Nanovid microscopy. Cell Motil Cytoskeleton 9: 30–47
de Brabander M, Nuydens R, Ishahara A, Holifield B, Jacobson K, Geertz H (1991) Lateral diffusion and retrograde movements of individual cell surface components on single motile cells observed with nanovid microscopy. J Cell Biol 112: 111–124
Devereux J, Haeberli P, Smithies O (1984) A comprehensive set of sequence analysis programs for the VAX. Nucleic Acids Res 12: 387–395
Dutrillaus B, Gerbault-Sereau M, Zafrani B (1990) A comparison of 30 paradiploid cases with few chromosome changes. Cancer Genet Cytogenet 49: 203–217
Eberle M, Traynor-Kaplan AE, Sklar LA, Norgauer J (1990) Is there a relationship between phosphatidylinositol triphosphate and F-actin polymerization in human neutrophils. J Biol Chem 265: 16725–16728
Eckstein DJ, Shur BD (1989) Laminin induces the stable expression of surface galactosyltransferase on lamellopodia of migrating cells. J Cell Biol 108: 2507–2517

Evans CP, Walsh DS, Kohn EC (1991) An autocrine motility factor secreted by the Dunning R-3327 rat prostatic adenocarcinoma cell subtype AT2.1. Int J Cancer 49: 109–113
Fernandez JLR, Geiger B, Saloman D, Ben-Ze'ev A (1993) Suppression of vinculin expression by antisense transfection confers changes in cell morphology, motility, and anchorage-dependent growth of 3T3 cells. J Cell Biol 122: 1285–1294
Fidler IJ (1990) Critical factors in the biology of human cancer metastasis. 28th GHA Clowes memorial award lecture. Cancer Res 50: 6130–6138
Fidler IJ, Gersten DM, Hart IR (1988) The biology of cancer invasion and metastasis. Adv Cancer Res 28: 149–250
Garcia-Castro I, Mato JM, Vasanthakumar G, Wiesmann W, Schiffmann E, Chiang PK (1983) Paradoxical effects of adenosine on neutrophil chemotaxis. J Biol Chem 258: 4345–4349
Geiger B, Volk T, Raz A (1985) Cell contacts and cytoskeletal organization of metastatic cell variants. Exp Biol Med 10: 39–53
Gherardi E, Gray J, Stoker M, Perryman M, Furlong R (1989) Purification of Scatter factor, a fibroblast-derived basic protein that modulates epithelial interactions and movement. Proc Natl Acad Sci USA 86: 5844–5848
Gilman AG (1987) G proteins: transducers of receptor-generated signals. Annu Rev Biochem 56: 615–649
Grey AM, Schor AM, Rushton G, Ellis I, Schor SL (1989) Purification of the migration stimulating factor produced by fetal and breast cancer patient fibroblasts. Proc Natl Acad Sci USA 86: 2438–2442
Grossi IM, Fitzgerald LA, Umbarger LA, Nelson KK, Diglio CA, Taylor JD, Honn KV (1989) Bidirectional control of membrane expression and/or activation of the tumor cell IRGpIIb/IIIa receptor and tumor cell adhesion by lipoxygenase products of arachidonic acid and linoleic acid. Cancer Res 49: 1029–1037
Grotendorst GR (1984) Alteration of the chemotactic response of NIH/3T3 cells to PDGF by growth factors, transformation, and tumor promoters. Cell 36: 279–285
Guirguis R, Margulies I, Taraboletti G, Schiffmann E, Liotta L (1987) Cytokine-induced pseudopodial protrusion is coupled to tumor cell migration. Nature 329: 261–263
Guirguis R, Schiffman E, Lui B, Birkbeck D, Engel J, Liotta LA (1988) Detection of autocrine motility factor in urine as a marker of bladder cancer. J Natl Cancer Inst 80: 1203–1211
Hall CL, Wang C, Lange LA, Turley EA (1994) Hyaluronan and the hyaluronan receptor RHAMM promote focal adhesion turnover and transient tyrosine kinase activity. J Cell Biol 126: 575–588
Hartmann G, Weidner KM, Schwartz H, Birchmeier W (1994) The motility signal of scatter factor/hepatocyte growth factor mediated through the receptor tyrosine kinase met requires intracellular action ras. J Biol Chem 269: 21936–21939
Hanks SK, Quinn AM, Hunter T (1988) The protein kinase family: conserved features and deduced phylogeny of the catalytic domains. Science 241: 42–52
Heffernan M, Yousefi S, Dennis JW (1989) Molecular characterization of P2B/LAMP-1, a major protein target of a metastasis-associated oligosaccharide structure. Cancer Res 49: 6077–6084
Hendrix MJC, Wood WR, Seftor EA, Lotan D, Nakiyama M, Misiorowski RL, Seftor REB, Stetler-Stevenson WG, Bevacqua SJ, Liotta L, Sobel ME, Raz A, Lotan R (1990) Retinoic acid inhibition of human melanoma cell invasion through a reconstituted basement membrane and its relation to decreases in the expression of proteolytic enzymes and motility factor receptor. Cancer Res 50: 4121–4130
Herskowitz I, Marsh L (1987) Conservation of a receptor/signal transduction system. Cell 50: 995–996
Heuser J (1989) Changes in lysosome shape and distribution correlated with changes in cytoplasmic pH. J Cell Biol 108: 855–864
Hopkins CR, Gibson A, Shipman M, Miller K (1990) Movement of internalized ligand receptor complexes along a continuous endosomal reticulum. Nature 346: 335–339
Hotta H, Ross AH, Huebner K, Isobe M, Wendeborn S, Chao MV, Ricciardi RP, Tsujimoto Y, Croce CM, Koprowski H (1988) Molecular cloning and characterization of an antigen associated with early stages of melanoma turnor progression. Cancer Res 48: 2955–2962
Huang B, Xie Y, Raz A (1995) Identification of an upstream region that controls the transcription of the human autocrine mobility factor receptor. Biochem Biophys Res Commun 212: 727–742
Ikezu T, Okamoto T, Ogata E, Nishimoto I (1992) Amino acids 356–372 constitute a G_1-activator sequence of the α_2-adrenergic receptor and have a Phe substitute in the G protein-activator sequence motif. FEBS Lett 311: 29–32
Ishihara A, Holifield B, Jacobson K (1988) Analysis of lateral redistribution of a monoclonal antibody complex plasma membrane glycoprotein which occurs during cell locomotion. J Cell Biol 106: 329–343

Ishisaki A, Oida S, Momose F, Amagasa T, Rikimaru K, Ichijo H (1994) Identification and characterization of autocrine-motility-factor-like activity in oral squamous-cell-carcinoma cells. Int J Cancer 59: 783–788

Kanbe K, Chigira M, Watanabe H (1994) Effects of protein kinase inhibitors on the cell motility stimulated by autocrine motility factor. Biochim Biophys Acta 1222: 395–399

Klominek J, Robert K-H, Sundqvist KG (1993) Chemotaxis and haptotaxis of human malignant mesothelioma cells: effects of fibronectin, laminin, type IV collagen, and an autocrine motility factor-like substance. Cancer Res 53: 4376–4382

Kohn EC, Liotta LA, Schiffmann E (1990) Autocrine motility factor stimulates a three-fold increase in inositol trisphosphate in human melanoma cells. Biochem Biophys Res Comm 166: 757–764

Kornfeld S, Mellman I (1989) The biogenesis of lysosomes. Annu Rev Cell Biol 5: 483–525

Kucik DF, Elson EL, Sheetz MP (1990) Cell migration does not produce membrane flow. J Cell Biol 111: 1617–1622

Kucik DF, Kuo SC, Elson EL, Sheetz MP (1991) Preferential attachment of membrane glycoproteins to the cytoskeleton at the leading edge of lamella. J Cell Biol 114: 1029–1036

Lam WC, Delikatny J, Orr FR, Wass J, Varani J, Wand PA (1981) The chemotactic response of tumor cells. A model for cancer metastasis. Am J Pathol 104: 69–76

Larsson C, Bystrom C, Skoog L, Rotstein S, Nordenskjold M (1990) Genomic alterations in human breast carcinomas. Genes Chromosom Cancer 2: 191–197

Liotta LA, Mandler R, Murano G, Katz DA, Gordon RK, Chiang PK, Schiffmann E (1986) Tumor cell autocrine motility factor. Proc Natl Acad Sci USA 83: 3302–3306

Lippincott-Schwartz J, Fambrough DM (1986) Lysosomal membrane dynamics: Structure and interorganelle movement of a major lysosomal membrane glycoprotein. J Cell Biol 102: 1593–1605

Lippincott-Schwartz J, Fambrough DM (1987) Cycling of the integral membrane glycoprotein, LEP100, between plasma membrane and lysosomes: kinetic and morphological analysis. Cell 49: 669–677

Lippman SM, Kessler JF, Meyskens FL (1987) Retinoids as preventive and therapeutic anticancer agents. Cancer Treat Rep 71: 391–405

Loetscher H, Pan YCE, Lahm HW, Gentz R, Brockhaus M, Tabuchi H, Lesslauen W (1990) Molecular cloning and expression of the human 55 KDa tumor necrosis factor receptor. Cell 61: 351–359

Lotan R (1980) Effects of vitamin A and its analogs (retinoids) on normal and neoplastic cells. Biochim Biophys Acta 605: 33–91

Lotan R, Amos B, Watanabe H, Raz A (1992) Suppression of motility factor receptor expression by retinoic acid. Cancer Res 52: 4878–4884

Mansouri A, Spurr N, Goodfellow PN, Kemler R (1988) Characterization and chromosomal localization of the gene encoding the human cell adhesion molecule uvomorulin. Differentiation 38: 67–71

Marchalonis JJ, Schlutter SF (1989) Evolution of variable and constant domains and joining segments of rearranging immunoglobulins. FASEB J 3: 2469–2479

Marcus PI (1962) Dynamics of surface modification in myxovirus-infected cells. Cold Spring Harbor Symp Quant Biol XXVII: 351–365

Matsumoto K, Hashimoto K, Yoshikawa K, Nakamura T (1991) Marked stimulation of growth and motility of human keratinocytes by hepatocyte growth factor. Exp Cell Res 196: 114–120

Matteoni R, Kreis TE (1987) Translocation and clustering of endosomes and lysosomes depends on microtubules. J Cell Biol 105: 1253–1265

McCarthy JB, Palm SL, Furcht LT (1983) Migration by haptotaxis of a schwann cell tumor line to the basement membrane glycoprotein laminin. J Cell Biol 97: 772–777

McCarthy JB, Basara ML, Palm SL, Sas F, Furcht TL (1985) The role of cell adhesion proteins laminin and fibronectin in the movement of malgnant and metastatic cells. Cancer Metastasis Rev 4: 125–152

McCarthy JB, Hagen ST, Furcht LT (1986) Human fibronectin contains distinct adhesion- and motility-promoting domains for metastatic melanoma cells. J Cell Biol 102: 179–188

Metzelaar MJ, Wijngaard PLJ, Peters PJ, Sixma JJ, Nieuwenhuis HK, Clevers HC (1991) CD63 antigen. J Biol Chem 266: 3239–3245

Mohler JL, Partin AW, Isaacs JT, Coffey DS (1987) Time lapse videomicroscopic identification of Dunning R-2237 adenocarcinoma and normal rat prostate cells. J Urol 137: 544–547

Mohler JL, Partin AW, Coffey DS (1988) Metastatic potential prediction by a visual grading system of cell motility: prospective validation in the Dunning R-3327 prostatic adenocarcinoma model. Cancer Res 48: 4312–4317

Morton RA, Ewing CM, Nagafuchi A, Tsukita S, Isaacs WB (1993) Reduction of E-cadherin levels and deletion of the α-catenin gene in human prostate cancer cells. Cancer Res 53: 3585–3590

Nabeshima K, Kataska H, Koona M (1986) Enhanced migration of tumor cells in response to collagen degradation products and tumor cell collagenolytic activity. Invasion Metastasis 6: 270–286
Nabi IR, Raz A (1987) Cell Shape modulation alters glycosylation of a metastatic melanoma cell surface antigen. Int J Cancer 40: 396–401
Nabi IR, Raz A (1988) Loss of metastatic responsiveness to cell shape modulation in a newly characterized B16 melanoma adhesive variant. Cancer Res 48: 1258–1264
Nabi IR, Watanabe H, Raz A (1990) Identification of B16-F1 melanoma autocrine motility-like factor receptor. Cancer Res 50: 409–414
Nabi IR, Le Bivic A, Fambrough DF, Rodriguez-Boulan E (1991) An endogenous MDCK membrane lysosomal glycoprotein is targeted basolaterally before delivery to lysosomes. J Cell Biol 115: 1573–1584
Nabi IR, Watanabe H, Raz A (1992) Autocrine motility factor and its receptor: role in cell locomotion and metastasis. Cancer Metastasis Rev 11: 5–20
Nicolson GL (1988) Organ specificity of tumor metastasis: role of preferential adhesion, invasion and growth of malignant cells at secondary sites. Cancer Metastasis Rev 7: 143–188
Nakamori S, Watanabe H, Kameyama M, Imaoka S, Furukawa H, Ishikawa O, Sasaki Y, Kabuto T, Iwanaga T, Ishiguro S, Raz A (1994) Expression of autocrine motility factor receptor, gp78, in colorectal cancer: a predictor of disease recurrence. Cancer 74: 1855–1862
Ochieng J, Basolo F, Albinin A, Melchiori A, Watanabe H, Elliot J, Raz A, Parodi S, Russo J (1991) Increased invasive, chemotactic and locomotive abilities of c-Ha-ras-transformed human breast epithelial cells. Invasion Metastasis 11: 38–47
Okamoto T, Katada T, Murayama Y; Ui M, Ogata E, Nishimoto I (1990) A simple structure encodes G protein-activating function of the IGF-II/mannose 6-phosphate receptor. Cell 62: 709–717
Okamoto T, Nishimoto I (1991) Analysis of stimulation-G protein subunit coupling by using active insulin-like growth factor II receptor peptide. Proc Natl Acad Sci USA 88: 8020–8023
O'Malley B (1990) The steroid receptor superfamily: more excitement predicted for the future. Mol Endocrinol 4: 363–369
Otto T, Birchmeier W, Schmidt U, Hinke A, Schipper J, Rubben H, Raz A (1994) Inverse relation of E-cadherin and autocrine motility factor receptor expression as a prognostic factor in patients with bladder carcinomas. Cancer Res 54: 3120–3123
Parsons JT, Weber MJ (1989) Genetics of src: structure and functional organization of a protein tyrosine kinase. Curr Top Microbiol Immunol 147: 79–127
Partin AW, Schoeniger JS, Mohler JL, Coffey DS (1989) Fourier analysis of cell motility: correlation of motility with metastatic potential. Proc Natl Acad Sci USA 86: 1254–1258
Pastan IH, Willingham MC (1981) Journey to the center of the cell: role of the receptosome. Science 214: 504–509
Poste G (1982) Experimental systems for analysis of the malignant phenotype. Cancer Metastasis Rev 1: 141–199
Raghavan D, Shipley W, Garnick MB, Russell PJ, Richie JP (1990) Bioloogy and management of bladder cancer. N Engl J Med 322: 1129–1138
Raz A, Ben-Ze'ev A (1983) Modulation of the metastatic capability in B16 melanoma by cell shape. Science 221: 1307–1310
Raz A, Ben-Ze'ev A (1987) Cell-contact and architecture of malignant cells and their relationship to metastasis. Cancer Metastasis Rev 6: 3–21
Raz A, Geiger B (1982) Altered organization of cell-substrate contacts and membrane associated cytoskeleton in tumor cell variants exhibiting different metastatic capabilities. Cancer Res 42: 5183–5190
Rogalski AA, Bergmann JE, Singer SJ (1984) Effect of microtubule status on the intracellular processing and surface expression of an integral protein of the plasma membrane. J Cell Biol 99: 1101–1109
Rosen EM, Meromsky L, Setter E, Vinter DW, Goldberg ID (1990) Quantitation of cytokine-stimulated migration of endothelium and epithelium by a new assay using microcarrier beads. Exp Cell Res 186: 22–31
Rotrosen D, Gallin JI, Spiegel AM, Malech HI (1988) Subcellular localization of $G_{i\alpha}$ in human neutrophils. J Biol Chem 263: 10958–10964
Ryan GB, Borysenko JZ, Karnovsky MJ (1974) Factors affecting the redistribution of surface-bound concanavalin A on human polymorphonuclearr leukocytes. J Cell Biol 62: 351–365
Samuels M, Ezzell RM, Cardozo TJ, Critchley DR, Coll J-L, Adamson ED (1993) Expression of chicken vinculin complements the adhesion-defective phenotype of mutant mouse F9 embryonal carcinoma cell. J Cell Biol. 121: 909–921

Schall TJ, Lewis M, Koller KJ, Lee A, Rice GC, Wong GHW, Gatanaga T, Granger GA, Lentz R, Raab H, Kohr WJ, Goeddel DV (1990) Molecular cloning and expression of a receptor for human tumor necrosis factor. Cell 61: 361–370

Schaller MD, Borgman CA, Cobb BS, Vines RR, Reynolds AB, Parsons JT (1992) pp125^{FAK}, a structurally distinctive protein-tyrosine kinase associated with focal adhesions. Proc Natl Acad Sci USA 89: 5192–5196

Schipper JH, Frixen UH, Behrens J, Unger A, Jahnke K, Birchmeier W (1991) E-cadherin expression in squamous cell carcinomas of the head and neck: inverse correlation with tumor dedifferentiation and lymph node metastasis. Cancer Res 51: 6328–6337

Schliwa M, Euteneuer U, Bulinski JC, Izant JG (1981) Calcium liability of cytoplasmic microtubules and its modulation by microtubule-associated proteins. Proc Natl Acad Sci USA 78: 1037–1041

Schor SL, Schor AM, Grey AM, Rushton G (1988) Foetal and cancer patient fibroblasts produce an autocrine migration stimulating factor not made by normal adult cells. J Cell Sci 90: 391–399

Schreiner GF, Braun J, Unanue ER (1976) Spontaneous redistribution of surface immunoglobulin in the motile B lymphocyte. J Exp Med 144: 1683–1688

Schwartz GK, Redwood M, Ohnuma T, Holland JF, Droller MJ, Liu BCS (1990) Inhibition of invasion human bladder carcinoma cells by protein kinase C inhibitor staurosporine. J Natl Cancer Inst 82: 1753–1756

Seiki M, Sato H, Liotta LA, Schiffmann E (1991) Comparison of autocrine mechanisms promoting motility in two metastatic cell lines: human melanoma and ras-transfected NIH-3T3 cells. Int J Cancer 49: 717–720

Shefcyk J, Yassin R, Volpi M, Molski TFP, Naccache PH, Munoz JJ, Becken EL, Feinstein MB, Sha'afi RI (1985) Pertussis toxin but not cholera toxin inhibits the stimulated increase in actin association with the cytoskeleton in rabbit neutrophils: role of the "G proteins" in stimulation-response coupling. Biochem Biophys Res Comm 126: 174–181

Shermen ME (ed) (1986) Retinoids and cell differentiation. CRC Press, Boca Raton

Silletti S, Raz A (1993) Autocrine motility factor is a growth factor. Biochem Biophys Res Commun 194: 446–457

Silletti S, Watanabe H, Hogan V, Nabi IR, Raz A (1991) Purification of B16-F1 melanoma autocrine motility factor and its receptor. Cancer Res 51: 3507–3511

Silletti S, Yao J, Sanford J, Mohammed AN, Otto T, Wolman SR, Raz A (1993) Autocrine motility factor receptor in human bladder carcinoma: gene expression, loss of cell-contact regulation and chromosomal mapping. Int J Oncology 3: 801–807

Silletti S, Timar J, Honn KV, Raz A (1994) Autocrine motility factor induces differential 12-lipoxygenase expression and activity in high and low-metastatic K-1735 melanoma cell variants. Cancer Res 54: 5752–5756

Silletti S, Paku S, Raz A (1995) Focal adhesion, cytoskeletal and intracellular signaling proteins are differentially regulated during autocrine motility factor-induced angiosarcoma cell motile responses (submitted)

Singer SJ, Kupfer A (1986) The directed migration of eukaryotic cells. Annu Rev Cell Biol 2: 337–365

Small JW, Isenberg G, Celis JE (1978) The polarity of actin at the leading edge of cultured cells. Nature 272: 638–639

Smith CD, Cox CC, Snyderman R (1986) Receptor coupled activation of phosphoinositide-specific phospholipase C by a G protein. Science 232: 97–100

Soussi T, de Fromental CC, May P (1990) Structural aspects of the p53 protein in relation to gene evolution. Oncogene 5: 945–952

Sporn MB, Roberts AB, Goodman DS (eds) (1984) The retinoids. Academic, New York

Steinman RM, Mellman IS, Muller WA, Cohn ZA (1983) Endocytosis and the recycling of plasma membrane. J Cell Biol 96: 1–27

Stoker M, Gherardi E, Oerryman M, Gray J (1987) Scatter factor is a fibroblast-derived modulator of epithelial cell mobility. Nature 327: 239–242

Stracke ML, Guirguis R, Liotta LA, Schiffmann E (1987) Pertusis toxin inhibits stimulated motility independently of the adenylate cyclase pathway in human melanoma cells. Biochem Biophys Res Commun 164: 339–345

Stracke ML, Krutzsch HC, Unsworth EJ, Arestad A, Cioce V, Schiffmann E, Liotta LA (1992) Identification, purification, and partial sequence analysis of autotaxin, a novel motility-stimulating protein. J Biol Chem 267: 2524–2529

Strauli P, Weiss L (1977) Cell locomotion and tumor penetration. Eur J Cancer 13: 1–12

Swanson J, Bushnell A, Silverstein SC (1987) Tubular lysosome morphology and distribution within macrophages depend on integrity of cytoplasmic microtubules. Proc Natl Acad Sci USA 84: 1921–1925

Taraboletti G, Roberts DD, Liotta L (1987) Thrombospondin-induced tumor cell migration: haptotaxis and chemotaxis are mediated by different molecular domains. J Cell Biol 105: 2409–2415
Theriot JA, Mitchison TJ (1991) Actin microfilment dynamics in locomoting cells. Nature 352: 126–131
Timar J, Silletti S, Bazaz R, Raz A, Honn KV (1993) Regulation of melanoma cell motility by the lipoxygenase metabolite 12-(*S*)-HETE. Int J Cancer 55: 1003–1010
Ullrich A, Schlessinger J (1990) Signal transduction by receptors with tyrosine kinase activity. Cell 61: 203–212
Umbas R, Schalken JA, Aalders TW, Cater BS, Karthaus HFM, Schaafsma HE, Debruyne FMJ, Isaacs WB (1992) Expression of the cellular adhesion molecule E-cadherin is reduced or absent in high-grade prostate cancer. Cancer Res 52: 5104–5109
Van Roy FV, Mareel M (1992) Tumor invasion: effects of cell adhesion and motility. Trends Cell Biol 2: 163–169
Verghes MW, Smithy CD, Snyderman R (1985) Potential role for a guanine nucleotide regulatory protein in chemoattractant receptor mediated polyphosphoinositide metabolism, Ca^{++} mobilization and cellular responses in leukocytes. Biochem Biophys Res Commun 127: 450–457
Volk T, Geiger B, Raz A (1984) Motility of high and low metastatic neoplastic cells. Cancer Res 44: 811–824
von Heijne G (1986) A new method for predicting signal sequence cleavage sites. Nucleic Acids Res 14: 4683–4690
Watanabe H, Carmi P, Hogan V, Raz T, Silletti S, Nabi IR, Raz A (1991a) Purification of human tumor cell autocrine motility factor and molecular cloning of its receptor. J Biol Chem 266: 13442–13448
Watanabe H, Nabi IR, Raz A (1991b) The relationship between motility factor receptor internalization and the lung colonization capacity of murine melanoma cells. Cancer Res 51: 2699–2705
Watanabe H, Kanbe K, Chigara M (1994) Differential purification of autocrine motility factor derived from a murine protein-free fibrosarcoma. Clin Exp Metastasis 12: 155–163
Weidner KM, Behrens J, Vanderkerchove J, Birchmeier W (1990) Scatter factor: molecular characteristics and effect on the invasiveness of epithelial cells. J Cell Biol 111: 2097–2108
Weidner KM, Arakai N, Hartmann G, Vanderkerchove J, Weingart S, Reider H, Fonatsch C, Tsubouchi H, Hishida T, Daikuhara Y, Birchmeier W (1991) Evidence for the identity of human scatter factor and human hepatocyte growth factor. Proc Natl Acad Sci USA 88: 7001–7005
Weinberg RA (1991) Tumor suppressor genes. Science 254: 1138–1146
Woodgett JR, Gould KL, Hunter T (1986) Substrate specificity of protein kinases C. Use the synthetic peptides corresponding to physiological sites as probes for substrate recognition requirements. Eur J Biochem 161: 177–184
Yoshida K, Ozaki T, Ushijina K, Hayashi H (1970) Studies on the mechanisms of invasion in cancer. 1. Isolation and purification of a factor chemotactic for cancer cells. Int J Cancer 6: 123–132
Young MR, Newby M, Meunier J (1985) Relationships between morphology, dissemination, migration, and prostaglandin E_2 secretion by cloned variants of Lewis lung carcinoma. Cancer Res 45: 3918–3923
Zvibel I, Raz A (1985) The establishment and characterization of a new Balb/c angiosarcoma tumor system. Int J Cancer 36: 261–272

Metastasis-Related *mts1* Gene

E.M. LUKANIDIN[1] and G.P. GEORGIEV[2]

1 Introduction 171
2 Isolation 173
3 Transcription in Mouse Normal and Tumor Cells 173
4 Structural Organization 175
5 Regulation of *mts1* Gene Expression 180
6 Transfection 183
7 mts1 Protein and Other Ca^{2+} Binding Proteins of the S100 Subfamily 184
8 Conclusion 190
References 190

1 Introduction

Metastasis is a multistep process involving detachment of cells from the primary tumor, invasion and penetration of the basement membranes of blood vessels, survival in the blood stream, extravasation, and growth of secondary tumors in distant organs (LIOTTA et al. 1983, 1991; SHIRRMACHER 1985; NICOLSON 1988; HART and SAINI 1992). Many proteins have been identified which may play a role in the different stages of tumor progression and metastasis and which distinguish metastatic and nonmetastatic cell populations (STETLER-STEVENSON et al 1993). Somatic cell hybridization (Tuck et al. 1991; RAMSHAW et al. 1983) and transfection of DNA from metastatic cells into nonmetastatic cells suggested a genetic involvement in the metastatic process (THORGEIRSSON et al. 1985; BERNSTEIN and WEINBERG 1985).

The expression of several genes was shown to be important during tumor progression (NICOLSON 1991; MUSCHEL and MCKENNA 1989; CHAMBERS and TUCK 1993). For example, the expression of certain oncogenes was found to correlate with the metastatic phenotype in some cases (MUSCHEL and MCKENNA 1989;

[1]Department of Molecular Cancer Biology, Division of Cancer Biology, Danish Cancer Society, Strandboulevarden 49, Bldg. 4.3, 2100 Copenhagen, Denmark
[2]Institute of Gene Biology, Russian Academy of Sciences, Vavilov Str. 34/5, Moscow, Russia

YOKOTA et al. 1986; CHAMBERS and TUCK 1988), and direct DNA transfer implicated oncogenes in tumor progression. However, in other systems oncogene transfection had no effect upon metastasis or other parameters of tumor progression (NICOLSON et al 1990). A high level of oncogene expression alone is usually not sufficient to change the metastatic phenotype of tumor cells (KLEIN and KLEIN 1986). It is more likely that oncogene expressioon regulates the expression of some other genes that are required for tumor progression (DE VOUGE and MUKHERJEE 1992).

Using differential or subtractive hybridization of cDNA libraries, several groups have identified genes whose expression correlates with tumor progression (STEEG et al. 1988; DEAR et al. 1988, 1989; PHILLIPS et al. 1990; BASSET et al. 1990; PENCIL et al. 1993). Some of these genes are specifically expressed in metastatic cells, while the others are preferentially expressed in nonmetastatic ones. Thus, several genes suspected of controlling metastatic process have been described.

For example, the *pGM21* gene is characterized by strong overexpression in a highly metastatic rat mammary adenocarcinoma cell line (PHILLIPS et al. 1990). The protein product of the *pGM21* gene has a homology to human elongation factor-1α (EF-1α). This product plays an important role in the motility of metastatic cells by binding to actin filaments (YANG et al. 1990). A good correlation was also found between the levels of *EF-1α* gene expression and metastatic ability in *fos*-transfected rat fibroblasts and in a mouse melanoma cell line (TANIGUCHI et al. 1992).

The transin gene was be induced in rat embryo fibroblast transformed with polyoma virus (MATRISIAN et al. 1985) and was overexpressed in invasive squamous cell carcinomas (MATRISIAN et al. 1986).

The *Mta-1* gene was isolated from a rat mammary adenocarcinoma metastatic system. The expreession of *Mta-1* was found to be four times higher in metastatic cells than in related non-metastatic cells (TOH et al. 1994).

The *Meta-1* gene, the spliced variant of a *CD44* gene, was found to be specifically expressed in different metastatic tumors and cell lines (HEIDER et al. 1993a,b; HOFMANN et al. 1991). Transfection of *Meta-1* into nonmetastatic cell lines converts these cells into rapidly spreading metastatic tumor cells (GÜNTHERT et al. 1991).

The combination of two experimental approaches, retroviral insertional mutagenesis and in vitro selection for invasive cell variant, enabled the isolation of a gene named *Tiam-1*. Transfection with the *Tiam-1* gene induced invasiveness of noninvasive cells. Invasive clones produced experimental metastases in nude mic (HABETS et al. 1994).

In addition to the genes that induce metastasis, suppressor genes have been discovered. One member of this group is an *nm23* gene which has been isolated by the method of differential hybridization as a gene that is selectively expressed in low-metastatic murine melanoma cells (STEEG et al. 1988). Amino acid sequence analysis shows that the *nm23* gene encodes nucleoside diphosphate kinase (GILLES et al. 1991). Transfection of the *nm23* gene into metastatic cells reduces the metastatic ability of transfectants (LEONE et al. 1991).

The substractive hybridization approach made it possible to isolate two other metastasis suppressor genes, *WDNM1* and *WDNM2*, whose transcription was down regulated in rat metastatic mammary adenocarcinoma cells (DEAR et al. 1988, 1989). Sequence analysis showed no homology to known genes.

Thus, although the biological function and role of metastasis-related genes in tumor progression and metastasis is still unknown, it is clear that phenotypic changes associated with metastasis reflect chages in the activity of many different genes. It is probable that many new genes which are involved in tumor progression will be isolated in the future and study of these genes will lead to the discovery of regulatory mechanisms which are associated with the progression and metastasis of human cancer. In this review we will summarize the data on one such gene, *mts1*, which seems to be involved in the regulation of the malignancy of tumor cells.

2 Isolation

To identify genes that are differentially expressed in metastatic and nonmetastatic tumor cells, we used two related murine cell lines. Originally, these cell lines were obtained from spontaneous mouse mammary (SENIN et al. 1983). One of these cell lines, CSML-100, was characterized by a high ability to metastasize into the lungs of subcutaneously injected animals. Its counterpart is the CSML-0 cell line, which is unable to metastasize after subcutaneous transplantation (SENIN et al. 1984). To enrich the CSML-100 cDNA with molecules specific for the metastatic cell line, it was hybridized with the excess of a driver poly(A) RNA isolated from related nonmetastatic CSML-0 cells and the hybrids were removed. More than 95% of the CSML-100 cDNA hybridized with CSML-0 mRNA, while the remaining cDNA formed double-stranded (ds)-cDNA molecules. These molecules were used to obtain a cDNA library. After differential screening of the obtained library with cDNA probes derived from the mRNA of CSML-100 and CSML-0 cells, several clones were isolated which hybridized only with CSML-100 cDNA. One of them, designated *mts1*, hybridized only with cDNA trancribed from the mRNA of the CSML-100 cell line and was taken for further analysis (EBRALIDZE et al. 1989).

3 Transcription in Mouse Normal and Tumor Cells

We investigated the correlation of transcription of the *mts1* gene with the metastatic phenotype of different mouse tumors and cell lines. All the analyzed metastatic cell lines, except one, contained a unique 0.55-kb transcript which hybridized with a *mts1* probe in northern blot analysis. The level of *mts1* expression correlated to some extent with the metastatic potential of tumor cells: the cells that had a high level of *mts1* expression demonstrated a high degree of metastasis. The absence of *mts1* expression in one of the highly metastatic cell lines suggested another regulatory mechanism for switching on metastasis in this particular cell line (Table 1).

The *mts1* gene expression was also detected in normal cells. Different mouse organs were analyzed for *mts1* expression. A rather high level of 0.55-kb mRNA was detected in the spleen, thymus, bone marrow, and bloodstream lymphocytes of mouse. Recently, we found that the *mts1* gene was also highly expressed in the basal layer of proliferating skin keratinocytes, lung, colon, uterus, and ovary. Low but detectable levels of *mts1* expression were found in the kidney, testis, and skeletal muscle. The *mts1* gene expression could not be detected in mouse brain, liver, or heart (Fig. 1).

Table 1. mts1 RNA transcription in different mouse tumor cell lines

Source of RNA	mts1RNA transcription	Spontaneous metastases	Target organs
Mammary carcinosarcoma			
CSML-0	–	Nonmetastatic	–
CSML-50	+	50%	Lung
CSML-100	++	High[a]	Lung
Mammary solid carcinoma			
HMC-0	–	Low [b]	Liver
HMC-Lr	–	High [a]	Liver
Teratocarcinoma cell line			
PCC4-B	–	Nonmetastatic	–
PCC4-P	–	Nonmetastatic	–
PCC4-107	–	Nonmetastatic	–
C12-	–	Nonmetastatic	–
Embryocarcinoma, T-36	++	50%	Lymph node
Cell line derived from T-36			
T-36	+	50%	Lymph node
Embryocarcinoma, LMEC	+	High [a]	Lymph node
Teratocarcinoma, T-9	–	Low [a]	Lymph node
Colon adenocarcinoma			
Acatol	–	Nonmetastatic	–
Melanoma, B-16	–	Low [b]	Lung
Lung carcinoma, RL-67	+	High [a]	Lung
Lewis lung carcinoma, LLC	++	High [a]	Lung

[a]Multiple metastases in 100% of target organs.
[b]Solitary metastases in 20% of injected mice.

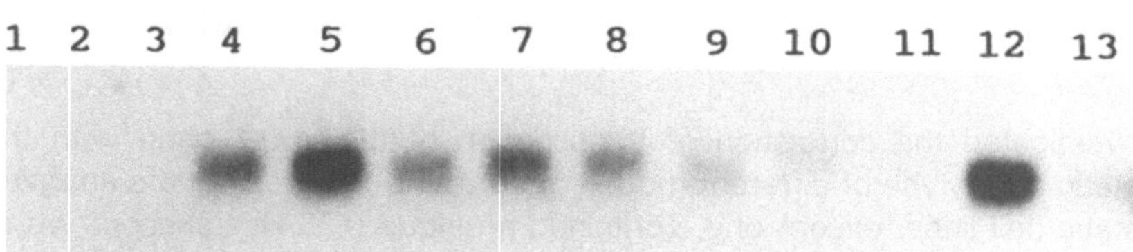

Fig. 1. Northern blot analysis of *mts1* mRNA expression in different mouse organs. Total cellular RNA was isolated from the following: *1* liver; *2* heart; *3* brain; *4* thymus; *5* spleen; *6* lung; *7* colon; *8* uterus; *9* mammary gland; *10* kidney; *11* testis; *12* ovary; *13* skeletal muscle. The RNA were loaded onto formaldehyde containing a agarose gel, blotted to a nitrocellulose membrane, and the membrane was hybridized to a *mts1* DNA probe

The data on the levels of *mts1* mRNA expression in different tissues are somewhat conflicting. For example, some groups found that the *mts1* gene was expressed in the heart, but not in the kidney and uterus (TAKENAGA et al. 1994a, b). The dicrepancy may be due to different conditions of cell growth or the sensitivity of methods used for detection of *mts1*-specific RNA. Indeed, we found that after passaging cells in vitro, the level of *mts1*-specific mRNA in the T36 carcinoma cell line was much lower than in the parental tumor (Table 1).

The expression of *mts1* in hematopoietic cells was analyzed in more detail. The level of *mts1* expression was quite high in T lymphocytes, granulocytes, macrophages, and corresponding cell lines. We could not find any expression of *mts1*-specific RNA in human B lymphocytes, mouse myeloma cell line Sp2/0, or in erythroid cells K562 and HEL92. Even polymerase chain reaction (PCR) amplification of Sp2/0 cDNA did not detect *mts1*-specific transcript (GRIGORIAN et al. 1994).

Interestingly, the hematopoietic cells that express the *mts1* gene mimic the properties of metastatic cells, such as invasiveness, high motility, and degradation of extracellular matrix components. Agents that modulate the funcitonal activity of macrophages also affect *mts1* gene expression. We found that lipopolysaccharide (LPS), tumor necrosis factor (TNF)-γ, concanavalin A (ConA), and modulation of Ca^{2+} ion concentration in the cells may increase or decrease *mts1* mRNA levels. The level of *mts1* expression seems to correlate with the invasive phenotype of hematopoietic cells; cells such as T lymphocytes, neutrophils, and macrophages demonstrate the highest level of *mts1* expression. It can be concluded that *mts1* gene expression also has an important role in the control of certain functions of normal cells.

4 Structural Organization

The *mts1* gene is present in the mouse haploid genome as a single copy. The genomic copies of mouse and human *mts1* gene were cloned and sequenced. Mouse *mts1* gene consists of three exons and two introns. Exon 1 (38 bp long) represents the nontranslated region of the gene. The second exon (156 bp) contains the start codon, which is localized 15 bp downstream from its 5'-end. The stop codon is localized in the middle of the third exon (302 bp). The third exon carries a polyadenylation signal and stretch of the GT and T which were shown to be essential for the formation of 3'-termini of mRNA (BIRNSTIEL et al. 1985). Acceptor and donor splicing sites of the introns coincide with the consensus sequences of these regions (MOUNT 1982).

To compare mouse and human *mts1* sequences, several human *mts1* cDNA-containing plasmids were isolated from the human osteosarcoma cDNA library. Two *mts1*-specific clones with different lengths of the cDNA insert were chosen for detailed analyses. The coding sequences of the isolated clones are almost identical, as are their 3'-nontranslated regions, and correspond to the structure of the mouse *mts1* gene. However, the 5'-ends of the molecules are different. The hu-*mts1* (variant) molecule differs from the hu-*mts1* mRNA because of the insertion of a 49-nt fragment (Fig. 2).

In order to clarify the origin of this 49-nt fragment and to determine the position of the exons of the hu-*mts1* gene, we cloned a genomic copy of the human *mts1* gene. A comparison of the sequence of genomic fragment with the sequences of hu-*mts1* and hu-*mts1* (var) cDNA enabled us to find the difference in intron/exon organization of human and mouse *mts1* gene.

The human *mts1* gene is composed of four exons separated by three introns of 232, 657, and 720 bp. The first and second exons, 54 and 49 bp long, are noncoding. The third and fourth exons, 156 and 302 bp long, contain the coding sequences of the *mts1* gene. Thus the structural organization of the human *mts1* gene coincides with that recently proposed for another memeber of the S100 subfamily of Ca^{2+} -binding proteins, the S100D gene (Engelkamp et al. 1993). The second noncoding exon in clone hu-*mts1* (var) is realized in cells only as an alternatively spliced variant.

Analysis of the distribution of the alternatively spliced variants of hu-*mts1* RNA in different human organs demonstrates a significant variation in its expression. The hu-*mts1* (var) is predominantly expressed in the colon and is almost completely absent in the liver. The hu-*mts1*(var) was not found in peripheral blood cells, i.e. leukocytes, neutrophils, macrophages, and lymphocytes. In the other tissues, where the expression of the *mts1* gene was detected, the variations are less prominent (Fig.3). The hu-*mts1*(var) retains the open reading frame (ORF) for the Mts-1 protein and does not contain any additional long ORF. Both mRNA were found in association with a polysomal fraction, suggesting that both spliced variants of the human *mts1* gene are translated in the cells. The physiological significance of the alternative splicing of the *mts1* gene in human cells has not been established.

In addition, the hu-*mts1*(var) is present in varying amounts in all human tumor cell lines expressing the *mts1* gene. In the human osteosarcoma cell line both spliced variants are present in equal amounts; in the human adrenal carcinoma cell line and in HeLa cells, the amount of the hu-*mts1*(var) is higher, whereas in human breast carcinoma and lung carcinoma the hu-*mts1* RNA is predominant (Fig.4). Changing of splicing specificity in tumor cells during progression was also shown for other genes such as fibronectin, tenascin (Oyama et al. 1989; Carnemolla et al. 1989; Sriramara and Bourdon 1993), and *CD44* (Günthert et al. 1991).

The *mts1* gene was independently cloned by several other groups under different names: *pEL98*, *18A2*, *42A*, *p9K*a, and *CAPL*. Clones *42A* and *p9Ka* represent the rat homologue of the *mts1* gene. *p9Ka* expression was found to be 15-fold more abundant in myoepithelial-like cells than in parental cuboidal epithelial stem cells (Barreclough et al. 1987). The *42A* gene was cloned by differential hybridization of a cDNA library from rat pheochromocytoma PC12 cells. The level of 42A RNA in PC12 cells treated with nerve growth factor (NGF) was elevated after 7 h, reached maximum by 24 h, and remained at a high level for 2 days (Masiakowski and Shooter 1988). Genes *pEL98* and *18A2* were cloned from

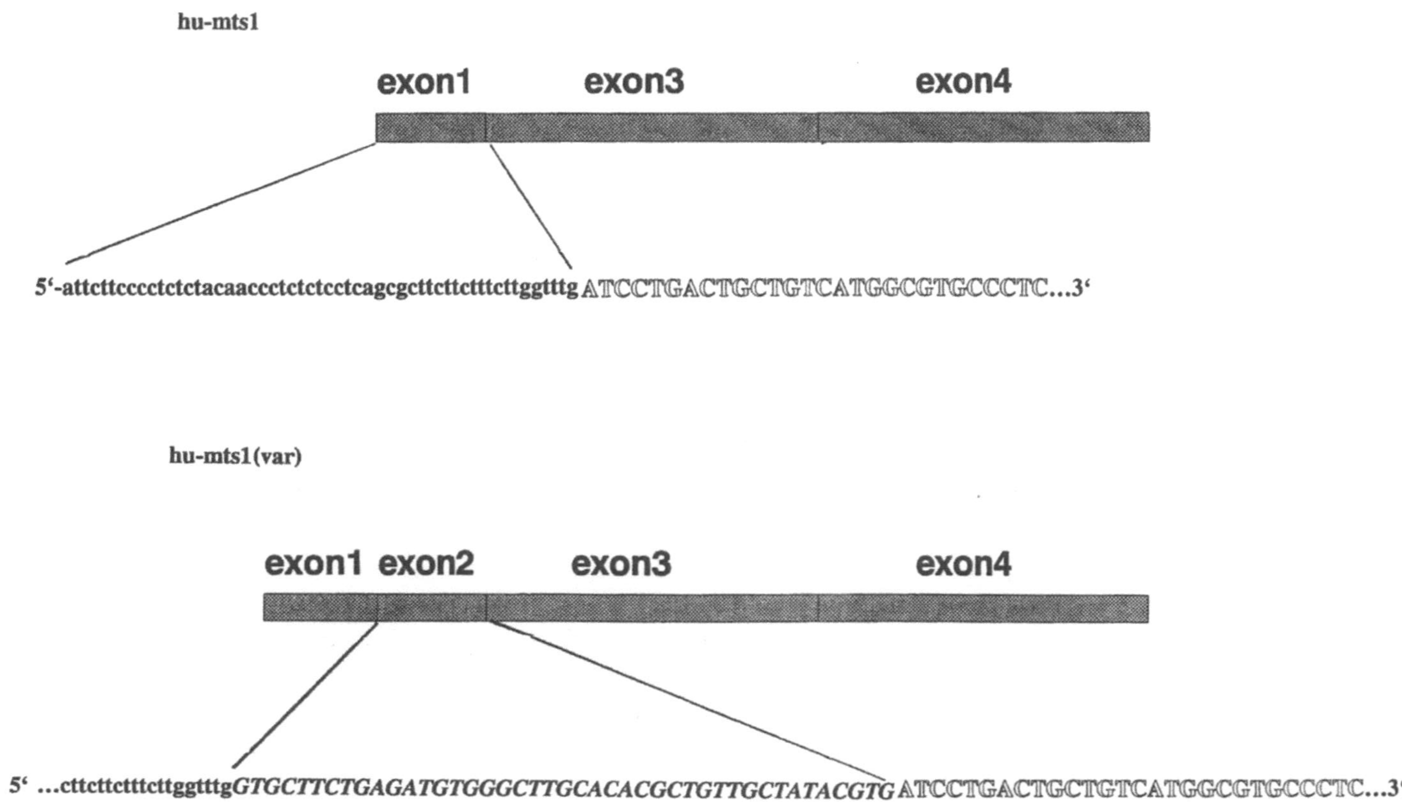

Fig. 2. Structural organization of two *mts1* specific mRNA. The sequences of the exon 1 and exon 3 are shown in *lower* and *upper case letters* respectively. The sequence of the exon 2 is shown in *italic upper case letters*

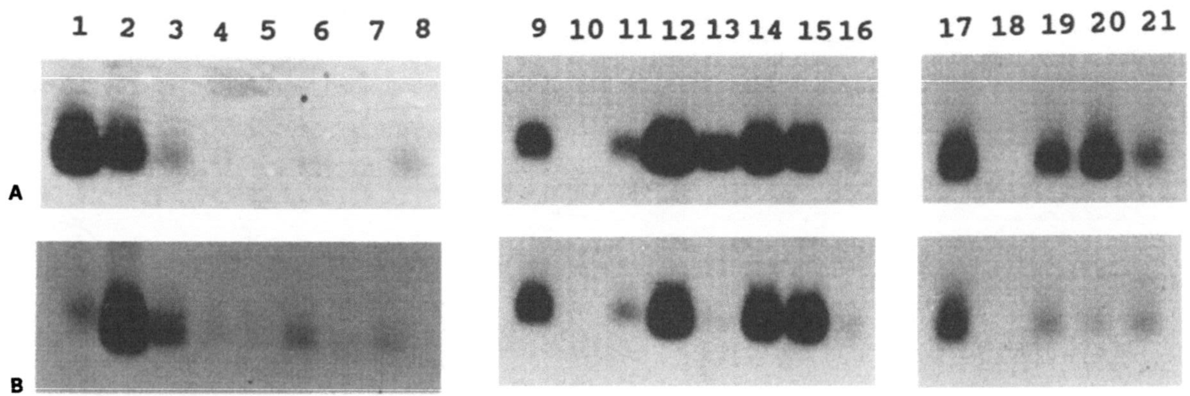

Fig. 3A,B. Detection of splice variant of *mts1* RNA in human organs. Northern blot analysis with the polymerase chain reaction (PCR)-amplified probes specific to **A** exon 1 and **B** exon 2. *1*, Peripheral blood leukocytes; *2*, colon; *3*, small intestine; *4*, ovary; *5*, testis; *6*, prostate; *7*, thymus; *8*, spleen; *9*, heart; *10*, brain; *11*, placenta; *12* lung; *13*, liver; *14*, skeletal muscle; *15*, kidney; *16*, pancreas; *17*, fetal heart; *18*, fetal brain; *19*, fetal lung; *20*, fetal liver; *21*, fetal kidney

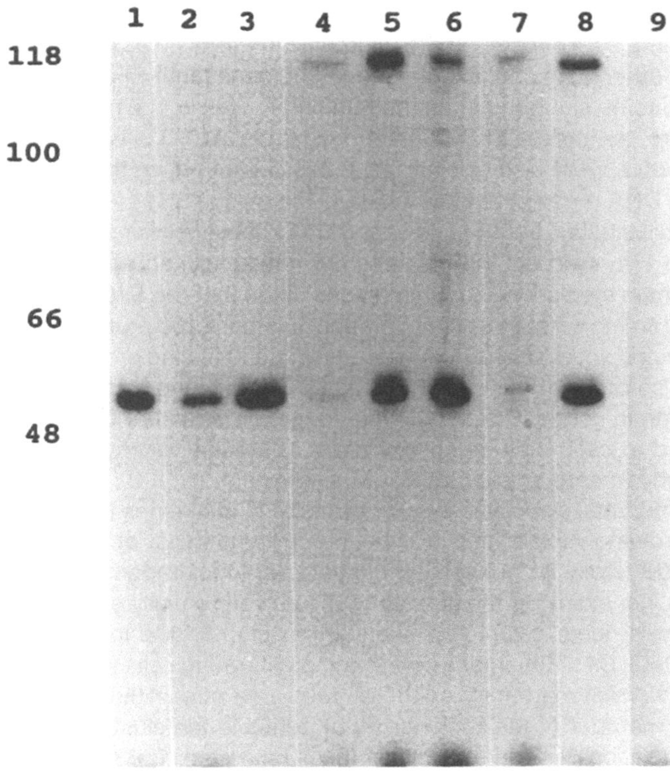

Fig. 4. Detection of spliced variants of *mts1* mRNA in human tumor cell lines and in normal hemopoietic cells. RNase protection assay of total RNA isolated from: *1*, neutrophils; *2*, lymphocytes activated with interleukin-2; *3*, macrophages; and from the following human tumor cell lines: *4*, TCAR (adrenal carcinoma); *5* OHS (osteosarcoma); *6* SK-BR-3 (breast carcinoma); *7*, HeLa; *8*, A549 (lung carcinoma); *9*, yeast RNA (negative control). The positions of the marker are indicated

mouse cells either as mRNA related to cell immortalization (*pEL98*) (GOTO et al. 1988) or as a serum-inducible mRNA in BALB/c 3T3 cells (*18A2*) (JACKSON GRUSBY et al. 1987). cDNA of the human homologue of the *mts1*, *CAPL*, was cloned from human heart using the polymerase chain reaction (ENGELKAMP et al. 1992). *CAPL* cDNA clone showed high (79%) sequence homology to the mouse *mts1* cDNA. An especially high degree of sequence homology (89%) was found in the translated part of cDNA. *CAPL* sequences of all three exons were identical to exon sequences of hu-*mts1* cDNA clone isolated from the osteosarcoma cDNA library. However, to our surprise, *CAPL* cDNA contained an additional 66 nt flanking the first exon which were not found in the genomic 650-bp fragment upstream of the hu-*mts1* gene. One explanation is that the variant of *CAPL* mRNA present in the heart is a result of a *trans*-splicing event. Clones *pEL98* and *18A2* are identical to the *mts1* gene, but the rat (*42A*) and human (*CAPL*) homologues of the *mts1* gene encode proteins which have four and seven amino acid substitutions, respectively.

The *mts1* gene was mapped on human chromosome 1 (ENGELKAMP et al. 1993) and mouse chromosome 3 (DORIN et al. 1990). Human *CAPL/mts1* was found in a cluster of at least five other genes belonging to the S100 gene family. Four out of six genes were found to be clustered in the human chromosome 1q21 region in the following head to tail order: 5'-*S100E-CAPL-S100D-CACY*-3'. Two other genes, S100L and S100α, were also colocalised in this cluster within the 450-kb fragment of genomic DNA (ENGELKAMP et al. 1993).

Three genes of this cluster, *S100E*, *CAPL*, and *CACY*, have similar structure, i.e., three exons and two introns. The first exon is not translated and coding sequences are present in the second and third exons. Exon 2 of the *S100D* gene encodes 13 additional amino acids that are not present in other S100 proteins, and the third and fourth exons encoded the body of a *S100D* Ca^{2+}-binding protein. The hu-*mts1* gene may also contain four exons and be structurally the same as the *S100D* gene. However, in contrast to *S100D*, the four-exon structure of the hu-*mts1* gene is realized in cells only as an alternatively spliced variant and the additional exon does not create a new ORF for Mts-1 protein.

In contrast to some other gene clusters, for example globin genes (KRUMLAUF 1992), which are coordinately expressed in a specific cell type (HIGGS et al. 1989), members of the *S100* family of genes demonstrate an independent specific pattern of expression. For example, the hu-*mts1* gene is preferentially expressed in peripheral blood cells, lung, colon, and skeletal muscle; *S100α* in neurons (SELINGFREUND et al. 1990); *CACY* in organs containing proliferating cells (GUO et al. 1990); and *S100L* in normal mammary epithelial cells (Lee et al. 1992). It was suggested that each member of the S100 family of genes in the cluster has its own independent regulation of gene expression (ENGELKAMP et al. 1993).

The long arm of chromosome 1 is characterized by a high degree of anomaly in human hematological and solid cancers. Duplication of the chromosomal segments of 1q11 and 1q32 was found to be associated with advantages in proliferation and metastasis formation in human leukemia B cell clones (GHOSE et al. 1990). Human gastric adenocarcinomas frequently undergo a loss of heterozygosity on chromosome 1q (SANO et al. 1991). Genetic alterations in breast cancer are often associated with chromosomal locus 1q21 (DEVILEE et al. 1991), 1q23–32 (CHEN et al. 1989), and 1q42–43 (GENDLER et al. 1990). These may have a prognostic value (WOLTMAN et al. 1992). PEDROCCHI et al.(1994a) examined expression of four members of the S100 family of genes which are clustered on human chromosome 1q21 in different human breast cancer cell lines and human breast cancer tissues. It was found that only expression of *CAPL/mts1* gene was specifically correlated with the tumorigenesis of human mammary cells.

5 Regulation of *mts1* Gene Expression

Several different approaches have been used for analysis of *mts1* transcriptional control. The first is computer analysis. Comparison of the 5'-flanking region of mouse and human genes shows a high degree of homology in the 135-bp region upstream of the *mts1* transcription. This region contains sequences homologous to the SV-40 enhancer and the promoter of the human prothrombin gene. In addition, the region of homology to the T cell specific enhancer of the CD3 gene is present in the first intron of *mts1*. The first intron also contains the sequences reminiscent of AP-1 and NFkB-binding sites, but differing from the typical sites.

The regions containing these sequences have been tested for binding of nuclear proteins and for induction of marker gene expression. By footprinting and gel retardation analysis, we found that, in vitro, the fragment with coordinates from –41 to –135 yields DNA/protein complexes with nuclear extracts from both metastatic and nonmetastatic cells. However, in chloramphenicol acetyltransferase (CAT) assay, only the 41-bp segment upstream of the cap site is sufficient for maximal levels of transcription from the *mts1* gene promoter in both metastatic and nonmetastatic cells (TULCHINSKY et al. 1992). Moreover, in vivo footprinting does not confirm the presence of proteins in the above-mentioned regions of the *mts1* gene.

Gel retardation shows that the fragment of the first intron containing homology with the CD3 enhancer also binds nuclear proteins from CSML-100 and CSML-0 cells. A 16-bp-long sequence in this fragment is protected by proteins, as shown by DNAaseI footprinting and methylation interference. CAT assay shows that the same 16-bp sequence possesses the properties of an enhancer. Again, it works in both CSML-100 and CSML-0 cells, confirming the presence of activating proteins in both types of cells.

However, in vivo footprinting demonstrates a difference between CSML-100 and CSML-0 cells. In vivo 16-bp oligonucleotide is protected by proteins in metastatic cell line. Thus an activating nuclear factor is present in both cell types, but only in CSML-100 binds with 16-bp oligonucleotide.

To explain this, the differences in the methylation state of CSML-100 and CSML-0 DNA have been analyzed. The *mts1* gene in CSML-0 cells is heavily methylated, while in CSML-100 cells it is not. Thus, binding of the nuclear factor that recognises the 16-bp sequence may depend on DNA methylation in its neighborhood. Molt-4, CEM, and Jurkat, which express different levels of *mts1* gene, were used for detail analysis of the influence of DNA methylation on the *mts1* gene expression in human lymphoma cell lines. Molt-4 cells show the highest levels of transcription, CEM cells show an intermediate level, and in the Jurkat cells the *mts1* gene-specific transcripts are not detectable. The DNA methylation status of the *mts1* gene in these cell lines was analyzed by hydrolysis of *mts1* DNA with methylation-sensitive and nonsensitive restriction endonucleases. We found that the methylation level of the regulatory region of the *mts1* gene (first intron) correlates with gene expression. For example in Jurkat cells, in which the *mts1* gene is not expressed, it is highly methylated (Table 2). It is possible to demethylate DNA by treating cells with 5'-azadeoxycytidine (Azad-C), an inhibitor of eukariotic DNA-methylase. When CEM or Jurkat cells are treated with Azad-C, the level of *mts1* gene transcription significantly increases (Fig. 5). Thus, DNA methylation may serve as an improtant mechanism in the regulation of

Table 2. The methylation status of the *HhaI* and *HpaII* sites spanning the *mts1* gene

Coordinates	MOLT-4	CEM	JURKAT	Lymphocytes	Macrophages	Neutrophils
HpaII (–4000)	?	?	+/–	+/–	+	+
HhaI (–3400)	–/+	+	+/–	–/+	–/+	–/+
HpaII (–2500)	?	–	+	+/–	+	+
HpaII (–800)	–	+	+	+/–	+	+
HhaI (+ 35)	–/+	+/–	+/–	–	–	–
HhaI (+ 386)	–/+	–/+	+/–	–	–	–
HhaI (+ 777)	–/+	–/+	+/–	–	–/+	–/+
HpaII (+1125)	–	?	?	+/–	+/–	+/–
HpaII (+1436)	+	+	+	+	+	+
HpaII (~+5000)	+	+	+	?	?	?

All sites are methylated; +/–, more than 50% of the sites are methylated; –/+, less than 50% of the sites are methylated; –, none of the sites are methylated; ?, the methylation status has not been studied.

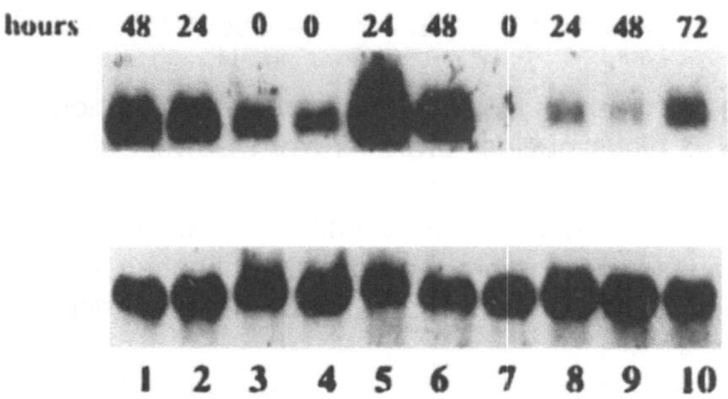

Fig. 5. Northern analysis of total cellular RNA isolated from MOL-4 (*lanes 1–3*), CEM (*lanes 4–6*), and Jurkat (*lanes 7–10*) cells. The cells were treated with 5'–azadeoxy-cytidine before RNA isolation for different periods, as indicated in the *top panel*. The filter was hybridized to an exon 2 *mts1* specific probe. The *bottom panel* show the reprobing of the same blot with human β-actin

mts1 gene transcription. There are many examples in which DNA methylation reduces or blocks transcription of different genes (Doerfler 1983). The mechanism of transcriptional silencing has been explained on the basis of either an indirect model in which densely methylated DNA is recognized by proteins that may displace crucial transcription factors (Sekler 1990) or a direct model in which binding of a single transcription factor is prevented by the presence of a methylated CpG dinucleotide localized in the sensitive region of a DNA motif (Iguchi-Ariga and Schaffner 1989).

Recently, the role of other sequences located within the first intron has been elucidated.

The first, AP-1-like binding site, has one base substitution compared with the consensus one that creates a CpG sequence. When methylated, it binds AP-1 proteins, while in a nonmethylated state it binds other proteins, as shown by the competition experiments. In the CAT assay the methylated AP-1 acts as a silencer.

The next site, a nontypical NFkB site, binds both NFkB and some other nuclear factor. This site acts as an enhancer element, and both proteins interacting with the NFkB-like binding site are involved in transcriptional activation.

The second AP-1 binding site seems to be a typical enhancer. In vivo footprinting shows that it is protected only in metastatic CSML-100 cells. CAT assay shows that deletion of NFkB sequences significantly reduces transcription, while deletion of both NFkB and AP-1 sequences completely abolishes it. The data suggest that NFkB and AP-1 sequences positively regulate expression of the *mts1* gene, but probably only AP-1 sequences are involved in differential regulation of *mts1* gene expression in metastatic cells.

Thus the complex regulatory mechanism is present in the first intron of the *mts1* gene, which includes enhancer and silencer elements that interact with well-known AP-1 and NFkB factors as well as with several novel regulatory proteins. In addition, DNA methylation plays an important role in transcription control.

6 Transfection

To demonstrate the association of the *mts1* gene with metastasis, DNA transfection experiments followed by in vivo spontaneous metastasis assay were performed. Two constructs containing the *mts1* gene in sense and antisense orientations were used. The transcription was controlled by strong non-specific elements such as Molony sarcoma virus long terminal repeat (LTR), cytomegalovirus enhancer, or hydroxymethyluracil (HMG)-protein enhancer. Clones obtained after transfection of metastatic CSML-100 cells with the antisense construct of the *mts1* gene, actively expressing antisense *mts1* RNA, demonstrated marked reduction of metastatic ability (Grigorian et al. 1993).

The sense constructions were tested with two cell lines, CSML-0 and Line-1. Transfection of CSML-0 cells with sense constructs enabled us to obtain the clones producing high amount of *mts1* mRNA. However, not all of them metastasize after subcutaneous transplantation to syngeneic mice. Surprisingly, neither clone produces Mts-1 protein, in spite of the presence of *mts1* mRNA.

The data suggest that the mechanism of downregulation of Mts1 protein translation exists in CSML-0 cells.

Line-1 is a highly metastatic cell line which expresses the *mts1* gene. However, after dimethylsulfoxide (DMSO) treatment, the level of *mts1* RNA is reduced as well as the metastatic ability of treated cells. In clones transfected with the sense construct of the *mts1* gene, DMSO treatment does not block transcription of exogeneously added *mts1* sequences, as this is controlled by viral LTR promoter. Exogenous *mts1* gene is highly expressed even in the presence of DMSO, and cells retain their high metastatic potential (Grigorian et al. 1993).

Another approach is to use *mts1*-specific ribozyme RNA molecules. Ribozyme was constructed that cuts *mts1* transcript in the second exon. Ribozyme construct was transfected into the metastatic human osteosarcoma cell line OHS. Several clones were examined for the presence of ribozyme-specific RNA and levels of *mts1* expression. The clones expressing ribozyme have a reduced level of *mts1* RNA expression. Their metastatic potential was examined by intracardiac injection in nude rats. Parental OHS cells usually cause bone marrow metastasis 20–25 days after injection. In contrast, the clones expressing *mts1* ribozyme demonstrated significant suppression of metastatic potential. Thus, expression of *mts1*-specific ribozyme can suppress the metastasis of OHS tumor cells (MAELANDSMO et al. 1995).

DAVIES et al. transfected the rat mammary epithelial Rama37 cell line with the *mts1* gene (DAVIES et al. 1993). Rama37 cells yield benign, nonmetastasizing tumors in rats and do not express the *mts1* gene. The clones expressiog high levels of the *mts1* mRNA and Mts-1 protein are able to produce tumors which metastasize into lung and lymph nodes. Taken together, these experiments suggest that the *mts1* gene is one of the key genes in the regulation of the metastatic phenotype of several tumor cells.

7 Mts1 Protein and Other Ca^{2+} Binding Proteins of the S100 Subfamily

The *mts1* gene encodes a 101-amino acid protein belonging to the S100 subfamily of Ca^{2+}-binding proteins, which are reported to be involved in the regulation of cell growth and differentiation by conversion of Ca^{2+} message signals into intracellular response. S100 subfamily proteins are cytosolic, acidic proteins that belong to the Ca^{2+}-binding protein family of the EF-hand type (KRETSINGER et al. 1991; HEIZMANN and HUNZIKER 1991). Two S100 proteins that have been more extensively characterized structurally are S100α and S100β polypeptides. Both peptides are approximately 10 kDa in size and share 58% amino acid and sequence indentity. S100α and S100β can form homodimeric and heterodimeric structures. The proteins obtained from different species and tissues have been characterized. The common feature of S100 proteins is the presence of a Ca^{2+}-binding domain which consists of two so-called E and F α-helices joined by a Ca^{2+}-binding loop. Each S100 protein contains two EF-hand Ca^{2+}-binding loops, but one of the loops contains 14 amino acids instead of 12, which is specific for S100 proteins (PERSECHINI et al. 1988).

The function of S100 proteins has not yet been established. S100α has been claimed to be involved in regulation of the pH and water balance in certain tissues (MOLIN et al. 1985) or regulation of the phosphorylation of some cytoskeletal proteins (HAGIVARA et al. 1988).

S100β was found to be identical by amino acid sequence analysis to a neurite extension factor (KLIGMAN and MARSHAK 1985). Some forms of S100β may be secreted and function as neurite growth factor in a primary culture of neurons and in a neuroblastoma cell line (KLIGMAN and HSIEH 1987). In the nervous system, S100 proteins are abundant in glial cells (DONATO 1991). The selective inhibition of S100 protein expression in a glial cell line by antisense techniques resulted in a more organized microfilament network (SELINGFREUND et al. 1990).

It has been suggested that S100 proteins play a general role in the Ca^{2+}-dependent regulation of the cytoskeleton (DONATO 1991). S100 proteins interact with two cytoskeletal proteins, caldesmon and calpactin, and regulate activities of these proteins (PRITCHARD and MARTSON 1991; BIANCHI et al. 1992). The binding of S100 protein to glial fibrillary acidic protein (GFAP), a major cytoskeleton constituent in glial cells, inhibits the polymerization of GFAP in a Ca^{2+} dependent manner (BIANCHI et al. 1993).

Recently, new additional members of the S100 subfamily of proteins have been described (HILT and KLIGMAN 1991). These demonstrate amino acid sequences of cDNA structures similar to S100 proteins, but have different cellular distribution and biochemical functions. For example, migration-inhibitory factor-related proteins MRP8 and MRP14 (HESSIAN et al. 1993) are expresssed during differentiation of granulocytes and monocytes. MRP8/MRP14 heterodimers modulate phosphorylation by inhibiting casein kinase (MURAO et al. 1989) and associate with membrane structure and intermediate filaments in a Ca^{2+}-dependent manner (BURWINKEL et al. 1994). The *CaN19* gene is preferentially expressed in the culture of human mammary epithelial cells as compared with tumor-derived mammary epithelial cells. The *CaN19* gene was mapped to the same region of chromosome 1 as some other *S100*-related genes. CaN19 has 63% homology to the Mts1 protein in amino acid sequence (LEE 1992) and is identical to S100L protein isolated from bovine lung. S100L protein is expressed at a high level in bovine kidney and lung tissue, at an intermediate level in muscle, and at a low level in brain and intestine (GLENNEY et al. 1989).

Another new member of the S100 family is p11 protein (also called p10, light chain of the cytoskeletal protein calpactin, and 42C protein). It has 50% sequence indentity with S100α. Unique freatures of p11 are deletions and substitutions in both EF-hand motifs which inactivate Ca^{2+}-binding sites. p11 can form a complex with another Ca^{2+}-binding protein, membrane phospholipid–annexin II, which is a cellular substrate of the *src* oncogene tyrosine kinase. The p11-annexin complex can modulate aggregation and fusion of phospholipid vesicles (HARDER et al. 1992).

The calcyclin gene was characterized as a serum-inducible gene in BHK cells and was found to be expressed at a high level in lymphocytes from patients with chronic myelogenic leukemia (CALABRETTA et al. 1985). Calcyclin-specific mRNA is expressed in a cell cycle and growth factor-dependent manner (HIRSCHHORN et al. 1984). Two proteins, annexin II and the glycolytic enzyme glyceraldehyde-3-phosphate dehydrogenase from murine Ehrlich ascite tumor cells, interact in vitro with the immobilized calcyclin in a Ca^{2+}-dependent manner (FILIPEK et al. 1991). Expression of calcyclin was found to be related to the metastasis of a human melanoma cell line in nude mice (WETERMAN et al. 1992).

An accumulating body of evidence indicates that each member of the S100 subfamily has a function specific to a particular species, tissue, developmental or proliferative state, or pathological process. It is difficult to make any general conclusion about their significance.

Native and recombinant Mts1 protein was isolated and analyzed in different groups under different names. Under the name of CAPL or calvasculin, the Mts1 protein was isolated from human heart and platelet (PEDROCCHI 1993; TOMIDA et al. 1992). It was found that CAPL is able to interact with phenylsepharose in the presence of Ca^{2+} and can be eluted when Ca^{2+} is absent. Using this property, CAPL was isolated from human heart and analyzed. The molecular mass of CAPL was determined by electrospray-ionization mass spectrometry (ESIMS) and calculated as 11 641 kDa. It was suggested that N-terminal methionine is cleaved off during translation and processing, and the next residue, an alanine, is blocked by acetylation. Similar processing probably occurs in bovine Mts1 protein isolated from retina (POLANS et al. 1994). It was shown that CAPL protein can form homodimers via cysteine bridges (PEDROCCHI et al. 1994b).

Using affinity chromatography with isoquinoline-sulfonamide as a ligand, (WATANABE et al.; WATANABE et al. 1992a) have purified 11-kDa Ca^{2+}-binding protein from bovine aorta. The protein, which was called calvasculin, showed about 90% amino acid sequence homology to mouse Mts1 protein. Calvasculin can form dimers even in the presence of 10 m*M* 2-mercaptoethanol, suggesting that not only disulfide bridges, but also protein conformation may be improtant for dimerization. In the absence of 2-mercaptoethanol, calvasculin can polymerize to multimeric complexes. Isoelectric focusing analysis of calvasculin demonstrates numerous isoelectric bands, suggesting the existence of several isoelectric variants or post-translational modification forms of calvasculin. It was shown by immunoblotting that, in contrast to the Mts1 protein, calvasculin was present in abundance in bovine aorta, where it was localized to the cytoplasmic granules. Such granular distribution of calvasculin is typical of secreted protein. Indeed it was shown that about 30% of newly synthesized calvasculin was secreted by rat embryo fibroblast cells (WATANABE et al. 1992b). The secretion of calvasculin might be induced by cell injury or by lowering the concentration of Ca^{2+}. Calvasculin can also bind one of the extracellular matrix proteins of bovine aorta, 36-kDa microfibril-associated glycoprotein (MAP). The binding is Ca^{2+}-dependent, and in the absence of Ca^{2+} calvasculin does not interact with MAP. The function of MAP protein is not known at present, but it has been shown that MAP can associate with cellular membranes and that this association is Ca^{2+}-dependent (WATANABE et al. 1992b). One can speculate that MAP somehow participates in the transloca-tion of calvasculin from cytoplasm through the membrane into the extracellular space.

It was shown that cytoplasmic granules containing calvasculin were distrib-uted in the cytoplasm along actin filaments (WATANABE et al. 1992b). Binding of calvasculin to actin filaments in vitro was demonstrated by a high-speed and low-speed centrifugation method. Calvasculin was found to coprecipitate with F-actin in the presence of Ca^{2+}. The ratio of calvasculin to actin molecules was 3:1 (WATANABE et al. 1993). The data indicate that calvasculin may be involved in the formation of microfilament bundles.

Several attempts were made to find the protein that interacts with the Mts-1 protein. One cytoskeletal protein which revealed binding activity to the pEL98/ Mts-l protein was mouse fibroblast tropomyosin (TM) (TAKENAGA et al. 1994b). TM belong to the family of actin-binding proteins (MATSUMURA et al. 1983), and there are two groups of TM: high and low molecular weight TM (TALBOT and MACLEOD 1983). pEL98/Mts-1 protein was found to interact with high molecular weight TM. The functions of TM in non muscle cells are still not known, but it was suggested that TM are associated with actin in microfilaments and regulate cell motility and division. In muscle cells TM fromed complexes with troponins along the actin filaments. Nonmuscle cells do not contain troponin complexes. However, in these cells TM interacts with actin, and this reaction is regulated by other proteins such as villin, non muscle caldesmon, or tropomodulin. Binding of pEL98/Mts1 to TM was specific and dependent on the presence of Ca^{2+}. To localize pEL98/Mts1 binding sites on TM, a series of deletion mutants of TM were analyzed in binding assay. The binding site was mapped in the region of TM between residues 39 and 107 (TAKENAGA et al. 1994b). It is still not clear whether the pEL98/Mts1 protein is associated with TM in vivo. Double-labeled immunofluorescence microscopy analysis suggests that some pEL98/Mts1 protein may associate with TM, but most of the protein exists diffusely in the cytoplasm of the cells (TAKENAGA et al. 1994a, b). It is also not clear whether TM can be coprecipitated with pEL98/Mts1 protein in immunoprecipitation assay.

In order search for intracellular molecules that can be associated with Mts1 protein, we developed a panel of monoclonal antibodies against the different synthetic peptides or recombinant Mts1 protein. Specificity of the monoclonal antibodies was examined by immunoblotting. The antibodies reacted with Mts1 protein with varying efficiency and specificity (CARDENAS et al. 1993). Two mAb which strongly and specifically reacted with Mts1 protein were used to look for intracelluar targets for Mts1 protein. These were mAb/HM-4 against the Mts1 carboxyterminal peptide and mAb-A8/7 against recombinant Mts1 protein.

Cell lysate from [^{35}S] methionine-labeled CSM-100 (metastatic, Mts1-positive) and CSML-0 (nonmetastatic, Mts1-negative) cells were prepared to identify Mts1-associated proteins. The proteins were immunoprecipitated and analyzed using sodium dodecyl sulfate polyacrylamide gel electrophoresis. Four major specific bands were detected in immunoprecipitates from metastatic cells. Besides the 11-kDa band of Mts1 protein, three other bands had molecular masses of approximately 200, 20, and 17 kDa, corresponding to one heavy and two light chains of myosin (Fig. 6).

The specificity of coprecipitation of Mts1 protein with heavy and light chains of myosin was confirmed by immunoprecipitation with antibodies against different epitopes of Mts1 protein as well as with antibodies against myosin. No precipitation of the myosin complex with anti-Mtsl antibodies was observed with lysate of nonmetastatic cells, where Mts1 protein was not expressed. However, myosin complex was easily detected on SDS-PAGE when lysate of nonmetastatic cells was mixed with recombinant Mts1 protein and immunoprecipitated with anti-Mts1 antibodies (Fig. 6). Thus, the precipitation of myosin by anti-Mts1 antibodies does indeed depend on the presence of Mts1 protein in cell lysate.

Sucrose gradient centrifugation confirmed the association between Mts1 and myosin. The Mts1 protein cosedimented with myosin complex.

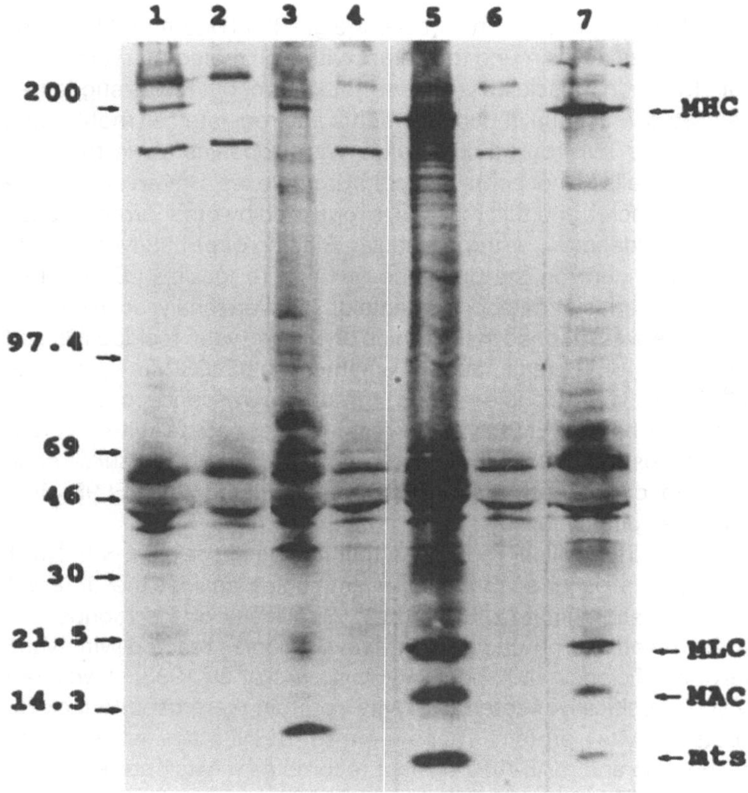

Fig. 6. Coimmunoprecipitation of the Mts1–myosin complex. Immunoprecipitation from CSML-0 (*lanes 1–4*) and CSML-100 (*lanes 5–7*) [^{35}S] methionine-labeled cell lysates. *1* Immunoprecipitation with anti-Mts1 monoclonal antibody (mAb) HM-4, obtained from CSML-0 cell extracts mixed with recombinant Mts1 protein; *2 ,5,* anti-Mts-1 mAb HM-4 (IgG_1); *3, 7,* rabbit antiserum against myosin heavy chain; *4, 6,* mouse immunoglobulin G_1 (IgG_1)

Direct evidence for an association between Mts1 protein and myosin complex was obtained using blot overlay technique. Recombinant Mts1 protein bound to the 200-kDa heavy chain of myosin in a Ca^{2+}-dependent manner (Fig. 7).

The analyzed cells contain two types of myosin: nonmuscle and smooth muscle myosins. Mts1 protein binds only to the heavy chain of nonmuscle myosin (Fig. 8).

Cellular localization of the Mts1 protein in metastatic cells was examined by indirect immunofluorescence microscopy. Mts1 protein appeared to be diffusely distributed throughout the cytoplasm and displayed enhanced immunofluorescence in the perinuclear region. Double staining with antibodies against Mts1 and myosin confirmed cytoplasmic colocalization of myosin and Mts1 protein.

Thus at least four proteins were described as possible targets for Mts1 protein:

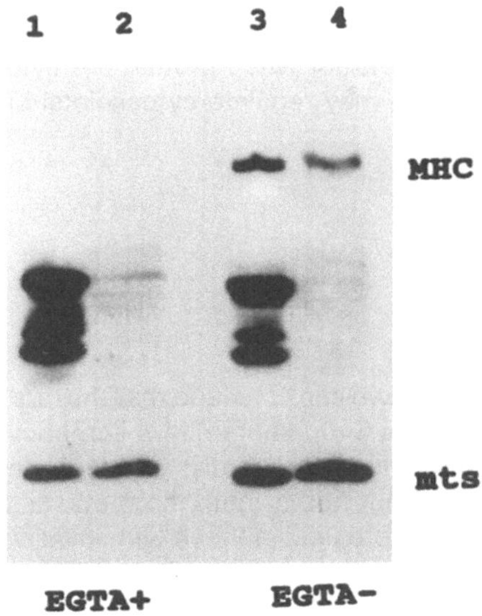

Fig. 7. Blot overlay analysis of the interaction of the Mts1 protein with heavy chain of myosin. Proteins immunoprecipitated with anti-Mts1 monoclonal antibody (mAb) HM-4 (*lanes 1,3*) or cytosolic extracts from CSML-100 cells (*lanes 2,4*) were separated in two-step sodium dodecyl sulfate polyacrylamide gel electrophoresis (SDS-PAGE) (6% and 15%) and blotted. Membranes were incubated with recombinant Mts1 protein in the presence of 1m*M* ethyleneglycoltetraacetic acid (EGTA; (*lanes 1,2*) or $CaCl_2$ (lanes 3,4))

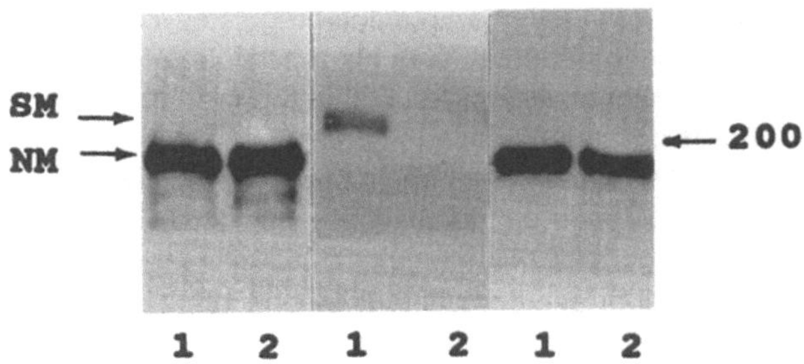

Fig. 8. Interaction between Mts1 protein and the heavy chain of nonmuscle myosin. Proteins were separated in 5% sodium dodecyl sulfate polacrylamide gel electrophoresis (SDS-PAGE) and blotted. The filter was probed with anti-nonmuscle myrosin antibodies (*left*) stripped, and reprobed with anti-smooth muscle myosin antibodies (*middle*) stripped again, and the blot overlay procedure was performed with recombinant Mts1 protein in the presence of 1 m*M* $CaCl_2$ (*right*) *1* Whole-cell lysate of CSML-100 cells; *2*, proteins immunoprecipitated with anti-Mts monoclonal antibody (mAb) from CSML-100 cells. *SM*, heavy chain of smooth muscle myosin; *NM*, heavy chain of nonmuscle myosin

1. A rat homologue of Mts1, p9Ka, as well as mouse pEL98/Mts1 protein were colocalized in fibers containing F-actin, and it was suggested that these proteins may interact with actin (Watanabe et al. 1993; Gibbs et al. 1994).
2. pEL98/Mts1 was also associated with TM and stained along actin filaments (Takenaga et al. 1994).
3. Calvasculin, the bovine homologue of Mts1, was found to be secreted and associated with extracellular matrix protein p36 (Takenaga et al. 1992b).
4. We demonstrated that Mts1 protein is associated with the heavy chain of myosin as a component of the actin–myosin complex (Kriajevska et al. 1994). The precise biochemical mechanism and functional role of the interaction

between Mts1 and its potential targets remains to be determined. However, the fact that three out of four possible targets for Mts1 protein are cytoskeletal proteins suggests that Mts1 protein may regulate cytoskeletal organization.

8 Conclusion

Summarizing the data presented in this review, we can conclude that the *mts1* gene encoding the Ca^{2+} -binding proteins of the S100 subfamily is somehow involved in the control of tumor metastasis. It is frequently, although not always, overexpressed in metastatic tumors, but not in their non metastatic counterparts. In normal cells, *mts1* is usually expressed in those with invasive behavior.

Transfection experiments show that, at least in some model systems, overexpression of *mts1* can strongly enhance the metastatic phenotype, while the expression of antisense *mts1* RNA or anti-*mts1* ribozyme downregulates metastasis.

The control of *mts1* expression is realized through a complex system of positive and negative *cis*-regulatory elements located within the first intron. DNA methylation is also involved in this control.

Mts1 protein seems to interact with the heavy chain of myosin and with TM. One can speculate that it might influence cell motility and, as a result, tumor metastasis through this interaction.

We believe that Mts1 is an important component in one of the multiple pathways leading to tumor progression and metastasis.

References

Barraclough R, Savin J, Dube SK, Rudland PS (1987) Molecular cloning and sequence of the gene for p9Ka, a cultured myoepithelial cell protein with strong homology to S100, a calcium-binding protein. J Mol Biol 198: 13–20

Basset P, Bellocq JP, Wolf C, Stoll I, Hutin P, Limacher JM, Podhajcer OL, Chenard MP, Rio MC, Chambon P (1990) A novel metalloproteinase gene specifically expressed in stromal cells of breast carcinomas. Nature 438: 699–704

Bernstein SC, Weinber RA (1985) Expression of the metastatic phenotype in cells tranfected with human metastatic tumor DNA. Proc Natl Acad Sci USA 82: 1726–1730

Bianchi R, Pula G, Ceccarelli P, Giambanco I, Donato R (1992) S100 protein binds to annexin II and p11, the heavy and light chains of calpactin I. Biochim Biophys Acta 1160: 67–75

Bianchi R, Giambanco I, Donato R (1993) S-100 protein, but not calmodulin, binds to the glial fibrillary acidic protein and inhibits its polymerization in a Ca^{2+}-dependent manner. J Biol Chem 268: 12669–12674

Birnstiel ML, Busslinger M, Strub K (1985) Transcription termination and 3' processing: the end is in site. Cell 41: 349–361

Burwinkel F, Roth J, Goebeler M, Bitter U, Wrocklage V, Vollmer E, Roessner A, Sorg C, Bocker W (1994) Ultrastructural localization of the S-100-like proteins MRP8 and MRP14 in monocytes is calcium-dependent. Histochemistry 101: 113–120

Calabretta B, Kaczmarek L, Mars W, Ochoa D, Gibson CW, Hirshhorn RR, Baserga R (1985) Cell cycle-specific genes differentially expressed in human leukemias. Proc Natl Acad Sci USA 82: 44663–44671

Cardenas MN, Vorovich MF, Ebralidze AK, Chertov OYu, Lukanidin EM (1993) Production of recombinant metastasin and its characterization. Bioorg Khim 19: 193–258

Carnemolla B, Balza E, Siri A, Zardi L, Nicotia MR, Bigotti A, Natali PG (1989) A tumor associated fibronectin generated by alternative splicing of messenger RNA precursor. J Cell Biol 108: 1139–1148

Cedar H (1988) DNA methylation and gene activity. Cell 34: 5503–5513

Chambers AF, Tuck AB (1988) Oncogene transformation and the metastatic phenotype. Anticancer Res 8: 861–872

Chambers AF, Tuck AB (1993) Ras-responsive genes and tumor metastasis. Crit Rev Oncog 4: 95–114

Chen L-C, Dollbaum C, Smith H (1989) Loss of heterozygosity on chromosome 1q in human breast cancer. Proc Natl Acad Sci USA 86: 7204–7207

Davies BR, Davies MPA, Gibbs FEM, Barraclough R, Philip S, Rudland PS (1993) Induction of metastatic phenotype by transfection of a benign rat mammary epithelial cell line with the gene for p9Ka, a rat calcium-binding protein, but not with the oncogene EJ-ras-1. Oncogene 8: 999–1008

De Vouge MW, Mukherjee BB (1992) Transformation of normal rat kidney cells by v-K-ras enhances expression of transin 2 and an S-100-related calcium-binding protein. Oncogene 7: 109–119

Dear TN, Ramshaw IA, Kefford RF (1988) Differential expression of a novel gene WDNM1, in nonmetastatic rat mammary adenocarcinoma cells. Cancer Res 48:5203–5209

Dear TN, Ramshaw IA, Kefford RF (1989) Transcriptional down regulation of a rat gene, WMDNM2, in metastatic DMBA–8 cells. Cancer Res 49: 5323–5328

Devilee P, van Vliet M, Bardoel A, Kievits T, Kuipers-Dijkshoorn, Pearson PL, Cornelisse CJ (1991) Frequent somatic imbalance of marker alleles for chromosome 1 in human primary breast carcinoma. Cancer Res 51: 1020–1025

Doerfler W (1983) DNA methylation and gene activity. Annu Rev Biochem 1983, 52: 93–124

Donato R (1988) calcium-independent, pH-regulated effects of S-100 proteins on assembly disassembly of brain microtubule protein in vitro. J Biol Chem 263: 106–110

Donato R (1991) Perspectives in S-100 protein biology. Cell Calcium 12: 713–726

Dorin JR, Emslie E, van Heyningen V (1990) Related calcium-binding proteins map to the same subregion of chromosome 1q and to extended region of synteny on mouse chromosome 3. Genomics 8: 420–426

Ebralize A, Tulchinsky E, Grigorian M, Afanasyeva A, Senin V, Revazova E, Lukanidin E (1989) Isolation and characterization of a gene specifically expressed in different metastatic cells and and whose deduced gene product has a high degree of homology to a Ca^{2+}-binding protein family. Genes Dev 3: 1086–1093

Engelkamp D, Schafer BW, Erne P, Heizmann CW (1992) S100a, CAPL, and CACY: Molecular cloning and expression analysis of three clacium-binding proteins from human heart. Biochemistry 31: 10258–10264

Engelkamp D, Schafer BW, Mattei MG, Erne P, Heizmann CW (1993) Six S100 genes are clustered on human chromosome 1q21: identification of two genes coding for the two previously unreported calcium-binding proteins S100D and S100E. Proc Natl Acad Sci USA 90: 6547–6551

Feinberg AP, Vogelstein B (1983) Hypomethylation distinguished genes of some human cancers from their normal counterparts. Nature 301: 89–92

Filipek A, Gerke V, Weber K, Kuznicki J (1991) Characterization of the cell-cycle-regulated protein calcyclin from Ehrlich ascites tumor cells. Eur J Biochem 195: 795–800

Gendler SJ, Cohen EO, Craston A, Duhig T, Johnstone G, Barnes D (1990) The locus of the polymorphic epithelial mucin (PEM) tumor antigen on chromosome 1q21 shows a high frequency of alteration in primary human breast tumors. Int J Cancer 45: 431–435

Ghose T, Lee CYL, Fernandez LA, Lee SHS, Raman R, Colp P (1990) Role of 1q trisomy in tumorigenicity, and metastasis of human leukemic B-cell clones in nude mice. Cancer Res 50: 3737–3742

Gibbs FEM, Wilkinson MC, Rudland PS, Barraclough R (1994) Interactions in vitro of p9Ka, the rat S-100-related, metastasis-inducing, calcium-binding protein, J Biol Chem 269: 18992–18999

Gilles A, Presecan E, Vonica A, Lascu I (1991) Nucleoside diphosphate kinase from human erythrocytes. J Biol Chem 266: 8784–8789
Glenney JR, Kindy MS, Zokas L (1989) Isolation of a new member of the S100 protein family: amino acid sequence, tissue, and subcellular distribution. J Cell Biol 108: 569–578
Goto K, Endo H, Fujiyoshi T (1988) Cloning of the sequences expressed abundantly in established cell lines: indentification of a cDNA clone highly homologous to S-100, a calcium binding protein. J Biochem 103: 48–53
Grigorian M, Tulchinsky E, Zain S, Ebralidze A, Kramerov D, Kriajevska M, Georgiev G, Lukanidin E (1993) The *mts1* and control of tumor metastasis. Gene 135: 229–238
Grigorian M, Tulchinsky E, Burrone O, Tarabykina S, Georgiev G, Lukanidin E (1994) Modulation of mts1 expression in mouse and human normal and tumor cells. Electrophoresis 15 463–468
Guo X, Chambers AF, Parfett CLJ, Waterhouse P, Murphy LC, Reid RE, Craig AM, Edwards DR, Denhardt DT (1990) Identification of a serum-inducible messenger RNA (5B10) as the mouse homologue of calcyclin: tisssue distribution and expression in metastatic, ras-transformed NIH 3T3 cells. Cell Growth Differ 1: 333–338
Günthert U, Hofmann M, Rudy W, Reber S, Zoller M, Haussmann I, Matzku S, Wenzel A, Ponta H, Herrlich P (1991) A new variant of glycoprotein CD44 confers metastatic potential to rat carcinoma cells. Cell 65: 13–24
Habets GGM, Scholtes EHM, Zuydgeest D, van der Kammen RA, Stam JC, Berns A, Collard GJ (1994) Identification of an invasion-including gene, Tiam-1, that encodes a protein with homology to GDP-GTP exchangers for Rho-like proteins. Cell 77: 537–549
Hagivara M, Ochiai M, Owada K, Tanaka T, Hidaka H (1988) Modulation of tyrosine phosphorylation of p36 and other substrates by the S100 protein J Biol Chem 263: 6438–6441
Harder T, Kube E, Gerke V (1992) Cloning and characterization of the human gene encoding p11: structural similarity to other members of the S-100 gene family. Gene 113: 269–274
Hart IR, Saini A (1992) Biology of tumor metastasis. Lancet 339: 1453–1457
Heider KH, Dammrich J, Skroch-Angel P, Müller-Hermelink HK, Vollmers HP, Herrlich P, Ponta H (1993a) Tumor type specific expression of metastasis-related variants of CD44 in human gastric cancer. Cancer Res 53: 4197–4203
Heider KH, Hofmann M, Horst E, van den Berg F, Ponta H, Herrlich P, Pals ST (1993b) A human homologue of the tat metastasis-associated variant of cd44 is expressed in colorectal carcinomas and adenomatous polyps. J Cell Biol 120: 227–233
Heizmann CW, Hunziker W (1991) Intracellular calcium-binding proteins: more sites than insights. Trends Biochem Sci 16: 98–103
Hessian PA, Edgeworth J, Hogg N (1993) MRP8 and MRP14, two abundant calcium-binding proteins of neutrophils and monocytes. Leukocyte Biol 53: 197–204
Higgs DR, Vickers MA, Wilkie AOM, Pretorius IM, Jarman AP, Weatherall DJ (1989) A review of the molecular genetics of the human alpha-globin gene cluster. Blood 73: 1081–1104
Hilt DC, Kilgman D (1991) The S-100 protein family: a biochemical and functional overview. In: Heizmann CW (ed) Novel calcium-binding proteins; fundamental and clinical implications. Springer, Berlin Heidelberg New York, pp 65–104
Hirschhorn RR, Allen P, Yuan Z, Gibson CW, Baserga R (1984) Cell-cycle-specific cDNAs from mammalian cells temperature sensitive for growth. Proc Natl Acad Sci USA 81: 6004–6008
Hoffmann M, Rudy W, Zoller M, Tolg C, Ponta H, Herrlich P, Günther U (1991) CD44 splice variants confer metastatic behaviour in rats: homologous sequences are expressed in human tumor cell lines. Cancer Res 51: 5292–5297
Iguchi-Ariga SM, Schaffner W (1989) CpG methylation of the cAMP-responsive enhancer/promoter sequence TGACGTCA abolishes selective factor binding as well as transcriptional activation. Genes Dev 3: 612–619
Jackson Grusby LL, Swiergiel J, Linzer IH (1987) a growth-related mRNA in cultured mouse cells encodes a placental calcium binding protein. Nucleic Acids Res 15: 6677–6690
Klein G, Klein E (1986) Conditioned tumorigenicity of activated oncogenes Cancer Res 46: 3211–3224
Kligman D, Marshak DR (1985) Purification and characterization of a neurite extension factor from bovine brain. Proc Natl Acad Sci USA 82: 7136–7139
Kligman D, Hsieh LS (1987) Neurite extension factor induces rapid morphological differentiation of mouse neuroblastoma cells in defined medium. Dev Brain Res 33: 296–300
Kretsinger RH, Tolbert D, Nakayama S, Pearson W (1991) The EF-hand, homologs and analogs. In: Heizmann CW (ed) Novel calcium-binding proteins: fundamentals and clinical implications. Springer, Berlin Heidelberg New York, pp17–37

Kriajevska MV, Cardenas MN, Grigorian MS, Ambartsumian NS, Georgiev GP, Lukanidin EM (1994) Non-muscle myosin heavy chain as a possible target for protein encoded by metastasis-related mts-1 gene. J Biol Chem 269: 19679–19682

Krumlauf R (1992) evolution of the vertebrate Hox homeobox genes. Bioessays 14: 245–252

Lee SW, Tomasetto C, Swisshelm K, Keyomarsi K, Sager R (1992) Down-regulation of a member of the S100 gene family in mammary carcinoma cells and reexpression by azadeoxycytidine treatment. Proc Natl Acad Sci USA 89: 2504–2508

Leone A, Flatow U, King C, Sandeen MA, Margulies I, Liotta L, Steeg P (1991) Reduced tumor incidence, metastatic potential, and cytokine responsiveness of nm23-transfected melanoma cells. Cell 65: 25–35

Liotta L, Rao C, Barsky S (1983) Tumor invasion and the extracellular matrix. Lab Invest 49: 636–649

Liotta LA, Steeg PS, Stetler-Stevenson WG (1991) Cancer metastasis and angiogenesis: an imbalance of positive and negative regulation. Cell 64: 327–336

Masiakowski P, Shooter EM (1988) Nerve growth factor induces the genes for two proteins related to a family of calcium binding proteins in PC12 cells. Proc Natl Acad Sci USA 85: 1277–1281

Matrisian LM, Glaichenhays N, Gesnel M-C, Breathnach R (1985) Epidermal growth factor and oncogenes induce transcription of the same cellular mRNA in rat fibroblasts. EMBO J 4: 1435–1440

Matrisian LM, Bowden GT, Krieg P, Furstenberger G, Briand JP, Leroy P, Breathnach R (1986) The mRNA coding for the secreted protease transin is expressed more abundantly in malignant than in benign tumors. Proc Natl Acad Sci USA 83: 9413–9417

Matsumura F, Yamashiro-Matsumura S, Lin JJ-C (1983). Isolation and characterization of tropomyosin-containing microfilaments from cultured cells. J Biol Chem 258: 6636–6644

Molin SO, Rosengren L, Bandier J, Hamberger A, Haglid K (1985) S-100 alpha-like immunoreactivity in tubules of rat kidney. A clue to the function of a "brain-specific" protein. J Histochem Cytochem 33: 367–374

Mount SM (1982) A catalogue of splice junction sequences. Nucleic Acids Res 10: 459–572

Murao S, Collard FR, Huberman E (1989) A protein containing the cystic fibrosis antigen is an inhibitor of protein kinase. J Biol Chem 264: 8356–8360

Muschel RJ, McKenna WG (1989) Oncogenes and tumor progression. Anticancer Res 9: 1395–1406

Maelandsmo GM, Hovig E, Skrede M, Kashani-Sabet M, Engebråten O, Flørenes VA, Myklebost O, Lukanidin E, Grigorian M, Scanlon KJ, Fodstad Ø (1995) Reversal of the in vivo metastatic phenotype of human osteosarcoma cells by an anti-CAPL (mts-1) ribozyme. Proc Natl Acad Sci USA (submitted)

Nicolson GL (1988) Cancer metastasis: tumor cell and host organ properties important in metastasis to specific secondary sites. Biochim Biophys Acta 948: 175–224

Nicolson GL, Gallick GE, Dulski KM, Spohn WH, Lembo T, Tainsky MA (1990) Lack of correlation of intercellular junctional communication, $p21^{rasEJ}$ expression and sponteneous metastatic properties of rat mammary cells after transfection with c-H-ras^{EJ} or neo genes. Oncogene 5: 747–753

Nicolson GL (1991) Quantitative variations in gene expression: possible role in cellular diversification and tumor progression. J Cell Biochem 46: 277–283

Oyama F, Hirohashi S, Shimosato Y, Titani K, Sakiguchi K (1989) Deregulation of alternative splicing of fibronectin pre-mRNA in malignant human liver tumors. J Biol Chem 264: 10331–10334

Pedrocchi M, Hauer CR, Schafer BW, Erne P, Heizmann CW (1993) Analysis of Ca^{2+}-binding S100 proteins in human heart by HPLC-electrospray mass spectrometry. Biochem Biophys Res Commun 197: 529–535

Pedrocchi M, Schafer BW, Mueller H, Eppenberger U, Heizmann CW (1994a) Expression of Ca^{2+}-binding proteins of the S100 family in malignant human breast-cancer cell lines and biopsy samples. Int J Cancer 57: 684–690

Pedrocchi M, Schafer BW, Durussel I, Cox JA, Héizmann CW (1994b) Purification and characterization of the recombinant human calcium-binding S100 protein CAPL and CACY. Biochemistry 33: 6732–6738

Pencil SD, Toh Y, Nicolson GL (1993) Candidate metastasis-associated genes of the rat 13762NF mammary adenocarcinoma. Breast Cancer Res Treat 25: 165–174

Persechini A, Moncrief ND, Kretsinger RH (1988) the EF-hand family of calcium-modulated proteins. Trends Neurosci 12: 462–467

Phillips SM, Bendall AJ, Ramshaw IA (1990) Isolation of gene associated with high metastatic potential in rat mammary adenocarcinomas. J Natl Cancer Inst 82: 199–203

Polans AS, Palczewski K, Asson-Batres MA, Ohguro H, Witkowska D, Haley TL, Baizer L, Carbb JW (1994) Purification and primary structure of Capl, an S-100-related calcium-binding protein isolated from bovine retina. J Biol Chem 269: 6233–6240

Pritchard K, Martson SB (1991) Ca^{2+}-dependent regulation of vascular smooth-muscle caldesmon by S100 and related smooth-muscle proteins. Biochem J 277: 819–824

Ramshaw IA, Carlsen S, Wang HC, Badenoch-Jones P (1983) The use of cell fusion to analyze factors involved in tumor cell metastasis. Int J Cancer 32: 471–478

Sano T, Tsujino T, Yoshida K, Nakayama H, Haruma K, Ito H, Nakamura Y, Kajiyama G, Tahara E (1991) Frequent loss of heterozygosity on chromosomes 1q, 5q, and 17p in human gastric carcinomas. Cancer Res 51: 2926–2931

Schirrmacher V (1985) Cancer metastasis: experimental approaches, theoretical concepts, and impacts for treatment strategies. Adv Cancer Res 43: 1–73

Sekler EU (1990) DNA methylation and chromatin structure: a view from below. TIBS 15: 103–107

Selinfreund RH, Barger SW, Welsh MJ, Van Eldik LJ (1990) Antisense inhibition of glial S100 beta production results in alterations in cell morphology, cytoskeletal organization, and cell proliferation. J Cell Biol 111: 2021–2028

Senin VM, Buntsevich AM, Afanasyeva AV, Kiseleva NS (1983) A new line of metastatic murine carcinosarcoma. Exp Oncol USSR 5: 35–38

Senin VM, Ivanov AM, Afanasyeva AV, Buntsevich AM (1984) New organotropic-metastatic transplanted tumors of mice and their use for studying laser effect on dissemination. Vestnic USSR Acad Med Sci 5: 85–91

Sriramara P, Bourdon M (1993) A novel tenascin type III repeat is part of a complex of tenascin mRNA alternative splices. Nucleic Acids Res 21: 163–168

Steeg PS, Bevilacqua G, Kopper L, Thorgeirsson UP, Talmadge JE, Liotta LA, Sobel ME (1988) Evidence for a novel genes accociated with low tumor metastatic potential. J Natl Cancer Inst 80: 200–204

Stetler-Stevenson WG, Azavoorian S, Liotta LA (1993) Tumor cell interactions with the extracellular matrix during invasion and metastasis. Annu Rev Cell Biol 9: 541–573

Takenaga K, Nakamura Y, Sakiyama S (1994a) Cellular localization of pEL98 protein, an S100-related calcium binding protein, in fibroblasts and its tissue distribution analyzed by monoclonal antibodies. Cell Struct Funct 19: 133–141

Takenaga K, Nakamura Y, Sakiyama S, Hasegawa Y, Sato K, Endo H (1994b) Binding of pEL98 protein, an S100-related calcium-binding protein, to nonmuscle tropomyosin. J Cell Biol 124: 757–768

Talbot K,MacLeod AR (1983) Novel form of non-muscle tropomyosin in human fibroblasts. J Mol Biol 164: 159–174

Taniguchi S, Miyamoto S, Sadano H, Kobayashi H (1992) Rat elongation factor 1 alpha: sequence of cDNA from highly metastatic fos-transferred cell line. Nucleic Acids Res 19: 6949

Thorgeirsson UP, Turpeenniemi-Hujane T, Williams JE, Westin EH, Heilman CA, Talmade JE, Liotta LA (1985) NIH-3T3 cells transfected with human tumor DNA containing activated ras oncogenes express the metastatic phenotype in nude mice. Mol Cell Biol 5: 259–262

Toh Y, Pencil SD, Nicolson GL (1994) A novel candidate metastasis-associiated gene, mta1 differentially expressed in highly metastatic mammary adenocarcinoma cell lines. J Biol Chem 269: 22958–229663

Tomida Y, Terasawa M, Kobayashi R, Hidaka H (1992) Calcyclin and calvasculin exist in human platelet. Biochem Biophys Res Commun 189: 1310–1313

Tuck AB, Wilson SM, Sergovich FR, Chambers AF (1991) Gene expression and metastasis of somatic cell hybrids between murine fibroblast cell lines of different malignant potential. Somat Cell Mol Genet 17: 377–389

Tulchinsky E, Ford HL, Kramerov D, Reshetnyak E, Grigorian M, Zain S, Lukanidin E (1992) Transcriptional analysis of the mts1 gene with specific reference to 5' flanking sequences. Proc Natl Acad Sci USA 89: 9146–9150

Tulchinsky E, Kramerov D, Ford HL, Reshetnyak E, Lukanidin E, Zain S (1993) Characterization of a positive regulatory element in the mts1 gene. Oncogene 8: 79–86

Watanabe Y, Kobayashi, Ishikawa T, Hidaka H (1992a) Isolation and characterization of a calcium-binding protein derived from mRNA termed p9Ka, pEL-98, 18A2, or 42A by the newly synthesized vasorelaxant W-66 affinity chromatography. Arch Biochem Biophys 292: 563–569

Watanabe Y, Usuda N, Tsugane S, Kobayashi R, Hidaka H (1992b) Calvasculin, an encoded protein from mRNA termed pEL-98 18A2, 42A, or p9Ka, is secreted by smooth muscle cells in culture and exhibits Ca^{2+}-dependent binding to 36-kDa microfibril-associated glycoprotein. J Biol Chem 267: 17136–17140

Watanabe Y, Usada N, Minami H, Mrita T, Tsugane S, Ishikawa R, Kohama K, Tomida Y, Hidaka H (1993) Calvasculin, as a factor affecting the microfilment assemblies in rat fibroblasts transfected by src gene. FEBS Lett 324: 51–55

Weterman MAJ, Stoopen GM, van Muijen GNP, Kuznicki J, Ruiter DJ, Bloemers HPJ (1992) Expression of calcyclin in human melanoma cell lines correlates with metastatic behavior in nude mice. Cancer Res 52: 1291–1296

Woltman SR, Pauley RJ, Mohamed AN, Dawson PJ, Visscher DW, Sarkar FH (1992) Genetic markers as prognostic indicators in breast cancer. Cancer Suppl 70: 1765–1774

Yang F, Demma M, Warren V, Dharmawardhane S, Condeelis J (1990) Identification of an actin-binding protein from Dictyostelium as elongation factor 1a. Nature 347: 494–496

Yokota J, Tsunetsugu-Yokota Y, Battifora H, Le Pevre C, Cline MJ (1986) Alterations of myc, myb and rasHa proto-oncogenes in cancers frequent and show clinical correlation. Science 246: 1265–1275

Alterations in Cell Cycle Control During Tumor Progression: Effects on Apoptosis and the Response to Therapeutic Agents

R.J. Muschel[1] and W.G. McKenna[2]

1 Introduction . . . 197

2 Regulation of the Cell Cycle . . . 198

3 Cyclins in Tumors . . . 199

4 Cyclin–cdk Inhibitors . . . 200
4.1 WAF1/CIP1 . . . 200
4.2 GADD45 . . . 201
4.3 p15 and p16 . . . 202
4.4 Transforming Growth Factor-β and $p27^{Kip1}$ . . . 202

5 Regulation of Cell Proliferation and the Control of Apoptosis . . . 203
5.1 p53 and Apoptosis . . . 204
5.2 *myc* and Apoptosis . . . 205
5.3 *bcl-2* and Apoptosis . . . 206

6 Cell Cycle Response to DNA Damage . . . 207
6.1 G_2 Phase and the Response to DNA Damage . . . 208

References . . . 209

1 Introduction

It has long been apparent that alterations in the control of the cell cycle accompany the development of tumorigenicity. The mere observation that cell growth continues in tumors whereas normal cells for the most part cease dividing and retain a diploid amount of DNA suggests that the controls regulating the normal cell cycle have failed to function in the tumor cell. In recent years the biochemical mechanisms underlying the transitions through the cell cycle have begun to be identified, and some of the gene products that are mutated in cancer cells have been shown to interact directly with these genes in control of the cell cycle. These mutated genes, as well as the cell cycle regulators, have been shown to affect whether cells undergo programmed cell death and to influence how cells respond to DNA-damaging agents that are frequently used therapeutically. Thus the genetic

[1] Department of Pathology and Laboratory Medicine, University of Pennsylvania, Philadelphia, PA 19104–6082, USA

[2] Department of Radiation Oncology, University of Pennsylvania, Philadelphia, PA 19104–6082, USA

changes in the cell that affect cell cycle progression also have a bearing on the therapeutic outcome. While the description of the changes affecting the G_0 to G_1 to S phase transition is currently better understand as regards malignancy, the G_2/M to phase transition also has important implications for therapy.

2 Regulation of the Cell Cycle

A detailed understanding of the interaction of cyclins and cyclin-dependent kinases (cdk) is rapidly being pieced together and many reviews are available on these subjects (DOREE and GALAS 1994; HUNTER and PINES 1994; KING et al. 1994; MORGAN 1995; NORBURY and NURSE 1992; SHERR 1993, 1994). The cdks are a series of related kinases, each of which is only enzymatically active as a complex with a cyclin. Many of these cyclins can interact with several kinases under physiological conditions so that both cyclin A and cyclin B1 can activate p34 cdc2 or both cyclin E and cyclin A can interact with cdk2. These complexes must also undergo a series of complex phosphorylations that are best understood for p34 cdc2 and less well understood for the other family members. There are both inhibitory and facilitating phosphorylations of p34 cdc2. The phosphorylation of a threonine residue (161 in human p34 cdc2) occurs at the same time as cyclin complexes form (KREK et al. 1992; KREK and NIGG 1991) This is a facilitating phosphorylation which in some experimental situations is required for optimal cyclin B1–p34 cdc2 interactions (DESAI et al. 1995a; GOULD et al. 1991; KREK et al. 1992; 1992). Phosphorylation of threonine 14 and tyrosine 15 also occur at the time of p34 cdc2 complex formation and these are inhibitory phosphorylations. The activation of p34 cdc2–cyclin B1 complexes then occurs when these two phosphorylations are removed through the activation of the phosphatase cdc 25. The active kinase then triggers the events leading the cells through G_2/M (DOREE and GALAS 1994; HUNTER and PINES 1994; KING et al. 1994; MORGAN 1995; NORBURY and NURSE 1992; SHERR 1993, 1994). The subsequent dephosphorylation of threonine 161 on p34 cdc2 is required for the continuation of cycling in G_1 (LORCA et al. 1992). These dephosphorylations have also been shown to occur on cdk2 cyclin E or A complexes, suggesting that similar events will also regulate the other cyclin cdk complexes (DESAI et al. 1995b). A final crucial aspect to the regulation of these complexes is the existence of a series of inhibitors that bind to the complexes and inhibit kinase activity.

The cyclin cdk complexes are activated and deactivated in a precisely timed series of events throughout the cell cycle. At the transition between G_0 and G_1 or in the later stages of G_1 concurrent with the restriction point, cyclin D1–cdk4 is activated. Cyclin E–cdk2 complexes become active slightly later in late G_1 and into S, followed by activation of cyclin A–cdk2 complexes in S. Later, as S continues, cyclin A–cdc2 complexes form, and at the G_2 transition cyclin B–cdc2 complexes form. The activity of the cdc complexes then occurs at the G_2 to M transition, and the activity of the cyclin B–cdc2 complexes must be destroyed for cells to enter G_1. This destruction occurs via degradation of cyclin B, which occurs through ubiquitination in anaphase.

3 Cyclins in Tumors

This sequence of events might lead to the expectation that triggering the activation of cyclin D1–cdk4 complexes might contribute to the lack of appropriate control of G_1 in tumor cells. In fact cyclin D1 was first identified as *bcl-1*, a gene whose expression is brought under control of the immunoglobulin heavy chain promoter by a characteristic translocation in a subset of B cell lymphomas. Cyclin D1 expression is also altered in parathyroid adenomas, again through a translocation that places the cyclin D1 gene, which was called PRAD1 in this case, in proximity to the parathyroid hormone promoter (Motokura et al. 1991). Cyclin D1 expression is also augmented in squamous cell carcinomas of the head and neck due to amplification of the gene (Callender et al. 1994) and is overexpressed in some lymphomas (Bosch et al. 1994). Finally, in one instance in a breast cancer cell line, the cyclin D1 gene was altered through a truncation of the region coding for the 3'-untranslated portion of the mRNA, which led to an increased half-life of the mRNA, resulting in higher cyclin D1 mRNA levels (Lebwoh et al. 1994). While these changes are certainly suggestive of a role of increased cyclin D1 in contributing to the malignancy of the cells, the consequence of elevated cyclin D1 has yet to be established. Resnitzky et al. placed the cyclin D1 gene under an inducible promoter and introduced this vector into cells (Resnitzky et al. 1994). Induced expression of cyclin D1 resulted in a shortening of G_1 by several hours in these cells. It is interesting to note, however, that the overall length of the cell cycle did not change; S was correspondingly longer. Musgrove et al. altered cyclin D1 expression in T47D cells, a breast cancer cell line, using a similar approach and also found G_1 to be shortened (Musgrove et al. 1994). This group also found that induction of cyclin D1 could relieve growth inhibition triggered by serum starvation. It is likely that the overexpression of cyclin D1 must be accompanied by other changes which act independently to control G_1 in order to result in the fully transformed phenotype.

Cyclin A can also be overexpressed in carcinomas, and indeed it was cloned by Wang et al. because it was located at the integration site of hepatitis B in a hepatocellular carcinoma, although this is unusual in these tumors (Wang et al. 1990). Cyclin E has also occasionally been described as being overexpressed in cancers (Keyomarsi and Pardee 1993; Leach et al. 1993). So far overexpression of either cyclin E or A appears to be a rare event in cancers. In yeast, high levels of expression of one cyclin can often substitute for another, suggesting the possibility that these rarer instances of overexpression of cyclin A or E might have effects through interactions which do not occur at physiological concentrations.

Further evidence of the involvement of cell cycle regulators in the changes accompanying tumor progression is the evidence that the tumor suppressor Rb (retinoblastoma) gene is intimately involved in the action of cyclin D1. Rb binds to E2F, a transcription factor complex which includes its required partner DP-1 and which inhibits its activity (SHIRODKAR et al. 1992). Both cyclin D1– and cyclin E–cdk complexes can phosphorylate Rb, leading to its dissociation from the E2F–Dp-1 complexes and allowing transcriptional activity by that complex. The activity of E2F is required for the transcription of a number of genes whose function is required in S, including thymidine kinase, thymidylate synthatase, dihydrofolate reductase, DNA polymerase α, c-*myc*, cdc2, and even cyclin D1 itself. Thus, in cells with functional inactivation of Rb, cyclin D1 was found to be at low levels. It has also been reported that cyclin D1 can bind directly to Rb (KATO et al. 1993). If Rb is the major target of cyclin D1 complexes, inactivation of Rb might obviate the need for cyclin D1. Thus it might be expected that mutations in Rb might have the same effect as overexpression of cyclin D1. Whether overexpression of cyclin D1 can substitute for inactivation of Rb during tumor progression in tissues has yet to be demonstrated. Adding further complexity is the observation that cyclin A–cdk2 can bind to E2F under conditions in which E2F is overexpressed and this interaction may result in phosphorylation of the E2F–Dp-1 complex, inhibiting its ability to bind DNA (DYNLACHT et al. 1994, XU et al. 1994). Thus, the formation of active cyclin A–cdk2 complexes act as a negative feedback on the action of E2F–Dp-1 as a transcriptional activator later in S. Thus E2F activity is triggered in late G_1 by activation of cyclin D and E complexes and later in S is inhibited by cyclin A complexes. It is attractive to speculate that either inactivation of Rb or overexpression of cyclin D1 can remove one control point at which the normal cell is subject to controls limiting progression from G_1 past the restriction point and hence would be an expected target of mutations during tumor progression.

4 Cyclin–cdk Inhibitors

The naturally occurring inhibitors that associate with the cyclin–cdk complexes are also proving to be a fertile ground to explore for mutations that alter cell cycle control. Cyclin–cdk complexes were noted to coprecipitate with a series of proteins, including those with molecular weights of 27, 21, 16, and 15. Each of these molecules has now been isolated and cloned, and each has been shown to have distinct inhibitory activity on the kinase activity of specific cyclin–cdk complexes.

4.1 WAF1/CIP1

p21 WAF1/CIP1 first attracted interest when it was simultaneously isolated by HARPER et al. (1993) and EL-DEIRY et al. (1993). HARPER et al. identified this gene using a two-hybrid screen to detect proteins which bound to cdk2. They showed that this gene product normally coprecipitated with cyclin A– or cyclin E–cdk2 complexes and with cyclin D1– or cyclin D2–cdk4 complexes and in increased amounts could inhibit their activity.

The link to tumor progression came from an alternate strategy for the isolation of p21 by EL-DEIRY et al. (1993). KASTAN et al. had shown that DNA damage would lead to a G_1 arrest in cells containing wild-type p53, but in cells with mutant p53 there was no G_1 arrest after DNA damage (KASTAN et al. 1991; KUERBITZ et al. 1992). To attempt to isolate downstream intermediates which could mediate this effect, EL-DEIRY et al. conducted subtractive hybridization to isolate mRNA whose levels were increased after exposure of cells with intact p53 to DNA damage. This resulted in the isolation of a cDNA for p21 WAF1/CIP1, which is induced only in cells with wild-type p53 after exposure to DNA damage through ultraviolet (UV) radiation, by various chemotherapeutic agents or X-rays. Thus p21 WAF1/CIP1 is induced by DNA damage and then leads to growth arrest through its ability to inhibit the cyclin–cdk complexes. Moreover, p21 WAF1/CIP1 has other effects on cell cycle machinery. Proliferating-cell nuclear antigen (PCNA), a subunit of DNA polymerase-δ, is part of the complex of cyclin D and cdk4 or cyclin A and cdk2 (XIONG et al. 1993). p21 can be found as a component of these complexes. The binding of a single molecule of p21 to a cdk complex does not appear to inhibit its activity, but increased numbers do (ZHANG et al. 1994). When p21 WAF1/CIP1 was added to in vitro DNA replication systems that include DNA polymerase-δ, there was an inhibition of DNA synthesis in a fashion independent of its association with cyclin–cdk complexes. Most importantly, in these same in vitro systems the addition of p21 WAF1/CIP1 did not inhibit the repair capacity of UV-damaged DNA (R. LI et al. 1994). This data suggests that p21 WAF1/CIP1 can block cell cycle progression both in G_1 through interactions with cyclin–cdk complexes and in S through direct interaction with the replication complexes, and at the same time it can augment repair. p21 WAF1/CIP1 can be induced by both p53 and p53-independent mechanisms. The cyclin–cdk complexes in most cells with intact p53 contain p21 WAF1/CIP1. Cells with defective p53 may contain less p21 WAF1 CIP1 overall.

4.2 GADD45

The induction of p53 by DNA damage also results in the induction of another mRNA whose protein product can contribute to a G_1 arrest and to augmented repair capacity. GADD45 was first isolated by subtractive hybridization after DNA damage to cells (FORNACE et al. 1989a). It proved to be induced during either growth arrest induced by serum starvation or after DNA damage (FORNACE et al. 1989b). It, too, is p53 dependent after DNA damage and also inhibits in vitro replication while enhancing repair (KASTAN et al. 1992; SMITH et al. 1994). The effect of adding both p21 and GADD45 to these in vitro replication repair systems has not yet been reported, but the combined action of both has a greater effect on growth inhibition than either one alone. Thus in cells lacking wild-type p53 activity, a frequent mutation in many human cancers, this inhibitory mechanism is lacking, resulting in a loss of G_1 arrest after DNA damage which may also coincide with diminished repair capacity. Thus the suggestion has arisen that p53 mutations may lead to increased levels of mutations because of the lack of a G_1 checkpoint associated with lack of augmented repair in vitro. So far it has been demonstrated that DNA amplifications occur at many orders of magnitude greater frequency in cells with defective p53 pathways (LIVINGSTONE et al. 1992). The nature of other types of mutations which might be enhanced due to the lack of p21 and GADD45 remain to be established.

4.3 p15 and p16

At least three other inhibitors of cdk complexes have been identified. p15 and p16 both map closely to one another on the human 9p21, and both inhibit kinase activity of cdk4 complexes and probably cdk6 complexes (HANNON and BEACH 1994; JEN et al. 1994;OKAMOTO et al. 1994; SERRANO et al. 1993). Inherited p16 mutations have been found in a subset of families with syndromes of familial melanoma, and homozygous deletions of p16 are frequently found in cell lines derived from various cancers (HUSSUSSIAN et al. 1994; KAMB et al. 1994a,b; WEAVERFELDHAUS et al. 1994). Homozygous deletions of p16 have also been found in tissues from the actual tumors of malignant mesothelioma, esophageal cancer, lung cancer, leukemias, and gliomas as well as melanoma (CHENG et al. 1994; IGAKI et al. 1994; JEN et al. 1994; OGAWA et al. 1994, OKAMOTO et al. 1994; SCHMIDT et al. 1994). The genetics of p15 mutations are currently less well explored, but in gliomas both p16 and p15 are affected by homozygous deletion in many cases (JEN et al. 1994). In cell lines from gliomas lacking alterations in p16, HE et al. found that cdk4 was amplified, indicating that uncontrolled cdk4 activity must be important in glioma tumor and that this can be achieved via several routes (HE et al. 1994). Analogously, in lung cancer cell lines with wild-type Rb, p16 expression was universally absent, both in small cell and non-small-cell carcinomas, whereas those cells with mutant Rb still retained p16 expression (OTTERSON et al. 1994). This again suggests that there are several targets for disrupting the function of the pathways involving cdk4, cyclin D1, and Rb. Furthermore, the transcriptional expression of p16 may be repressed by Rb (LI et al. 1994).

4.4 Transforming Growth Factor-β and $p27^{Kip1}$

Transforming growth factor (TGF)-β was first discovered as a growth factor secreted by transformed cell lines. Subsequently, it was found to inhibit the growth of many normal cell types such as keratinocytes while promoting the growth of squamous cell carcinoma cells. TGF-β treated cells were found to contain an inhibitor $p27^{Kip1}$ in cdk complexes which was not present in the complexes from untreated cells (OSTEEN et al. 1994; POLYAK et al. 1994a, 1994b, SLINGERLAND et al. 1994; TOYOSHIMA and HUNTER 1994). $p27^{Kip1}$ binds to and preferentially inhibits the activity of cyclin E–cdc2 complexes. p15 has been found to be induced in TGF-β-treated cells, so that both of these inhibitors may contribute to the growth arrest initiated by TGF-β (HANNON and BEACH 1994). These results do not define the mechanism by which TGF-β acts as a positive growth factor in many transformed cells in contrast to its growth-inhibitory effects on many normal cells. One clue that transformed cells may behave very differently from normal cells with regard to their cdk complex formation comes from the work of XIONG et al. (1993). In diploid fibroblasts, cyclin D exists in a complex with p21, cdk, and PCNA, yet in transformed cells cdk4 is not found associated with cyclin D, PCNA, or p21, but is only found associated with p16. Cyclin B1–cdc2 complexes no longer contained PCNA or p21 after transformation, and cyclin A was found to be associated with an as yet uncharacterized protein, p19.

5 Regulation of Cell Proliferation and the Control of Apoptosis

There are several instances where control of apoptosis seems to be linked to some of the genes which also regulate cell cycle progression. Thymocytes undergo apoptosis at the G_0 to G_1 transition in response to glucocorticoids or X-rays (TELFORD et al. 1991; YAMADA and OHYAMA 1988). p53 expression is necessary for this effect (LOWE et al. 1993). Thymocytes from p53 null mice are extremely resistant to induction of apoptosis in response to X-rays. Thus the p53 gene, which is strongly implicated in human and animal carcinogenesis, can also be a significant regulator of the process of apoptosis. While aberrant regulation of cell proliferation, both by subversion of tumor suppressor gene activity and by the action of dominant oncogenes, is now recognized to be the hall mark of most, if not all, malignancies, recent work has demonstrated that these alterations in cell-proliferative activity are also intimately linked to the control of apoptosis. This linkage has been recognized both for the tumor suppressor genes p53 and Rb and for the dominant oncogene *myc*. Additionally, it has been recognized that many transforming viruses exert their transforming function in part by abrogating apoptotic functions in the cell. This has been most fully studied for SV40, the herpes viruses, and adenoviruses, but similar mechanisms clearly are also found in polyoma, Epstein-Barr virus, and the baculoviruses. In viruses these anti apoptotic functions presumably evolved to permit viral replication, which could be subverted by apoptosis, but they in turn promote the uncontrolled cellular proliferation which is similarly seen in both virally induced and spontaneous tumors (WHITE 1993).

5.1 p53 and Apoptosis

p53 mutations are now recognized to be the most common genetic changes in human cancer. Mutant p53 was first isolated as what was believed to be a dominant cooperating oncogene which behaved in many ways like *myc* or the E1A gene of adenovirus in that it would cooperate with *ras* in transformation assays of 3T3 cells. Subsequently, it was realized that the form of p53 which was first isolated yielded a mutant, but long-lived form of the protein which displaced the wild-type, but short-lived native form from its binding site and, because of this, appeared to act dominantly. p53 is now more properly thought of as a tumor suppressor gene. The action of p53 is complex. It binds to many important cellular proteins and is involved in the control of gene expression. The last 2 years have seen an intense focus on its roles in cell cycle delay in G_1 phase and in apoptosis. It is now recognized that p53 can regulate both cell proliferation and induce apoptosis depending on the circumstances and cellular background (LANE 1992, 1993; OREN 1992). An insight into its complex role can be derived from recent studies of the interaction of p53 with the Rb gene product (WHITE 1994). Several recent studies have demonstrated that, in the absence of retinoblastoma function, the apoptotic action of p53 can compensate for this loss and thus prevent malignant transformation (MORGENBESSER et al. 1994).

The Rb protein prevents cell cycle progression at G_1/S by inhibiting members of the E2F family of transcription factors, thereby inhibiting the expression of many genes implicated in S phase, including TK, *myc*, *myb*, DHFR, and DNA polymerase-α. Thus the normal expression of the Rb gene product in the developing lens causes the cells to undergo growth arrest and to terminally differentiate. Recent studies in Rb knockout mice have shown that in this case the cells continue to enter S phase; however, an active p53 gene product causes the cells to undergo apoptosis. Such mice show abnormal eye development (micropthalmia), but do not develop tumors. p53 knockout mice show normal eye development, since in these mice Rb functions normally (MORGENBESSER et al. 1994). Double-knockout mice cannot be studied, since they die in midgestation, but an insight into the effect of the absence of both genes can be gained from the study of mice transgenic for the human papilloma virus E6 and E7 genes, which target and inactivate p53 and Rb. Such transgenic mice develop eye tumors (PAN and GRIEP 1994). In this case, therefore, the apoptotic action of p53 is essential for its tumor suppressor activity in the absence of Rb. Similar results were found using transgenic mice with an SV40 T antigen fragment that inactivates Rb, but does not affect p53. Tumors develop slowly in these mice in comparison to their development in mice expressing this transgene that are p53 null, in which case rapid tumor development with minimal apoptosis is seen. It is of great interest that the functions of both Rb and p53 are tightly linked to cell cycle events in the G_1 phase of the cell cycle. The Rb gene product is linked to events at the G_0 to G_1 transition. p53 is also known to be able to induce G_1 arrest as well as apoptosis. X-rays or other DNA-damaging agents have been shown to increase the levels of p53 by a post-translational mechanism. The results of increased p53 expression were summarized above.

The contribution of p53 to tumor progression has also been seen in the development of squamous cell carcinoma after exposure of skin to UV radiation. BRASH and his laboratory have shown that p53 mutations have sequence patterns suggestive of repair of thymine dimers and hence can be supposed to result from repair after UV radiation. They have also shown that after UV irradiation there is a clonal expansion of cells containing p53 mutations, presumably because the radiation-damaged cells with wild-type p53 will undergo apoptosis, while those cells with DNA damage that now express a mutant p53 will survive and replace the dying cells (ZEIGLER et al. 1993, 1994). Thus the induction of a mutation in p53 can serve both an initiating and a promotion function in the language of the two-step models of carcinogenesis.

5.2 *myc* and Apoptosis

The *myc* oncogene has also been intensively studied for its roles in both cell proliferation and apoptosis. In nontransformed cells, *myc* expression is tightly linked to mitogenic stimuli and is a prerequisite for cell growth. Expression of *myc* has been shown to be both necessary and sufficient to cause G_0 fibroblasts to enter the cell cycle. Conversely, immortalized fibroblasts which constitutively express *myc* are unable to exit the cycle upon serum withdrawal, and they then undergo apoptosis. Similar results are seen in lymphocytes expressing *myc* (HARRINGTON et al. 1994a). The relationship of *myc*-induced apoptosis to cell cycle progression is somewhat unclear. The *myc* gene is an early response gene whose expression rises rapidly at the G_0 to G_1 transition and whose currently known functions are largely linked to G_1 and early S. Unlike most early response genes, however, *myc* expression is sustained throughout the cell cycle. In addition to its proliferative function, *myc* can be shown to induce apoptosis in G_0 and G_1 and EVAN has argued that it can also do so in S phase (EVAN et al. 1992), based on an experiment in which cells carrying an estrogen-*myc* chimeric protein, in which estrogen induces *myc* activity, were held for 48 h in high concentrations of thymidine to induce an S phase arrest. When estrogen was added, after 48 h 100% of the cells in the culture underwent apoptosis, from which EVAN concluded that *myc* could induce apoptosis in S phase. From this experiment EVAN has argued that the apoptotic action of *myc* is not cell cycle specific.

EVAN has argued also that *myc* is a gene whose proliferative function is not linked to any specific phase of the cell cycle. Its continued, unregulated expression throughout the cycle would tend to support this point of view. However, multiple experiments point to roles for *myc* at very specific points in the cell cycle. As was previously stated, *myc* expression can cause G_0 cells to enter the cycle (EILERS et al. 1991). In continuously proliferating cells, a role for *myc* can also be demonstrated in G_1. Serum withdrawal from proliferating fibroblasts leads to rapid disappearance of *myc* from all cells. Nevertheless, cells in S, G_2 and M at the time of serum withdrawal continue to move through the cycle and the entire population accumulates in G_1. These effects are not restricted to fibroblasts. Treatment of proliferating keratinocytes or hematopoietic cells with anti sense oligonucleotides to c-*myc* leads to an accumulation of cells in G_1 (HEIKKILA et al. 1987). There is thus an abundance of evidence that *myc* plays an essential role in G_1. Evidence for its role in other parts of the cycle is much more limited. SHIBUYA et al. studied BAF-B03 pro-B cells expressing a transfected epithelial growth factor (EGF) receptor (SHIBUYA et al. 1992). These cells did not express c-*myc* in response to EGF, but entered S normally, only to block before entering G_2. This led them to conclude that *myc* was necessary for the S/G_2 transition in these cells, although S/G_2 blocks were not noted in the other experiments described above. These experiments taken together provide a somewhat confusing view of the role of *myc* in relationship to the cell cycle. They do, however, suggest that the actions of the *myc* oncogene are certainly not random with respect to the cell cycle, but they may be very specific with regard to the system under study.

Several models have been proposed to explain the apparently contradictory roles of *myc* in both proliferation and apoptosis. EVAN originally proposed what he termed the "conflict model," in which *myc* is viewed as promoting proliferation. In the presence of mitogens, there is no conflict between the action of *myc* and the other signals the cells are receiving via other mitogens. When, however, mitogens are *absent*, as in serum withdrawal, or when a cytostatic drug is added, there is a "conflict" between the signal from *myc* to proliferate and the other signals the cells are receiving to cease proliferation. This conflict is viewed, in an unspecified manner, as leading to apoptosis. However, it is now thought that the functions of *myc* in promoting proliferation and apoptosis may be inseparable. *myc* is a transcriptional regulator acting via an amino-terminal transactivating domain which can be shown to be involved in both proliferation and apoptosis (EVAN and LITTLEWOOD 1993). Levels of *myc* expression correlate with both proliferative capacity and susceptibility to apoptosis. Finally, *myc* is known to form an active complex by heterologous dimerization with a partner protein named Max; *myc*–Max dimerization has been shown to be necessary for both proliferation and apoptosis (AMATI et al. 1993; AMATI and LAND 1994; EVAN and LITTLEWOOD 1993). As evidence has accumulated that *myc* is itself directly involved in the control of apoptosis, EVAN has more recently modified the conflict model by proposing what he now terms the "dual signal" model. In this model, *myc* simultaneously activates both a proliferative and an apoptotic pathway (HARRINGTON et al. 1994b). Mitogens then stimulate the proliferative pathway, while the apoptotic pathway is actively held in check by other antiapoptotic cellular factors. The most widely studied of these antiapoptotic factors is the oncogene *bcl-2*.

5.3 *bcl-2* and Apoptosis

The *bcl-2* gene was discovered as a gene whose expression was increased by chromosomal translocations in B cell malignancies (this subject has recently been extensively reviewed in REED 1994). It is found to be activated in the majority of follicular non-Hodgkin's lymphomas. It has been noted less commonly in other malignancies such as prostate cancer. Its activation has also been seen in some benign conditions such as follicular hypertrophy of lymph nodes and tonsils. In multiple systems, e.g. lymphocytes, fibroblasts, neurons, and hematopoietic cells, expression of *bcl-2* can be shown to be able to delay or even prevent apoptosis. Conversely, down regulation of *bcl-2* in many of these same systems can be shown to promote apoptosis. Its mechanism of action in this regard is poorly understood at this time. Although it has been proposed to act as an oxygen radical scavenger, there is evidence contradicting this hypothesis (HOCKENBERY et al. 1993; KANE et al. 1993; MUSCHEL et al. 1995).

The *bcl-2* gene specifically blocks the ability of c-*myc* to induce apoptosis, but does not affect the mitogenic properties of *myc*. The ability of *bcl-2* to block *myc*-induced apoptosis has been invoked to explain the ability of *bcl-2* and *myc* to function synergistically in tumor induction (NUNEZ et al. 1989; REED et al. 1990). Transgenic mice, engineered to express *bcl-2* and *myc*, develop B cell lymphomas at a much higher rate than mice engineered to express only one of the two genes (STRASSER et al. 1990). Two things should be noted, however. While *bcl-2* overexpression is common in spontaneous B cell malignancies in man, these tumors represent a small, albeit important, part of the cancer burden; *myc* deregulation, on the other hand, is a common event in human malignancies of many kinds. Thus, tumor progression is affected both through alterations is genes that regulate the G_1 to S transition and by genes that reduce the occurrence of apoptosis. In some cases the same genes appear to contribute to the regulation of both processes.

6 Cell Cycle Response to DNA Damage

The genes regulating apoptosis and cell cycle progression may also play a role in the response to tumors to chemotherapy or radiation therapy. Many chemotherapeutic agents as well as ionizing radiation result in damage to the cellular DNA. After damage of the DNA, the cell may undergo apoptosis, repair the DNA, or be unable to repair the DNA and go on to a mitotic or a necrotic death. Part of the cellular response to DNA damage includes perturbations of the cell cycle. After ionizing radiation, cells may prolong both the G_1 and the G_2 phase of the cycle. The G_1 delay is dependent upon the cells having wild-type p53 and is mediated through the transcriptional induction of WAF1/CIP1 as well as GADD45. Although both WAF1 and GADD45 can be induced by p53-independent stimuli, so far the G_1 response to DNA damage appears to be dependent upon p53 (JIANG et al. 1994; MICHIELI et al. 1994; SHEIKH et al. 1994). It has been proposed that the G_1 delay might alter repair and hence survival, and indeed both WAF1 and GADD45 have been found to alter some types of repair, but the effect upon apoptosis might be equally significant.

6.1 G_2 Phase and the Response to DNA Damage

The G_2 phase of the cell cycle is also prolonged after various forms of DNA damage, including those induced by ionizing radiation or topoisomerase I inhibitors, such as camptothecin, or other agents that interact directly with DNA, such as bleomycin. There is currently no evidence to suggest that the G_2 delay is directly influenced by p53. In HeLa cells, the G_2 arrest has been shown to correlate temporally with a decline in cyclin B1 protein levels, while cyclin A is unafffected (Muschel et al. 1991, 1993). Cyclin B1 usually increases in amount at the S/G_2 boundary, but after irradiation, although the cells remain viable and although they enter G_2, they do not show increased levels of cyclin B1 until much later as they resume cycling. The cyclin B1 mRNA levels show a similar pattern. Cyclin B1 mRNA levels vary by as much as 50-fold through the cell cycle, with little seen in G_1, more in S, and the maximum in G_2/M. Maity et al. have shown that cyclin B1 mRNA level stability also varies through the cell cycle, with a rapid half-life of 1-2 h in G_1 and a longer half-life of 13 h in G_2/M (Maity et al. 1994). Irradiation results in a reduction of the half-life of cyclin B1 mRNA to 2.5 h. Although irradiation also results in reduced cyclin B1 mRNA levels in rat embryo cells transformed by *ras* plus *myc*, it does not in several types of lymphocytic cell lines. In some cells hyperphosphorylation of p34 cdc2 seems to accompany the G_2 block (Lock 1992; O'Connor et al. 1992, 1993). O'Connor et al. have shown that inactivation of cdc25 via phosphorylation can accompany treatment of cells with nitrogen mustard, a DNA-damaging agent that results in a G_2 delay (O'Connor et al. 1993). Thus there may be several mechanisms underlying the G_2 delay induced by irradiation, depending upon the cell type.

There are suggestions that the G_2 delay may also contribute to survival. In yeast, mutations that abrogate the induction of a G_2 delay by DNA-damaging agents also render the cells markedly radiosensitive (Weinert and Hartwell 1988). In mammalian cells caffeine will greatly reduce the G_2 delay after irradiation and will also render the cells markedly more radiosensitive (Busse et al. 1977). In HeLa cells, treatment with caffeine at the time of irradiation also restores cyclin B1 mRNA levels to normal (Bernhard et al. 1994). Although the dose of caffeine and its congener pentoxyphylline would be too large to use in patients, these effects leave the hope that adjuvant treatments may be developed that could result in radiosensitization through altered cell cycle response to DNA damage.

In analogy to the G_1 checkpoint, events at the G_2 checkpoint may also affect the induction of apoptosis. Treatment of EL4 cells with caffeine after irradiation enhanced the level of apoptosis. These cells do, of course, undergo significant amounts of apoptosis after irradiation (Palayoor et al. 1995); however, we have found that HeLa cells will also undergo apoptosis after irradiation when treated with caffeine, while radiation itself results in predominantly necrotic and mitotic cell death (Bernhard et al. 1996).

Note added in Proof: Since writing this review much work has been published that bears on the concepts described here. Interested readers will note that transgenic mice lacking p21 WAF1/CIP1 have been generated and their cells have been shown to be defective in the G1 checkpoint (Deng et al. 1995). Cyclin D1 deficient mice have also been generated and are beginning to be evaluated for susceptibility to carcinogenesis (Sicinski et al. 1995). Finally data linking p53 expression to the G2 checkpoint has been published (Agarwal et al. 1995; Aloni-Grinstein et al. 1995; Guillouf et al. 1995; Stewart et al. 1995) as well as data of an indirect link based on the response to caffeine (Fan et al. 1995; Powell et al. 1995; Russell et al. 1995).

References

Agarwal ML, Agarwal A, Taylor WR, Stark GR (1995) p53 controls both the G2/M and the G1 cell cycle checkpoints and mediates reversible growth arrest in human fibroblasts. Proc Natl Acad Sci USA 92: 8493–8497

Aloni-Grinstein R, Schwartz D, Rotter V (1995) Accumulation of wild type p53 protein upon gamma irradiation induces a G2 arrest-dependent immunoglobulin kappa light chain gene expression. EMBO J 14: 1392–1401

Amati B, Land H (1994) Myc-Max-Mad: a transcription factor network controlling cell cycle progression, differentiation and death. Curr Biol 4: 102–108

Amati A, Littlewood TD, Evan GI, Land H (1993) The c-myc protein induces cell cycle progression and apoptosis through dimerization with Max. EMBO J 12: 5083–5087

Bernhard EJ, Maity A, Muschel RJ, McKenna WG (1994) Increased expression of cyclin B1 mRNA coincides wirth diminished G2-phase arrest in irradiated HeLa cells treated with staurosporine or caffeine. Radiat Res 140: 393–400

Bernhard EJ, Muschel RJ, Bakanauskas VJ, McKenna WG (1996) Reducing the radiation-induced G2 delay causes HeLa cells to undergo apoptosis instead of mitotic death. Int J Radiat Biol (in press)

Bosch F, Jares P, Campo E, Lopez-Guillermo A, Piris M, Villamor N, Tassies D, Jaffe E, Montserrat E, Rozman C, Cardesa A (1994) PRAD-1/Cyclin D1 gene overexpression in chronic lymphoproliferative disorders: a highly specific marker of mantle cell lymphoma. Blood 84: 2726–2732

Busse P, Bose S, Jones R, Tolmach L (1977) The action of caffeine on X-irradiated HeLa cells. II. Synergistic lethality. Radiat Res 71: 666–667

Callender T, el-Naggar A, Lee M, Frankenthaler R, Luna M, Batsakis J (1994) PRAD-1 (CCND1)/cyclin D1 oncogene amplification in primary head and neck squamous cell carcinoma. Cancer 74: 152–158

Cheng J, Jhanwar S, Klein W, Bell D, Lee W-C, Altomare D, Nobori T, Olopade O, Buckler A, Testa J (1994) p16 alterations and deletion mapping of 9p21-p22 in malignant mesothelioma. Cancer Res 54: 5547–5551

Deng G, Zhang P, Harper JW, Elledge SJ, Leder P (1995) Mice lacking p21 cip1/waf1 undergo normal development, but are defective in G1 checkpoint control. Cell 82: 675–684

Desai D, Wessling H, Fisher R, Morgan D (1995a) Effects of phosphorylation by CAK on cyclin binding by cdc2 and cdk2. Mol Cell Biol 15: 345–350

Desai D, Wessling H, Fisher R, Morgan D (!995b) Effects of phosphorylation by CAK on cyclin binding by CDC2 and CDK2. Mol Cell Biol 15: 345–350

Doree M, Galas S (1994) The cyclin-dependent protein kinases and the control of cell division. FASEB J 8: 1114–1121

Dynlacht B, Flores O, Lees J, Harlow E (1994) Differential regulation of E2F transactivation by cyclin/cdk2 complexes. Genes Dev 8: 1772–1786

Eilers M, Schirm S, Bishop JM (1991) The myc protein activates transcription of the alpha-prothymosin gene. EMBO J 10: 133–141

el-Deiry W, Tokino T, Velculescu V, Levy D, Parson R, Trent J, Lin D, Mercer W, Kinzler K, Vogelstein B (1993) WAF1, a potential mediator of p53 tumor supression. Cell 75: 817–825

Evan GI, Littlewood TD (1993) The role of c-myc in cell growth. Curr Opin Genet Dev 3: 44–49

Evan GI, Gilbert CS, Littlewood TD, Land H, Brooks M, Waters CM, Penn LZ, Hancock DC (1992) Induction of apoptosis in fibroblasts by c-myc protein. Cell 69: 119–128

Fan SJ, Smith ML, Rivet DJ, Duba D, Zhan QM, Kohn KW, Fornace AJ, O'Connor PM (1995) Disruption of p53 function sensitizes breast cancer MCF7 cells to cisplatin and pentoxifylline. Cancer Res 55: 1649–1654

Fornace A, Alamo I, Hollander M (1989a) DNA damage-inducible transcripts in mammalian cells. Proc Natl Acad Sci USA 85: 8800–8804

Fornace A, Nebert D, Hollander M, Leuthy J, Papathanasiou M, Fargnoli J, Holbrook N (1989b) Mammalian genes coordinately regulated by growth arrest signals and DNA-damaging agents. Mol Cell Biol 9: 4196–4203

Gould K, Moreno S, Owen D, Sazer S, Nurse P (1991) Phosphorylation at thr 167 is required for schizosaccharomyces pombe p34cdc2 function. EMBO J 10: 3297–3309

Guillouf C, Rosselli F, Krishnaraju K, Moustacchi E, Hoffman B, Liebermann DA (1995) p53 involvement in control of G2 exit of the cell cycle: role in DNA damage-induced apoptosis. Oncogene 11: 2263–2270

Hannon GJ, Beach D (1994) p15 (INK4B) is a potential effector of TGF-beta induced cell cycle arrest. Nature 371: 257–261

Harper JW, Adami GR, Wei N, Keyomarcki K, Elledge SJ (1993) The p21 cdk-interacting protein Cip1 is a potent inhibitor of G1 cyclin dependent kinases. Cell 75: 805–816

Harrington EA, Bennett MR, Fanidi A, Evan GI (1994a) c-Myc-induced apoptosis in fibroblasts in inhibited by specific cytokines. EMBO J 13: 3286–3295

Harrington EA, Fanidi A, Evan GI (1994b) Oncogenes and cell death. Curr Opin Gene Dev 4: 120–129

He J, Allen J, Collins V, Allalunisturner M, Godbout R, Day R, James C (1994) Cdk4 amplification is an alternative mechanism to p16 gene homozygous deletion in glioma cell-lines. Cancer Res 54: 5804

Heikkila R, Schwab G, Wickstrom E, Loke S, Pluznik D, Watt R, Neckers L (1987) A c-myc antisense oligodeoxynucleotide inhibits entry into S phase but not progress from G0 to G1. Nature 328: 445–449

Hockenbery DM, Oltvai ZN, Yin XM, Milliman CL, Korsmeyer SJ (1993) Bcl-2 functions in an antioxidant pathway to prevent apoptosis. Cell 75: 241–251

Hunter T, Pines J (1994) Cyclins and cancer II: cyclin D and CDK inhibitors come of age. Cell 79: 573–582

Hussussian C, Struewing J, Goldstein A, Higgins P, Ally D, Sheahan M, Clark W, Tucker M, Dracopoli N (1994) Germline p16 mutations in familial melanoma. Nature Genet 8: 15–21

Igaki H, Sasaki H, Kishi T, Sakamoto H, Tachimori Y, Kato H, Watanabe H, Sugimura T, Terada M (1994) Highly frequent homozygous deletion of the p16 gene in estophageal cancer cell lines. Biochem Biophys Res Commun 203: 1090–1095

Jen J, Harper J, Bigner S, Bigner D, Papadopoulos N, Markowitz S, Willson J, Kinzler K, Vogelstein B (1994) Deletion of *p16* and *p15* genes in brain tumors. Cancer Res 54: 6353–6358

Jiang H, Lin J, Su Z, Collart F, Huberman E, Fisher P (1994) Induction of differentiation in human promyelocytic HL-60 leukemia cell activates p21, WAF1/CIP1, expression in the absence of p53. Oncogene 9: 3397–3406

Kamb A, Gruis N, Weaverfeldhaus J et al. (1994a) A cell cycle regulator potentially involved in genesis of many tumor types. Science 264: 436–440

Kamb A, Shattuckeidens D, Eeles R et al. (1994b) Analysis of the p16 gene (cdkn2) as a candidate for the chromosome 9p melanoma susceptibility locus. Nature Genet 8: 22–26

Kane DJ, Safafian TA, Anton R, Hahn H, Gralla EB, Valentine JS, Ord T, Bredesen DE (1993) Bcl-2 inhibition of neural death: decreased generation of reactive oxygen species. Science 262: 1274–1277

Kastan M, Zhan Q, el-Deiry W, Carrier F, Jacks T, Walsh W, Plunkett B, Vogelstein B, Fornace A (1992) A mammalian cell cycle checkpoint pathway utilizing p53 and GADD45 is defective in ataxia-telangiectasia. Cell 71: 587–597

Kastan MB, Onyekwere O, Sidransky D, Vogelstein B, Craig RW (1991) Participation of p53 protein in the cellular response to DNA damage. Cancer Res 51: 6304–6311

Kato J-Y, Matusushime H, Hiebert S, Ewen M, Sherr C (1993) Direct binding of cyclin D to the retinoblastoma gene product (PRb) and pRb phosphorylation by the cyclin D-dependent kinase CDK4. Genes Dev 7: 331–342

Keyomarsi K, Pardee A (1993) Redundant cyclin overexpression and gene amplification in breast cancer cells. Proc Natl Acad Sci USA 90: 1112–1116

King RW, Jackson PK, Kirschner MW (1994) Mitosis in transition. Cell 79: 563–570

Krek W, Marks J, Schmitz N, Nigg E, Simanis V (1992) Vertebrate p34cdc2 phosphorylation site mutants: effects upon cell cycle progression in the fission yeast *Schizosaccharomyces pombe*. J Cell Sci 102: 43–53

Krek W, Nigg E (1991) Differential phosphorylation of vertebrate p34cdc2 kinase at the G1/S and G2/M transitions of the cell cycle: identification of major phosphorylation sites. EMBO J 10: 305–316

Kuerbitz S, Plunkett B, Walsh W, Kastan M (1992) Wild-type p53 is a cell cycle checkpoint determinant following irradiation. Proc Natl Acad Sci USA 89: 7491–7495
Lane DP (1992) p53, guardian of the genome. Nature 358: 15–16
Lane DP (1993) A death in the life of p53. Nature 362: 786
Leach R, Elledge S, Sherr C, Willson J, Markowitz S, Kinzler K, Vogenstein B (1993) Amplification of cyclin genes in colorectal carcinomas. Cancer Res 53: 1986–1989
Lebwoh D, Muise-Helmericks R, Sepp-Lorenzino L, Serve S, Timaul M, Bol R, Borgen P, Rosen N (1994) A truncated cyclin D1 gene encodes a stable mRNA in a human breast cancer cell line. Oncogene 9: 1925–1929
Li R, Waga S, Hannon G, Beach D, Stillman B (1994) Differential effects by the p21 inhibitor on PCNA-dependent DNA replication and repair. Nature 37: 534–537
Li Y, Nichols M, Shay J, Xiong Y (1994) Transcriptional repression of the D-type cyclin-dependent kinase inhibitor p16 by the retinoblastoma susceptibility gene product pRb. Cancer Res 54: 6078–6082
Livingstone LR, White A, Sprouse J, Livanos E, Jacks T, Tlsty TD (1992) Altered cell cycle arrest and gene amplification potential accompany loss of wild-type p53. Cell 70: 923–935
Lock R (1992) Inhibition of p34 cdc2 kinase activation, tyrosine dephosphorylation, and mitotic progression in Chinese hamster ovary cells exposed to etoposide. Cancer Res 52: 1817–1822
Lorca T, Labbe J, Devault A, Fesquet D, Capony J, Cavadore J, Bouffant FL, Doree M (1992) Dephosphorylation of cdc2 on threonine 161 is required for cdc2 kinase inactivation and normal anaphase. EMBO J 11: 2381–2390
Lowe S, Schmitt E, Smith S, Osborne B, Jacks T (1993) P53 is required for radiation-induced apoptosis in mouse thymocytes. Nature 362: 847–849
Maity A, McKenna WG, Muschel RJ (1994) Posttranscriptional regulation of cyclin B1 mRNA through the cell cycle and after irradiation. EMBO J 14: 603–609
Michieli P, Chedid M, Lin D, Pierce J, Mercer W, Givol D (1994) Induction of WAF1/CIP1 by a p53-independent pathway. Cancer Res 54: 3391–3395
Morgan D (1995) Principles of CDK regulation. Nature 374: 131
Morgenbesser SD, Williams BO, Jacks T, Depinho RA (1994) p53 dependent apoptosis produced by rb deficiency in the developing mouse lens. Nature 371: 72–74
Motokura T, Bloom T, Kim H, Juppner H, Ruderman J, Kronenberg H, Arnold A (1991) A novel cyclin encoded by a bcl1-linked candidate oncogene. Nature 350: 512–515
Muschel R, Zhang H, Iliakis G, McKenna W (1991) Cyclin B expression in HeLa cells during the G2 block induced by ionizing radiation. Cancer Res 51: 5113–5117
Muschel R, Zhang H, McKenna W (1993) Differential effect of ionizing radiation on the expression of cyclin A and cyclin B in Hela cells. Cancer Res 53: 1–8
Muschel R, Bernhard E, Garza L, McKenna W, Koch C (1995) Induction of apoptosis at different oxygen tensions:evidence that oxygen radicals do not mediate apoptotic signalling. Cancer Res 55: 995–998
Musgrove E, Lee C, Buckley M, Sutherland R (1994) Cyclin D1 induction in breast cancer cells shortens G_1 and is sufficient for cells arrested in G_1 to complete the cell cycle. Proc Natl Acad Sci USA 91: 8022–8026
Norbury C, Nurse P (1992) Animal cell cycles and their control. Annu Rev Biochem 61: 441–470
Nunez G, Seto M, Seremetis S, Ferrero D, Grignani F, Korsmeyer S, Favera RD (1989) Growth and tumor promoting effects of deregulated bcl2 in human B-lymphoblastoid cells. Proc Natl Acad Sci USA 86: 4589–4593
O'Connor PM, Ferris DK, White GA, Pines J, Hunter T, Longo DL, Kohn KW (1992) Relationships between cdc2 kinase, DNA cross-linking, and cell cycle perturbations induced by nitrogen mustard. Cell Growth Differ 3: 43–52
O'Connor PM, Ferris DK, Pagano M, Draetta G, Pines J, Hunter T, Longo DL, Kohn KW (1993) G2 delay induced by nitrogen mustard in human cells affects cyclin A/cdk2 and cyclin B1/cdc2-kinase complexes differently. J Biol Chem 268: 8298–8308
O'Connor PM, Ferris DK, Hoffman I, Jackman J, Draetta G, Kohn KW (1994) Role of the cdc25C phosphatase in G2 arrest induced by nitrogen mustard in human lymphoma cells. Proc Natl Acad Sci USA 91: 9480–9484
Ogawa S, Hirano N, Sato N, Takahashi T, Hangaishi A, Tanaka K, Kurokawa M, Tanaka T, Mitani K, Yazaki Y, Hirai H (1994) Homozygous loss of the cyclin-dependent kinase 4-inhibitor (P16) gene in human leukemias. Blood 84: 2431–2435
Okamoto A, Demetrick D, Spillare E, Hagiwara K, Hussain S, Bennett W, Forrester K, Gerwin B, Serrano M, Beach D, Harris C (1994) Mutations and altered expression of $p16^{INK4}$ in human cancer. Proc Natl Acad Sci USA 91: 11045–11049

Oren M (1992) The involvement of oncogenes and tumor suppressor genes in the control of apoptosis. Cancer Metastasis Rev 11: 141–148
Osteen K, Rodgers W, Gaire M, Hargrove J, Gorstein F, Matrisian L (1994) Stromalepithelial interaction mediatesteroidal regulation of metalloproteinase expression in human endometrium. Proc Natl Acad Sci USA 91: 10129–10133
Otterson G, Kratzke R, Coxon A, Kim Y, Kaye F (1994) Absence of p16ink4 protein is restricted to the subset of lung cancer lines that retains wildtype RB. Oncogene 9: 3375–3378
Palayoor S, Macklis R, Bump E, Coleman C (1995) Modulation of radiation-induced apoptosis and G(2)M block in murine T-lymphoma cells. Radiat Res 141: 235–243
Pan HC, Griep AE (1994) Altered cell cycle regulation in the lens of HPV-16 E6 or E7 transgenic mice-implications for tumor-suppressor gene function in development. Genes Dev 8: 1285–1299
Polyak K, Kato J-Y, Solomon M, Sherr C, Massague J, Roberts J, Koff A (1994a) $p27^{Kip1}$, a cyclin-Cdk inhibitor, links transforming growth factor-β and contact inhibition to cell cycle arrest. Genes Dev 8: 9–22
Polyak K, Lee M-H, Erdjument-Bromage H, Koff A, Roberts J, Tempst P, Massague J (1994b) Cloning of $p27^{Kip1}$, a cyclin-dependent kinase inhibitor and a potential mediator of extracellular antimitogenic signals. Cell 78: 59–66
Powell SN, Defrank JS, Connell P, Edgan M, Preffer F, Dombkowski D, Tang W, Friend S (1995) Differential sensitivity of p53– and p53+ cells to caffeine-induced radiosensitization and override of G (2) delay. Cancer Res 55: 1643–1648
Reed JC (1994) Mini-review: cellular mechanisms of disease series. J Cell Biol 124: 1–6
Reed J, Cuddy M, Haldar S, Croce C, Nowell P, Makover D, Bradley K (1990) Bcl-2-mediated tumorigenicity of a human T-lymphoid cell line: synergy with myc and inhibition by bc12 antisense. Proc Natl Acad Sci USA 87: 3660–3664
Resnitzky D, Gossen M, Bujard H, Reed S (1994) Acceleration of the G1/S phase transition by expression of cyclins D1 and E with an inducible system. Mol Cell Biol 14: 1669–1679
Russell KJ, Wiens LW, Demers GW, Galloway DA, Plon SE, Groudine M (1995) Abrogation of the G(2) checkpoint results in differential radiosensitization of G(1) checkpoint deficient and G(1) checkpoint-competent cells. Cancer Res 55: 1639–1642
Schmidt E, Ichimura K, Reifenberger G, Collins V (1994) CDKN2 (p16/MTS1) gene deletion of CDK4 amplification occurs in the majority of glioblastomas. Cancer Res 54: 6321–6324
Serrano M, Hannon G Beach D (1993) A new regulatory motif in cell cycle control causing specific inhibition of cyclin D/CDK4. Nature 366: 704
Sheikh M, Li X-S, Chen J-C, Shao Z-M, Ordonez J, Fontana J (1994) Mechanisms of regulation of WAF1/Cip1 gene expression in human breast carcinoma: role of p53-dependent and independent signal transduction pathways. Oncogene 9: 3407–3415
Sherr C (1993) Mammalian G_1 cyclins. Cell 73: 1059–1065
Sherr C (1994) G1 phase progression: cycling on cue. Cell 79: 551–555
Shibuya H, Yoneyama M, Ninomiya-Tsuji J, Matsumoto K, Taniguchi T(1992) IL-2 and EGF receptors stimulate the homatopoietic cell cycle via different signaling pathways: demonstration of a novel role for c-myc. Cell 70: 57–67
Shirodkar S, Ewen M, DeCapro J, Morgan J, Livingston D, Chittenden T (1992) The transcription factor E2F interacts with the retinoblastoma product and a p107-cyclin A complex in a cell cycle-regulated manner. Cell 68: 157–166
Sicinski P, Donaher JL, Parker SB, Li TS, Gardner H, Haslam SZ, Bronson RT, Elledge SJ, Weinberg RA (1995) Cyclin D1 provides a link between development and oncogenesis in the retina and breast. Cell 82: 621–630
Slingerland J, Hengst L, Pan G-H, Alexander D, Stampfer M, Reed S (1994) A novel inhibitor of cyclin-cdk activity detected in transforming growth factor β-arrested epithelial cells. Mol Cell Biol 14: 3683–3694
Smith M, Chen I, Zhan Q, Bae I, Chen C, Gilmer T, Kastan M, OConnor P, Fornace A (1994) Interaction of the p53-regulated protein GADD45 with proliferating cell nuclear antigen. Science 266: 1376–1380
Stewart N, Hicks GG, Paraskevas F, Mowat M (1995) Evidence for a second cell cycle block at G2/M by p53. Oncogene 10: 109–115
Strasser A, Harris A, Bath M, Cory S (1990) Novel primitive lymphoid tumours induced in transgenic mice by cooperation between myc and bcl-2. Nature 348: 331–333
Telford WG, King LE, Fraker PJ (1991) Evaluation of glucocorticoid-induced DNA fragmentation in mouse thymocytes by flow cytometry. Cell Prolif 24: 447–459

Toyoshima H, Hunter T (1994) p27, a novel inhibitor of G1 cyclin-cdk protein kinase activity, is related to p21. Cell 78: 67–74

Wang J, Chenivesse X, Henglein B, Brechot C (1990) Hepatitis B virus integration in a cyclin A gene in a hepatocellular carcinoma. Nature 343:555–557

Weaverfeldhaus J, Guirs N, Neuhausen S, Lepaslier D, Stockert E, Skolnick M, Kamb A (1994) Localization of a putative tumor-suppressor gene by using homozysgous deletions in melanomas. Proc Natl Acad Sci USA 1994: 7563–7567

Weinert TA, Hartwell LH (1988) The RAD9 gene controls the cell cycle response to DNA damage in *Saccharomyces cerevisiae*. Science 241: 317–322

White E (1993) Regulation of apoptosis by the transforming genes of the DNA tumor virus adenovirus. PSEBM 204: 30–39

White E (1994) P53, guardian of rb. Nature 371: 21–22

Xiong Y, Zhang H, Beach D (1993) Subunit rearrangement of the cyclin-dependent kinases is associated with cellular transformation. Genes Dev 7: 1572–1583

Xu M, Sheppard K-A, Peng C-Y, Yee A, Piwnica-Worms H (1994) Cyclin A/CDK2binds directly to E2F-1 and inhibits the dna N-binding activity of E2F-1/DP-1 by phosphorylation. Mol Cell Biol 14: 8420–8431

Yamada T, Ohyama H (1988) Radiation-induced interphase death of rat thymocytes is internally programmed (apoptosis). Int J Radiat Biol 53: 65–75

Zeigler A, Jonason A, Leffell D, Simon J, Sharma H, Kimmelman J, Remington L, Jacks T, Brash D (1994) Sunburn and p53 in the onset of skin cancer. Nature 372: 773–776

Zeigler A, Leffell D, Kunala S, Sharma H, Gailani M, Simon J, Halgerin A, Baden H, Shapiro P, Bale A, Brash D (1993) Mutation hotspots due to sunlight in the p53 gene of nonmelanoma skin cancers. Proc Natl Acad Sci USA 90: 4216–4220

Zhang H, Hannon G, Beach D (1994) p21-containing cyclin kinases exist in both active and inactive states. Genes Dev 8: 1750–1758

W[illegible]

Differential Gene Expression in Tumor Metastasis: Nm23

J.M.P. Freije, N.J. MacDonald, and P.S. Steeg

1 Identification and Characterization of Differentially Expressed Genes in Tumor Metastasis: Nm23 . . . 215
2 Transfection Experiments . . . 218
2.1 In Vivo Analysis . . . 218
2.2 In Vitro Analysis . . . 219
2.3 Implications from "Negative" Data . . . 221
3 Potential Functions in Normal Cellular Biology . . . 221
3.1 Differentiation . . . 221
3.2 Proliferation . . . 223
4 Potential Biochemical Mechanisms . . . 224
4.1 NDPK Activity . . . 224
4.2 Serine Phosphorylation . . . 226
4.3 Transcriptional Activation . . . 227
4.4 Other Activities . . . 229
References . . . 229

1 Identification and Characterization of Differentially Expressed Genes in Tumor Metastasis: Nm23

A central theme in metastasis research is the complex interplay between the tumor cell and its environment. Tumor cells reversibly adhere to each other and to the extracellular matrix (ECM), degrade ECM, move toward soluble and ECM-derived signals, arrest in host capillaries (sometimes in an organ-specific manner), colonize in response to host cytokines, and activate host angiogenic responses. A central hypothesis in metastasis research proposes that some of these phenotypic changes are mediated, at least in part, by increased or decreased tumor cell expression levels of specific genes.

Several techniques have been used to identify differentially expressed genes in the metastatic process. These include differential colony hybridization, subtractive hybridization, and differential display. All are founded on the analysis of mRNA from highly versus non- or low metastatic potential tumors to identify genes more

Women's Cancers Section, Laboratory of Pathology, National Cancer Institute, Bldg. 10, Rm. 2A33, Bethesda, MD 20892, USA abundant in one mRNA preparation than the other. Most of these investigations involve the use of cell lines from model systems to control for various aspects of normal cell metabolism and differentiation and to eliminate the potential artifactual influences of contaminating normal fibroblasts, endothelial cells, and immune cells found in tumors. It is critical to check the metastatic potential of each cell line by in vivo injection to ensure that its relative level of aggressiveness is stable.

We studied a series of murine melanoma cell lines derived from a single tumor, developed by Drs. Fidler and Kripke. Using a single low metastatic potential clonal cell line and a higher metastatic potential cell line, we used differential colony hybridization to identify 24 clones that were differentially expressed over three rounds of screening from a library of approximately 40 000 clones (Steeg et al. 1988a). While it is an enticing hypothesis that most or all of these cDNA clones are differentially expressed between low and highly metastatic melanoma cell lines, it is also possible that the two cell lines utilized differ in other aspects of their biology that are not understood at present. To identify those cDNA clones best correlated with differences in tumor metastatic potential we determined their relative expression levels on northern blots among seven related K-1735 murine melanoma cell lines of known in vivo metastatic potentials. Only one of the cDNA clones, *nm23*, exhibited a pattern of expression that was consistent with the tumor metastatic potential of all seven cell lines. Nm23 RNA levels were approximately ten fold higher in the two low metastatic potential K-1735 murine melanoma cell lines than among five related but more highly metastatic K-1735 cell lines (Steeg et al. 1988a). These data are typical of differential gene expression experiments using any of the above-mentioned techniques, and it is strongly suggested to utilize a model system with more than one cell line available of each phenotype.

Differential cloning approaches have identified genes thought to be involved in metastatic tumor progression. In addition, previously unidentified genes have been identified, leading to new hypotheses on the mechanism and/or regulation of the tumor metastatic process. Such genes include *nm23*, *WDNM1*, *WDNM2*, and *mta-1*. Once identified, we can ask (a) what proteins these genes encode, (b) Whether differential gene expression is correlated with or, alternatively, causally involved in metastatic progression, and (c) how the protein products function in metastasis.

Subsequent cloning has revealed that *nm23* encodes a family of at least two genes in most species. In humans *nm23-H1* and *nm23-H2* are both localized to chromosome 17q21 and encode 17-kDa proteins approximately 90% identical in amino acid sequence (Rosengard et al. 1989; Stahl et al. 1991; Backer et al. 1993; Kelsell et al. 1993). Nm23 genes have been independently cloned and named *awd* (abnormal wing discs) in *Drosophila* and nucleoside diphosphate kinases (NDPK) in several species.

Validation of *nm23* has proceeded by the examination of its expression patterns in other metastasis model systems. Reduced *nm23* expression, either at the RNA or protein level, was correlated with high tumor metastatic potential in

N-nitrosomethylurea induced rat mammary carcinomas (STEEG et al. 1988a), mouse mammary tumor virus induced mammary tumors (CALIGO et al. 1992), *ras* or *ras*+adenovirus 2 *E1a* transfected rat embryo fibroblasts (STEEG et al. 1988b) and B16 murine melanomas (LAKSHMI et al. 1993). In other systems no correlation was observed between *nm23* expression and tumor metastatic potential. This could signify that *nm23* is irrelevant to the regulation of tumor metastatic potential among these cell lines or, alternatively, that *nm23* is deregulated by means other than its simple reduced expression.

The potential relevance of *nm23* to human cancer metastasis has been further investigated by examining *nm23* expression levels in tumor cohorts. Metastatic potential is assessed by patient follow-up, (disease-free or overall survival), the presence of lymph node metastases, or other histopathological criteria of aggressive clinical behavior. In general, a significant correlation of low *nm23* expression and high metastatic potential has been observed in cohorts of breast, hepatocellular, gastric, ovarian, and cervical carcinomas and melanoma (reviewed, in DE LA ROSA et al. 1995).

Data on breast carcinoma cohorts are summarized on Table 1. Using RNA or five different antibodies, decreased Nm23 expression has been significantly correlated with either decreased patient survival or the presence of lympth node metastases. In two other reports using a single antibody, reduced Nm23 expression was not significantly correlated with lymph node status or survival but did exhibit a significant correlation with histopathological criteria of high aggressiveness. In three other reports either no trend was observed in *nm23* expres-

Table 1. Cohort studies of Nm23 expression patterns in human breast carcinoma

Reference	Indicator [a]	Trend	*N*	*P*
Correlations with lymph node status or patient survival				
BEVILACQUA et al. 1989	RNA	D	31	–
HENNESSTY et al. 1991	RNA	D	71	0.003
BARNES et al. 1991	Ab Nm23-PII	D	39	0.05
TOKUNAGA et al. 1993	mAb-Nm23-HI	D	116	0.014
	mAB-Nm23-H2	D	116	0.07
ROYDS et al. 1993	Ab Nm23-PII	D	127	0.0001
NOGUCHI et al. 1994a	Ab-Nm23-P	D	124	0.04
CROPP et al. 1994	mAb-301	D	47	0.012
NOGUCHI et al. 1994b	mAb-Nm23-HI	D	144	0.035
NOGUCHI et al. 1994c	Ab Nm23 P	D	112	0.04
Correlations with histopathological criteria				
HIRAYAMA et al. 1991	Ab NDPK	D	29	0.05
YAMASHITA et al. 1993	Ab NDPK	D	78	0.01
No significant correlation				
SASTRE-GARAU et al. 1992	Ab NDPKA	NC[a]	111	–
SAWAN et al. 1994	Ab NDPKA	NC	197	–
GOODALL et al. 1994	RNA	D	121	0.10

[a] RNA, northern blot, or in situ hybridization analysis. Ab, antibody; mAb,monoclonal antibody:
D, decreased expression in aggressive cases; NC, no change in expression.

sion, or a trend of reduced expression was observed but failed to reach statistical significance. Possible reasons for these discordant data include the specificity of the antibodies used (Steeg et al. 1993) and/or the methodology involved in the analysis of differential expression levels. Factors such as heterogeneous expression and intensity of expression have been shown to be critical for other molecular markers and may be important. Although most of the published papers report a significant association of low *nm23* expression and high metastatic potential, they do not confirm *nm23* as an independent prognostic factor. Rather, they indicate that the trend of differential gene expression observed in model systems is also seen in actual human tumors and suggest the hypothesis that *nm23* exerts a metastasis regulatory function in these tumor cell types.

In other tumor cohorts, such as lung, colon, prostate, kidney, etc., no trend of *nm23* expression and tumor metastatic potential has been evident (reviewed in De La Rosa et al. 1995). As in model system data, these cohort studies may indicate that *nm23* is irrelevant to tumor progression in these cancer cell types, or that it is deregulated by alternate means. In colorectal carcinoma and human neuroblastoma, mutations in *nm23* have been observed in aggressive tumors (Hailat et al. 1991; Leone et al. 1993b; Chang et al. 1994). Given the complexity of the metastatic process, and the number of redundant mechanistic alternatives to accomplish each phenotypic change (numerous proteases for invasion, for example), it is expected that alterations in the expression of a single gene are not uniformly relevant.

2 Transfection Experiments

2.1 In Vivo Analysis

Multiple research approaches can be used to implicate a gene functionally as a classical tumor suppressor gene, including alterations in familial diseases, allelic deletion/mutation analyses in sporadic tumors, and transfection studies. In metastasis, where changes in gene expression levels predominate over mutational events, many of these traditional research avenues are closed, and transfection studies remain the best accepted functional evidence. Given the instability of the metastatic phenotype it is imperative that transfection experiments compare the gene of interest with side-by-side control transfectants which received the same plasmid or viral vector minus the cDNA insert. Multiple clones should also be analyzed of both the control and experimental transfectants. In vivo data are essential, but a thorough in vitro analysis can confirm and extend the in vivo metastasis data. Given the complexity and redundancy of the metastatic process it is considered unlikely that a single gene exerts metastasis suppressive activity in all cell lines studied.

Transfection of *nm23* cDNAs has now been conducted by four independent laboratories, for which the in vivo data is summarized on Table 2. Cell lines of murine, rat, and human origin have been utilized, including melanoma and mammary carcinomas. In each case transfection of *nm23* cDNA resulted in a reduction of in vivo metastatic potential, using experimental (tail vein) or spontaneous (mammary fat pad) injection routes (LEONE et al. 1991, 1993; BABA et al. 1995; PARHAR et al. 1995; FUKUDA et al., submitted for publication). The data confirm *nm23* as a metastasis suppressor gene in a variety of model systems. In no case did *nm23* overexpression completely suppress in vivo tumor metastatic potential, indicating the participation of additional independent mechanisms in this process. A consistent lack of effect of *nm23* transfection was also reported on primary tumor size upon subcutaneous or mammary fat pad injection.

2.2 In Vitro Analysis

In vitro analysis reveals *nm23* to be multifunctional in metastasis suppression and permits the development of mechanistic hypotheses. A summary of in vitro phenotypic data on *nm23* transfectants is presented in Table 3. Multiple aspects of the tumor metastatic process are implicated, including motility, invasiveness, cell adhesion, immunosensitivity, and colonization.

Table 2. Summary of transfection experiments reporting an in vivo metastasis suppressor activity for nm23

cDNA	Cell line	primary tumor size[a]	Metastasis assay	Metastasis suppression[b]	Reference
nm23-M1	Murine K-1735 TK melanoma	Same	Experimental	58%–96%	LEONE et al. (1991)
nm23-H1	Human MDA-MB-435 breast carcinoma	Same	Spontaneous	65%–90%	LEONE et al. (1993a)
nm23-M1	Murine BI6F10 melanoma	NR	Experimental	93%	PARHAR et al. (1995)
nm23-M1	Murine BI6 FE7 melanoma	Same	Experimental	80%–87%	BABA et al. (1995)
nm23-M2	Murine B1 FE7 melanoma	Same	Experimental	78%–86%	BABA et al. (1995)
nm23-M1 and nm23-M2	Murine B16 FE7 melanoma	Same	Experimental	65%–81%	BABA et al. (1995)
nm23-a	Rat MTLn3 mammary carcinoma	Same	Spontaneous[c]	44%–52%	FUKUDA et al. (submitted)

NR, Not reported.
[a] Based on subcutaneous or mammary fat pad injection, compared to control transfectants.
[b] Where multiple clonal lines were reported, data are shown as the range of percentage inhibition, compared to the mean of the control transfectants.
[c] Lung metastases.

Table 3. In vitro phenotypes observed in nm23 tranfectants

Phenotype	Model system	Effect of nm23 transfection	Reference
Motility	K-1735 melanoma MDA-MB-435 breast ca.	Decreased in response to serum, IGF Decreased in response to serum, IGF	KANTOR et al. (1993)
Invasiveness	B16F10 melanoma	Decreased in response to complete medium	PARHAR et al. (1995)
Cell adhesion	B16F10 melanoma	Decreased production of soluble ICAM-1	PARHAR et al. (1995)
	HT29-HD3 colon ca.	Decreased adherence using anti-sense nm23 oligos	HSU et al. (1994)
Immunosensitivity	B16F10 melanoma	More susceptible to LAK cytotoxicity	PARHAR et al. (1995)
Growth on plastic	B19FE7 melanoma	No change in 3% or 10% FCS	BABA et al. (1995)
	K-1735 melanoma	No change in 10% FCS	LEONE et al. (1995)
	MDA-MB-435 breast ca.	No change in thymdine labelling	HOWLETT et al. (1994)
Soft agar colonization	K-1735 melanoma	Decreased in response to TGF-β	LEONE et al. (1991)
	MDA-MD-435 Breast ca.	Decreased intrinsic colonization, Decreased in response to TGF-β	LEONE et al. (1993a)

For growth properties, data on tissue culture plastic were uniformly negative among three transfection experiments using different culture conditions and different assays.

A portion of the in vitro data suggest that Nm23 acts to shut off the signal transduction process. For instance, in the K-1735 TK melanoma series, *nm23-1* transfection had no effect on unstimulated (random) motility or intrinsic colonization in soft agar but suppressed the stimulation of motility by serum or insulin-like growth factor (IGF), and transforming growth factor-β (TGF-β) induced colonization (LEONE et al. 1991; KANTOR et al. 1993). For the MDA-MB-435 breast carcinoma series intrinsic colonization in soft agar was somewhat reduced by *nm23-H1* transfection; however, we cannot rule out the possibility that this cell line produces autocrine growth factors which add a signal response element to an "unstimulated" assay. If Nm23 is inhibiting the signal transduction pathway, it is likely not at the level of a particular receptor, as Nm23 transfection has been implicated in responses to IGF, TGF-β, platelet-derived growth factor, and other factors.

Further consideration of the in vitro data suggests that particular facets of intracellular signaling may be involved in Nm23 function. Most of the in vitro assays demonstrating a functional effect of *nm23* transfection used cells plated either on ECM components or in three-dimensional cultures. These include motility on ECM-coated filters, invasiveness through ECM proteins, adherence to ECM protein coated dishes, and colonization in soft agar. A distinct lack of *nm23* function is noted on tissue culture plastic. These data suggest the hypothesis that *nm23* serves in the signaling process mediated by cell: ECM interactions such as integrin signaling, cell:cell interactions found in three-dimensional culture, pathways in which ECM and growth factor signaling converge, transcriptional regulatory pathways regulating these response, etc.

2.3 Implications from "Negative" Data

Three transfection studies have reported particular *nm23* transfection constructs that failed to suppress in vivo metastasis, while side-by-side transfections of other cDNA were metastasis suppressive. First, we reported that bulk transfectants of *nm23-H1*, more than *nm23-H2*, suppress in vivo metastastic potential of MDA-MB-435 breast carcinoma cells (LEONE et al. 1993a). While these data have been interpreted to mean that *nm23-H2* is not metastasis suppressive, we cautioned in the original article that this conclusion is premature until stable, high-expression clonal transfectants are analyzed.

FUKUDA et al. (submitted for publication) have reported that rat *nm23a*, but not *nm23b*, suppressed spontaneous metastatic potential of rat mammary carcinoma cells. These constructs differ by 17 amino acids, clustered in two regions of the protein, including seven amino acids located between positions 37 and 50, and six amino acids between positions 130 and 150 (the remaining changes at positions 4, 62, 69, and 89). Analysis of these regions for known protein motifs failed to reveal potential functions. The data, however, suggest the hypothesis that these regions are important to the metastasis-suppressive activity of Nm23.

Even more perplexing are data from BABA et al. (1995), who found that both *nm23-M1* and *-M2* suppress the tumor metastatic potential of murine B16 FE7 melanoma cells, but that *nm23-H1* and *-H2* fail to do so. Sequence analysis shows that there are no amino acids maintained in both Nm23-M1 and -M2 but altered in both Nm23-H1 and -H2. Confirmation of this apparent species specificity requires repetition in other model systems. If repeatable, it suggests the hypothesis that the protein interacting with Nm23 exhibits species' differences in a functional amino acid(s).

3 Potential Functions in Normal Cellular Biology

3.1 Differentiation

Several lines of evidence suggest that *nm23* has a role in the normal development and differentiation process. This role was first suggested by the cloning of *awd*, the *Drosophila* homolog of *nm23* (DEAROLF et al. 1988a,b). Reduced expression of *awd* resulted in normal fly development through metamorphosis. Afterward, abnormalities occur in (a) the morphology of the wing discs, brain, and proventriculus, (b) the differentiation of the wing, leg, and eye-antennae discs and ovaries, and (c) the cell viability of the wing discs. Developmental abnormalities were characterized by a tremendous degree of heterogeneity, to the extent that two wing discs from the same larva could be different. On a cellular basis imaginal neuroblasts from mutant *awd* larvae exhibited condensed chromosomes (BIGGS et al. 1990). The killer of prune mutation of *awd*, *awd*$^{k\text{-}pn}$, in which proline 97 is mutated to a serine, results in no obvious phenotype but exhibits the reduced/null *awd* phenotypes when expressed in combination with the *pn* (prune) eye color gene mutation. The *pn* gene has been cloned, but its biochemical function remains unknown (see TENG et al. 1991 and comments thereafter).

We first looked for evidence of a relationship between Nm23 expression, at the protein level, and differentiation of mammalian organisms by immunohistochemical staining of mouse embryos through each day of embryonic development (LAKSO et al. 1992). Nm23 expression was uniform and low through days 1–10 of embryonic development. Afterwards increased Nm23 immunostaining was observed in the embryonic heart and brain, the first two organs to differentiate. Increased Nm23 expression was also observed coincident with functional epithelial differentiation, in the liver, kidney, stomach, skin, intestine, and adrenals. "Negative" data also bears consideration: Nm23 expression was not increased with the functional differentiation of lung. Furthermore, while Nm23 expression increased with fundamental differentiation in embryogenesis in many epithelial tissues, it was not universally maintained in adult differentiated cells, such as colonic cells. We have hypothesized that Nm23 is relevant to the differentiation process of certain epithelial cells but is not necessary for the maintenance of the differentiated state.

Functional data linking *nm23* to differentiation have been published in two mammalian model systems. In collaboration with Dr. Mina Bissell's laboratory we examined morphologic and biosynthetic aspects of mammary differentiation using the MDA-MB-435 breast carcinoma cell line (HOWLETT et al. 1994). Dr. Bissell pioneered the three-dimensional culture of normal human mammary epithelial cells within reconstituted basement membrane components, resulting in the formation of the duct-like ascini, deposition of basement membrane components, and production of sialomucin. A series of breast carcinoma cell lines and primary cultures failed to recapitulate this process upon three-dimensional culture (PETERSEN et al. 1992). Using control transfectants and *nm23-H1* transfectants of the MDA-MB-435 human breast carcinoma cell line, we observed no evidence of differentiation when cultured on tissue culture plastic, but ascinar formation, expression of basement membrane components, sialomucin production, and growth arrest of the *nm23-H1* transfectants upon three-dimensional culture (HOWLETT et al. 1994). While *nm23-H1* was clearly implicated in the observed differentiation pattern, it was noted that other high Nm23 expressing breast carcinoma cell lines such as MCF-7 failed to differentiate in this system, arguing that other genetic events are also critically involved.

Dr. Eileen Friedman's laboratory independently reported a functional association of *nm23* expression and differentiation of one of two related colon carcinoma cell lines (HSU et al. 1994). Two sublines of the HT29 colon carcinoma were studied, the HD3 line which exhibits the differentiation properties of growth arrest and cell adherence in response to TGF-β, and the U9 subline, which is more tumorigenic and invasive in vivo and responds to TGF-β by proliferation rather than differentiation. Antisense oligonucleotides to *nm23* blocked the adherence and growth inhibition of HD3 cells to TGF-β, but had no effect on U9 cells. The data indicate that *nm23* is functionally involved in TGF-β mediated colon differentiation but clearly indicate that *nm23* cannot reverse TGF-β stimulatory signal transduction pathways once cells have become undifferentiated.

A component of both differentiation studies is a growth-inhibitory effect of *nm23* coincident with differentiation. While transfection of *nm23* in several model systems has failed to influence either proliferation on tissue culture plastic or primary tumor size in vivo (see "Transfection Experiments"), growth of MDA-MB-435 breast carcinoma cells within ECM components and HD3 colon carcinoma cells on ECM proteins was altered. These data suggest that *nm23* inhibits proliferative responses which are ECM- dependent but not those which are ECM independent. The apparent lack of *nm23* inhibition of primary tumor size in spontaneous metastasis assays may have several explanations. First, *nm23* may actually be growth inhibitory for primary tumor formation for some cell types in vivo. Our inability to observe this may be due to the artificial nature of in vivo assays, where 10^5 or more cells are deposited simultaneously at a site. In these circumstances, cell interactions with the ECM may be significantly less functional than those in which a single transformed cell arose in contact with ECM. Second, protease production by tumor cells, or stromal cells stimulated by the tumor cells, may further hamper cell:ECM interactions.

Homologs of *nm23* were also cloned in *Dictyostelium*, the cellular slime mold, and reduced gip17 expression at the RNA level reported to be associated with starvation-induced cellular aggregation and differentiation to slug and fruiting body forms (Wallet et al. 1990). This differential expression was reported to be less pronounced at the protein level (Wallet et al. 1990), and overexpression of wild -type gip17 cDNA failed to influence the differentiated phenotype (Sellam et al. 1995). Thus the association of *nm23* with differentiation appears not to be functionally involved in organisms as far removed evolutionarily as *Dictyostelium*.

3.2 Proliferation

The enzymatic activity of Nm23 as a NDPK, which could influence the size of nucleotide pools, suggests a role for Nm23 in DNA replication and cellular proliferation. Several reports have documented a correlation between increased Nm23 and cell proliferation, principally in nonepithelial cells. Increased Nm23 was found in proliferating peripheral blood lymphocytes and leukemias (Keim et al. 1992), and microinjection of antibody to a Nm23 peptide inhibited fibroblast proliferation (Sorcher et al. 1993).

The complexity of the apparent growth-related function of *nm23* is noted by additional reports. Ohneda et al. (1994) found that increased Nm23 is correlated with proliferation in SV40 transfromed but not normal fibroblasts, while Yamashiro et al. (1994) found that Nm23 maintains the proliferative state of MEG-01 leukemia cells but may be involved in the early differentiation of K562 erythroleukemia cells. Overexpression or disruption of the *Saccharomyces cerevisiae* homolog of *nm23* failed to affect growth or spore formation (Fukuchi et al. 1993).

The apparent contrast of these data linking *nm23* to proliferation, and that previously presented linking *nm23* to differentiation and lack of proliferation, is unexplainable to date on a biochemical basis. Several hypotheses have emerged from the literature. First, most of the reports on *nm23* and differentiation utilized epithelial cells, which form multicellular structures bounded by ECM, i.e., mammary ducts and colonic villi. Many of the proliferation-related model systems are unicellular in origin or fibroblastic, i.e., without a high degree of cell:cell or cell:ECM interaction. It can be hypothesized that the *nm23* differentiative effects is linked to cell:cell or cell:ECM biochemical pathways. Second, Ohenda et al. (1994) noted that basal levels of Nm23 in lymphocytes are relatively low, while those of erythroleukemia cells, where *nm23* has a role in differentiation, are higher. If confirmed in additional studies, this hypothesis suggests that the overall level of Nm23 may regulate the cell's proliferation versus differentiation responses.

Whether the sole function of Nm23 in proliferation is the provision of nucleotide pools through its NDPK activity is not known, as nucleotide pool sizes have not been universally correlated with NDPK activity (Golden et al. 1993). Potential functions of Nm23 directly or indirectly in transcription and as a cytokine are reviewed elsewhere in this chapter.

4 Potential Biochemical Mechanisms

The biochemical mechanism of action of Nm23 suppression of tumor metastatic potential is unknown to date. The biochemical characterization of Nm23 function has been fraught with difficulties, owing to its abundant nature and nonspecific binding to many proteins. It should be noted that most of the proposed activities of Nm23 are not mutually exclusive of one another.

4.1 NDPK Activity

Nm23 possesses a nonspecific phosphotransferase activity as a NDPK (Wallet et al. 1990). This enzyme removes the terminal phosphate from a nucleoside triphosphate (such as ATP) and transfers it to a nucleoside diphosphate (such as GDP) via a high-energy Nm23-phosphohistidine intermediate. Other enzyme systems can serve the same phosphotransferase function (reviewed in De La Rosa et al. 1995).

Many functions have been postulated for NDPK, and Nm23 in turn (reviewed in De La Rosa et al. 1995). These include the maintenance of nucleotide triphosphate pools for DNA synthesis, transphosphorylation of microtubules, transphosphorylation of small G proteins such as *ras*, transphosphorylation of heterotrimeric G proteins in signal transduction. Considerable evidence has been published arguing against the transphosphorylation of small or large G proteins by NDPK (RANDAZZO et al. 1992), and its participation in microtubule polymerization is also open to controversy (MELKI et al. 1992).

Correlational data have cast doubt as to whether the NDPK activity of Nm23 is directly responsible for its biological suppression of metastasis. We noted that control and *nm23* transfected tumor cells did not exhibit significantly different NDPK activities either in whole cell or subcellular lysates (GOLDEN et al. 1993). Similarly, transfectants expressing the rat ndpka cDNA into a mammary carcinoma cell line, which exhibited suppressed in vivo metastatic potential, contained only 109%–141% of the NDPK activity that control transfectants did (but expressing higher Nm23/NDPK protein levels), while rat NDPKb cDNA transfectants contained 105%–163% of control NDPK activity but failed to suppress metastatic potential in vivo (FUKUDA et al. submitted).

Site-directed mutagenesis of the histidine in Nm23 responsible for the formation of the NDPK-phosphorylated intermediate has been attempted several times to determine the relationship between the NDPK activity and biological functions of Nm23. Transfectants expressing this construct have not been isolated (MACDONALD et al. 1993; BABA et al. 1995). We observed that mutant Nm23 proteins lacking a histidine can remove the terminal phosphate from a nucleoside triphosphate but, lacking the Nm23 histidine acceptor, release it as Pi (MACDONALD et al. 1993). Thus, these mutant Nm23s may act as uncontrolled ATPases, which would be toxic to cells. Even if such transfectants can be generated, it does not necessarily eliminate a role for NDPK activity. It could be hypothesized that a basal level of NDPK is required, provided by the endogenous low levels of the wild-type gene. For instance, this low level of wild-type Nm23 could transphosphorylate a serine residue on another Nm23 molecule (STRELKOV et al. 1995), with resultant biological effects due to the abundance of the transphosphorylated species.

Recently nucleotides other than nucleotide di- and triphosphates have been shown to bind NDPK. In vitro studies have reported cAMP to bind the NDPK/Nm23 protein and to act as a competitive inhibitor of NDPK activity (STRELKOV et al. 1995; WEBER et al. 1995). We reported conditions in which cAMP did not inhibit NDPK activity in vitro but inhibited a low-energy serine phosphorylation of Nm23 in vitro (see below; MACDONALD et al. 1993). The apparent discrepancy over whether cAMP is a competitive inhibitor of NDPK activity may be due to the varying concentrations used in the different reports. With nucleotide di- and triphosphates in excess, cAMP may exert effects on phosphoserine formation, while at equimolar concentrations effects on both NDPK activity of phosphoserine formation of Nm23 may be observed. The biological implications of cAMP interaction with NDPK/Nm23 protein are not yet understood but could potentially tie signal transduction with decreased NDPK activity or decreased transphosphorylation.

It is also possible that NDPK mediates other biochemical activities. For instance, the NDPK activity may represent the first part of a protein kinase pathway. Lacking the appropriate protein substrate in our in vitro reactions, the forward reaction whereby NDPK-phosphohistidine is formed, may move in the reverse direction, resulting in the reformation of a nucleoside triphosphate. The NDPK-phosphohistidine intermediate may also transphosphorylate serine residues on other NDPK/Nm23 molecules, discussed below.

4.2 Serine Phosphorylation

Autophosphorylation of Nm23 proteins in vitro has yielded the acid labile phosphohistidine intermediate of the NDPK activity as well as an acid stable phosphorylated form (Hemmerich and Pecht 1992; MacDonald et al. 1993; Muñoz-Dorado et al. 1993; Bominaar et al. 1994). Phosphoamino acid analysis has indicated phosphoserine to be the acid stable form. We protease-digested and acid-treated autophosphorylated recombinant Nm23-H1 protein, separated the peptides on HPLC, performed amino acid sequencing, and determined which peptides contained acid stable (serine) phosphorylation (MacDonald et al. 1993). Two protein peptides contained evidence of serine phosphorylation: peptide 1 contained only one serine residue at position 44 while peptide 2 contained less radioactivity, and three serine residues at positions 120, 122, and 125, of which only the serine 120 is conserved through evolution. In *Dictyostelium* site-directed mutagenesis of serine 124 to either glycine or cysteine resulted in acid stable autophosphorylation, indicating that multiple sites of phosphorylation also occur in this species (Bominaar et al. 1994). The phosphate bond energy of phosphoserine is approximately 2 kcal/mmol, far less than the 7 kcal/mmol phosphohistidine bond energy required for the transfer of phosphates in the NDPK activity. Serine phosphorylation is therefore not directly involved in the NDPK phosphotransferase activity and may signal a new biochemical function for Nm23 proteins.

We have shown the presence of acid-stable Nm23, immunoprecipitated from [^{32}P]-orthophosphate-labeled human breast carcinoma cells. Thus serine phosphorylated Nm23 occurs in vivo as well as in vitro. Using the murine melanoma transfection model, we used successive passages of control- and *nm23* transfectants to determine metastatic potential in vivo, *nm23* RNA expression levels, NDPK activity of cellular lysates, and [^{32}P]-orthophosphate-labeled, acid-stable Nm23 protein levels. We found that suppression of metastasis among the *nm23*-1 transfectants is accompained by RNA overexpression and increased in vivo phosphoserine-Nm23 but not by a difference in NDPK activity (MacDonald et al. 1993). The data suggest that Nm23-phosphoserine is correlated with, and may therefore be relevant, to its biological functions.

The relative abundance of Nm23 serine phosphorylation is under study. In *Myxococcus* phosphoserine-NDPK/Nm23 was relatively abundant (Muño-Dorado et al. 1993), while Bominaar et al. (1994) studied autophosphorylated proteins from several species and reported that the serine phosphorylated form represented less than 0.2% of total NDPK/Nm23 protein. Figure 1 presents data from the MCF-7 human breast carcinoma cell line in vivo. MCF-7 cells were sequentially labeled with ^{32}P-orthophosphate, Nm23 was immunoprecipitated from the lysate, resolved on a two-dimensional gel, and transferred to a nitrocellulose filter. The radioactive Nm23 was detected autoradiographically (Fig. 1a) and the proteins in the gel were subsequently detected immunologically as a Western blot (Fig. 1b). As can be seen, phosphoserine-Nm23 (Fig. 1a) represents the acidic, very minor proportion of total Nm23 (Fig. 1b) in this cell line.

Phosphorylation of serine residues on Nm23 may occur through multiple mechanisms. We reported that site-directed mutation of histidine-118 in murine Nm23-1, when produced and purified in recombinant form, autophosphorylated on serine residues (MacDonald et al. 1993). In similar studies using Nm23/NDPK from *Myxococcus* (Muñoz-Dorado et al. 1993) and *Dictyostellium* (Bominaar et al. 1994), histidine was required for acid stable (serine) autophosphorylation. For human Nm23-H1, histidine-118 is also required for acid stable phosphorylation (Freije, unpublished data). Serines located in the NDPK active channel, such as serine-120, could be phosphorylated by direct transfer from phosphohistidine. Other serines, such as serine-44, are localized to the surface of the protein and may be phosphorylated by intermolecular transfer. The phosphate donor could involve the phosphohistidine of another Nm23 molecule or an as yet unknown protein kinase. Muñoz-Dorado et al. (1993) have noted that serine-110 mutant Nm23/NDPK proteins from *Myxococcus* formed dimers as a result of disulfide bonds, evidence that intermolecular Nm23 associations can form.

4.3 Transcriptional Activation

Evidence of a transcriptional activating function for Nm23 proteins is discussed elsewhere in this volume, and briefly summarized here. Screening of a cDNA library for proteins binding a PuF recognition sequence GGGTGGG in the *myc* promoter identified Nm23-H2 (Postel et al. 1993). While recombinant Nm23-H2 activates *myc* transcription in in vitro assays, semipurified preparations from HeLa cell extracts function approximately 100-fold better (Postel et al. 1993). Furthermore, when recombinant Nm23-H2 was added to an immunodepleted, semipurified preparation of human PuF there was a significant increase in *myc* transcription activity. These observations suggest that a major part of the DNA binding and/or transcriptional activating function resides in an as yet unknown cofactor (Postel et al. 1993; Postel and Ferrone 1994). The NDPK activity of Nm23-H2 is not required for its transcriptional activating function (Postel and Ferrone 1994). Many questions remain concerning this interesting observation. Does Nm23-H1 possess similar activity? Is Myc involved in the suppression of metastasis or, more probably, the proliferative effects associated with Nm23 expression in lymphocytes? What other genes have the PuF sequence in their promoters? Do Nm23 proteins directly bind DNA and with what specificity?

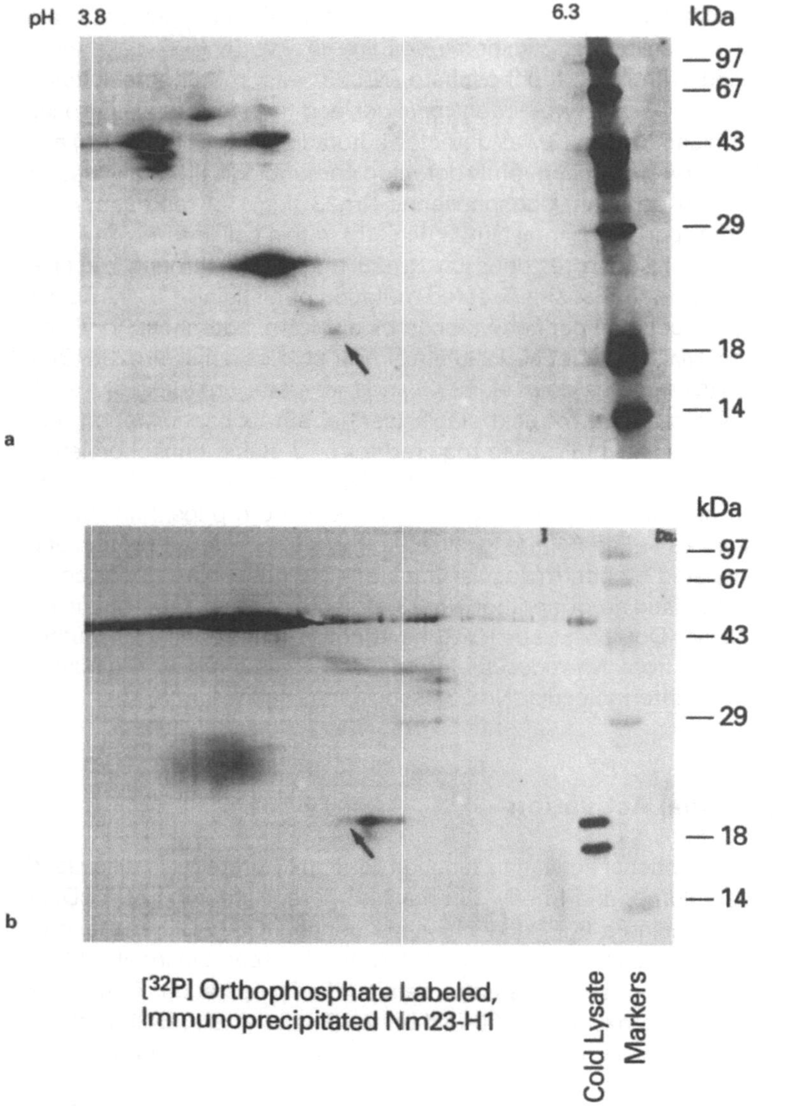

Fig. 1a, b. Nm23 protein phosphorylated in vivo. MCF-7 breast carcinoma cells were labeled with [^{32}P]orthophosphate for 3 hours. Nm23 protein was immunoprecipitated from 300 µg lysate and resolved by two-dimensional gel electrophoresis; 30 µg lysate from unlabeled MCF-7 cells was run in parallel in the second dimension (cold lysate), as were a mixture of prestained molecular weight markers. Proteins were transferred to a nitrocellulose membrane. **a** Detection of [^{32}P]orthophosphate-labeled Nm23 protein by autoradiography (*arrow*). **b** Western blot detection of total Nm23 protein on the same membrane using anti-Nm23 peptide 11 antibody

4.4 Other Activities

Nm23 has been identified as a cytokine involved in the inhibition of myeloid leukemia differentiation (Okabe-Kado et al. 1992). The NDPK activity of Nm23 is not required for this function (Okabe-Kado et al. 1995). Expression of Nm23 was also reported on the cell surface of 100% of myeloid and lymphoid cell lines and 37% of nonhematopoietic cell lines surveyed but not on peripheral blood lymphocytes or erythrocytes, and an internal RGD sequence was postulated to involved in signaling (Urano et al. 1993). We found no evidence of secreted Nm23 protein in concentrated culture supernatants or control or *nm23* transfected murine melanoma or human carcinoma cell lines (Steeg et al. 1993). The mechanism of Nm23 cell surface expression and/or secretion is unknown as Nm23 lacks a signal peptide. One possibility is that erythrocytes, which express extremely high levels of Nm23, leak this protein into the bloodstream as they die. These data may also explain in part the proliferative/differentiation inhibitory functions of Nm23 in lymphocytic and other cell lines.

References

Baba H, Urano T, Okada K, Furukawa K, Nakayama E, Tanaka H, Iwasaki K, Shiku H (1995) Two isotypes of murine *nm23*/nucleoside diphosphate Kinse, *nm23-M1* and *nm23-M2*, are involved in metastatic suppression of a murine melanoma line. Cancer Res 55: 1977–1981

Backer JM, Mendola CE, Kovesdi I, Fairhurst JL, O'Hara B Jr, Shows TB, Mathew S, Murty VVS, Chaganti RSK (1993) Chromosomal localization and nucleoside diphosphate kinase activity of human metastasis-suppressor *NM23-1* and *NM23-2*. Oncogene 8: 497–502

Barnes R, Masood S, Barker E, Rosengard AM, Coggin DJ, Crowell T, King CR, Porter-Jordan K, Wargotz E, Liotta LA, Steeg PS (1991) Low Nm23 protein expression in infiltrating ductal breast carcinomas correlates with reduced patient survival. Pathology 139: 245–250

Bevilacqua G, Sobel ME, Liotta LA, Steeg PS (1989) Association of low *nm23* RNA levels in human primary infiltrating ductal breast carcinomas with lymph node involvement and other histopatholgical indicators of high metastatic potential. Cancer 49: 5185–5190

Biggs J, Hersperger E, Steeg PS, Liotta LA, Shearn A (1990) A *Drosophila* gene that is homologous to a mammalian gene associated with tumor metastasis codes for a nucleoside diphosphate kinase. Cell 63: 933–940

Bominaar A, Tepper A, Veron M (1994) Autophosphorylation of nucleoside diphosphate kinase on non-histidine residues. FEBS Lett 353: 5–8

Caligo MA, Cipollini G, Valromita CD, Bistocchi M, Bevilacqua G (1992) Decreasing expression of *nm23* gene in metastatic murine mammary tumors of viral etiology. Anticancer Res 12: 969–973

Chang C, Zhu X-X, Thoraval D, Ungar D, Rawwas J, Hora N, Strahler J, Hansh S, Randany E (1994) Nm23-H1 mutation in neuroblastoma. Nature 370: 335–336

Cropp c, Lidereau R, Leone A, Liscia D, Cappa A, Campbell G, Barker E, Doussal VL, Steeg P, Callahan R (1994) NME1 protein expression and loss of heterozygosity mutations in primary human breast tumors. J Natl Cancer Inst 86: 1167–1169

De La Rosa A, Williams RL, Steeg PS (1995) Nm23/nucleoside diphosphate kinases: toward a structural and biochemical understanding of its biological functions. Bioessays 17: 53–62

Dearolf C, Hersperger E, Shearn A (1988a) Developmental consequences of *awdb3*, a cell autonomous lethal mutation of Drosophila induced by hybrid dysgenesis. Dev Biol 129: 159–168

Dearolf C, Tripoulas N, Biggs J, Shearn A (1988b) Molecular consequences of *awdb3*, A cell autonomous lethal mutation of *Drosophila* induced by hybrid dysgenesis. Dev Biol 129: 169–178

Fukuchi T, Nikawa J, Kimura N, Watanabe K (1993) Isolation, overexpression and disruption of a gene encoding nucleoside diphosphate kinase from *Saccharomyces cerevisiae*. Gene 129: 141–146

Fukuda M, Ishii A, Yasutomo Y, Shimada N, Ishikawa N, Hanai N, Nagara N, Irimura T, Nicolson G, Kimura N (submitted for publication) Metastatic potential of rat mammary adenocarcinoma cells associated with decreased expression of nucleoside diphosphate kinase/*nm23*: reduction by transfection of NDP Kinase a isoform, an *nm23-H2* gene homolog

Golden A, Benedict M, Shearn A, Kimura N, Leone A, Liotta LA, Steeg PS (1993) Nucleoside diphosphate kinase, *nm23* and tumor metastasis: possible biochemical mechanisms. In: Benz C, Liu E (eds) Oncogenes and tumor suppressor genes in human malignancies. Kluwer, Boston, pp 345–357

Goodall R, Dawkins H, Robbins P, Hahnel E, Sarna M, Hahnel R, Papadimitriou J, Harvey J, Sterrett G (1994) Evaluation of the expression levels of nm23-H1 RNA in primary breast cancer, benign breast disease, axillary lymph nodes and normal breast tissue. Pathology 26: 423–428

Hailat N, Keim DR, Melhem RF, Zhu X, Eckerstrom C, Brodeur GM, Reynolds CP, Seeger RC, Lottspeich F, Strahler JR, Hanash SM (1991) High levels of p19/*nm23* protein in neuroblastoma are associated with advanced disease and *N-myc* gene amplification. J Clin Invest 88: 341–345

Hemmerich S, Pecht I (1992) Oligomeric structures and autophosphorylation of nucleoside diphosphate kinase from rat mucosal mast cells. Biochemistry 31: 4580–4587

Hennessy C, Henry J, May FEB, Westly B, Angus B, Lennard TWJ (1991) Expression of the antimetastatic gene nm23 in human breast cancer: an association with good prognosis. J Natl Cancer Inst 83: 281–285

Hirayama R, Sawai S, Takagi Y, Mishima Y, Kimura N, Shimada N, Esaki Y, Kurashima C, Utsuyama M, Hirokawa K (1991) Positive relationship between expression of anti-metastatic factor (nm23 gene product or nucleoside diphosphate kinase) and good prognosis in human breast cancer. J Natl Cancer Inst 83: 1249–1250

Howlett AR, Petersen OW, Steeg PS, Bissell MJ (1994) A novel function for Nm23: overexpression in human breast carcinoma cells leads to the formative of basement membrane and growth arrest. J Natl Cancer Inst 86: 1838–1844

Hsu S, Huang F, Wang L, Banerjee S, Winawer S, Friedman E (1994) The role of *nm23* in transforming growth factor b1-mediated adherence and growth arrest. Cell Growth Differ 5: 909–917

Kantor JD, McCormick B, Steeg PS, Zetter BR (1993) Inhibition of cell motility after *nm23* transfection of human and murine tumor cells. Cancer Res 53: 1971–1973

Keim D, Hailat N, Melhem R, Zhu XX, Lascu I, Veron M, Strahler J (1992) Proliferation-related expression of p19/*nm23* nucleoside diphosphate kinase. J Clin Invest 89: 919–924

Kelsell DP, Black DM, Soloman E, Spurr NK (1993) Localization of a second NM23 gene, NME2, to chromosome 17q21–22. Genomics 17: 522–524

Lakshmi M, Parker C, Sherbert G (1993) Metastasis associated mts1 and *nm23* genes affect tubulin polymerisation in B16 melanomas: a possible mechanism of their regulation of metastatic behaviour of tumours. Anticancer Res 13: 299–304

Lakso M, Steeg PS, Westphal H (1992) Embryonic expression of Nm23 during mouse organogenesis. Cell Growth Differ 3: 873–879

Leone A, Flatow U, King CR, Sandeen MA, Margulies IMK, Liotta LA, Steeg PS (1991) Reduced tumor incidence, metastatic potential, and cytokine responsiveness of *nm23*-transfected melanoma cells. Cell 65: 25–35

Leone A, Flatow U, VanHoutte K, Steeg PS (1993a) Transfection of human *nm23-H1* into the human MDA-MB-435 breast carcinoma cell line: effects on tumor metastatic potential, colonization, and enzymatic activity. Oncogene 8: 2325–2333

Leone A, Seeger RC, Hong CM, Hu YY, Arboleda MJ, Brodeur GM, Stram D, Slamon DJ, Steeg PS (1993b) Evidence for *nm23* RNA overexpression, DNA amplification and mutation in aggressive childhood neuroblastomas. Oncogene 8: 855–865

MacDonald NJ, De La Rosa A, Benedict MA, Freije JMP, Krutsch H, Steeg PS (1993) A novel phosphorylation of Nm23, and not its nucleoside diphosphate kinase activity, correlates with suppression of tumor metastasis. J Biol Chem 269: 25780–25789

Melki R, Lascu I, Carlier MF, Veron M (1992) Nucleoside diphosphate kinase does not directly interact with tubulin or microtubules. Biochem Biophys Res Commun 187: 65–72

Muñoz-Dorado J, Almaula N, Inouye S, Inouye M (1993) Autophosphorylation of nucleoside diphosphate kinase from *Myxococcus xanthus*. J Bacteriol 175: 1176–1181

Noguchi M, Earashi M, Ohnishi I, Kinoshita K, Thomas M, Fusida S, Miyazaki I, Mizukami Y (1994a) Nm23 expression versus *Helix pomatia* lectin binding in human breast cancer metastases. Int J Oncol 4: 1353–1358

Noguchi M, Earashi M, Ohnishi I, Kitagawa H, Fusida S, Miyazaki I, Mizukami Y (1994b) Relationship between *nm23* expression and axillary and internal mammary lymph node metastases in invasive breast cancer. Oncol Rep 1: 795–799

Noguchi M, Earashi M, Ohnishi I, Kinoshita K, Thomas M, Fusida S, Miyazaki I, Mizukami Y (1994c) nm23 and c-erb-3-2 expression in invasive breast cancer. Oncol Rep 1: 523–528

Ohenda K, Kukuda M, Shimada N, Ishikawa N, Ichou T, Kaji K, Toyota T, Kimura N (1994) Increased expression of nucleoside diphosphate kinases/*nm23* in human diploid fibroblasts transformed by SV40 large T antigen or 60C irradiation. FEBS Lett 348: 273–277

Okabe-Kado J, Kasukabe T, Honma Y, Hayashi M, Henzel WJ, Hozumi M (1992) Identity of a differentiation inhibiting factor for mouse myeloid leukemia cells with Nm23/nucleoside diphosphate kinase. Biochem Biophys Res Commun 182: 897–994

Okabe-Kado J, Kasukabe T, Hozumi M, Honma Y, Kimura N, Baba H, Urano T, Shiku H (1995) A new function of Nm23/NDP kinase as a differentiation inhibitory factor, which does not require its kinase activity. FEBS Lett 363: 311–315

Parhar RS, Shi Y, Zou M, Farid NR, Ernst P, Al-Sedairy S (1995) Effects of cytokine mediated modulation of Nm23 expression on the invasion and metastatic behavior of B16F10 melanoma cells. Int J Cancer 60: 204–210

Petersen OW, Rønnov-Jessen L, Howlett AR, Bissel MJ (1992) Interaction with basement membrane serves to rapidly distinguish growth and differentiation patterns of normal and malignant human breast epithelial cells. Proc Natl Acad Sci USA 89: 9064–9068

Postel EH, Ferrone CA (1994) Nucleoside diphosphate kinase enzyme activity of Nm23-H2/PuF is not required for its DNA binding and in vitro transcriptional functions. J Biol Chem 269: 8627–8630

Postel EH, Berberich SJ, Flint SJ, Ferrone CA (1993) Human *c-myc* transcription factor PuF identified as Nm23-H2 nucleoside diphosphate kinase, a candidate suppressor of tumor metastasis. Science 261: 478–480

Randazzo PA, Northup JK, Kahn RA (1992) Regulatory GTP binding proteins (ARF, Gt and RAS) are not activated directly by nucleoside diphosphate kinase. J Biol Chem 267: 18182–18189

Rosengard AM, Krutzsch HC, Shearn A, Biggs JR, Barker E, Margulies IMK, King CR, Liotta LA, Steeg PS (1989) Reduced Nm23/Awd protein in tumor metastasis and aberrant Drosophila development. Nature 342: 177–180

Royds JA, Stephenson TJ, Rees RC, Shorthouse AJ, Silcocks PB (1993) Nm23 product expression in in situ and invasive human breast cancer. J Natl Cancer Inst 85: 727–731

Sastre-Garau X, Lacombe ML, Veron M, Magdelenat H (1992) Nucleoside diphosphate kinase/Nm23 expression in breast cancer: lack of correlation with lymph-node metastasis. Int Cancer 50: 533–538

Sawan A, Lascu I, Veron M, Anderson JJ, Wright C, Horne C, Angus B (1994) NDP-K/nm23 expression in human breast cancer in relation to relapse, survival and other prognostic factors: an immunohistochemical study. Pathology 172: 27–34

Sellam O, Veron M, Hildebrandt M (1995) Overexpression of wild-type and mutant NDP kinase in Dictyostelium discoideum. Mol Microbiol 16: 79–85

Sorcher SM, Steeg P, Feramisco JR, Buckmaster C, Boss GR, Meinkoth J (1993) Microinjection of an NM23 specific antibody inhibits cell division in rat embryo fibroblasts. Biochem Biophys Res Commun 195: 336–345

Stahl JA, Leone A, Rosengard AM, Porter L, King CR, Steeg PS (1991) Identification of a second human *nm23* gene, *nm23-H2*. Cancer Res 51: 445–449

Steeg PS, Bevilacqua G, Kopper L, Thorgeirsson UP, Talmadge JE, Liotta LA, Sobel ME (1988a) Evidence for a novel gene associated with low tumor metastatic potential. J Natl Cancer Inst 80: 200–204

Steeg PS, Bevilacqua G, Pozzatti R, Liotta LA, Sobel.ME (1988b) Altered expression of *nm23*, a gene associated with low tumor metastatic potential, during adenovirus 2 E1a inhibition of experimental metastasis. Cancer Res 48: 6550–6554

Steeg PS, De La Rosa A, Flatow U, MacDonald NJ, Benedict M, Leone A (1993) Nm23 and breast cancer metastasis. Breast Cancer Res Treat 25: 175–187

Strelkov S, Perisic O, Webb P, Williams R (1995) The 1.9 Å crystal structure of a nucleoside diphosphate kinase complex with adenosine 3',5' cyclic monophosphate: evidence for competitive inhibition. J Mol Biol 249: 665–674

Teng DHF, Engele CM, Venkatesh TR (1991) A product of the prune locus of *Drosophila* is similar to mammalian GTPase-activating protein. Nature 353: 437–440

Tokunaga Y, Urano T, Furakawa K, Kondo H, Kanematsu T, Shiku H (1993) Reduced expression of Nm23-H1, but not of Nm23-H2, is concordant with the frequency of lymphnode metastasis of human breast cancer. Int J Cancer 55: 66–71

Urano T, Furukawa K, Shiku H (1993) Expression of *nm23*/NDP kinase proteins on the cell surface. Oncogene 8: 1371–1376

Wallet V, Mutzel R, Troll H, Barzu O, Wurster N, Veron M, Lacombe ML (1990) *Dictyostelium* nucleoside diphosphate kinase highly homologous to Nm23 and awd proteins involved in mammalian tumor metastasis and *Drosophila* development. J Natl Cancer Inst 82: 1199–1202

Weber B, Weber E, Buck F, Hilz H (1995) Isolation of the *myc* transcription factor nucleoside diphosphate kinase and the multifunctional enzyme glyceraldephyde-3-phosphate dehydrogenase by cAMP affinity chromatography. Int J Biochem Cell Biol 27: 215–224

Yamashiro S, Urano T, Shiku H, Furukawa K (1994) Alternation of *nm23* gene expression during induced differentiation of human leukemia cell lines. Oncogene 2: 2461–2468

Yamashita H, Kobayashi S, Iwase H, Itoh Y, Kuzushima T, Iwata H, Itoh K, Naito A, Yamashita T, Masaoka A, Kimura N (1993) Analysis of oncogenes and tumor suppressor genes in human breast cancer. Jpn J Cancer Res 84: 871–878

NM23/Nucleoside Diphosphate Kinase as a Transcriptional Activator of c-*myc*

E.H. Postel

1 Introduction 233
1.1 NM23 in Normal Development and Cellular Proliferation 234
1.2 NM23 Is the Enzyme Nucleoside Diphosphate Kinase 235
1.3 NM23-H2 in Metastasis and Oncogenic Development 236
1.4 NM23-H2 Functions as a Transcription Factor 237

2 NM23-H2 Is the c-*myc* Transcription Factor PuF 238
2.1 Control of c-*myc* Expression in Human Cells 238
2.2 c-*myc* in Malignancy and Metastasis 238
2.3 Human c-*myc* Promoter Interacts with the PuF Transcription Factor 239
2.4 Interactions of PuF / NM23-H2 with Other Responsive Gene Promoters 241
2.5 Functional Cloning and Identification of PuF as NM23-H2/Nucleoside Diphosphate Kinase-β 241
2.6 Nucleoside Diphosphate Kinase Phosphotransferase Activity of NM23-H2/PuF Is Not Required for DNA Binding and Transcription In Vitro 242
2.7 NM23-H2 Transactivates c-*myc* in Cell Transfection Assays via the NHE Element 243
2.8 DNA Binding May Be a Function of NM23-H2 but not NM23-H1 244

3 How Does NM23-H2 Recognize DNA and Activate c-*myc* Transcription? 245
3.1 Structure and Function Relationships 245
3.2 A Model for Transcriptional Activation by PuF / NM23 / Nucleoside Diphosphate Kinase 245
3.3 The c-*myc* Connection 246

4 Conclusions and Prospects for the Future 247

References 248

1 Introduction

Expression of the *nm23* gene has been correlated with a variety of cellular mechanisms including tumor metastasis (Steeg et al. 1988; Nakayama et al. 1992, 1993), neoplastic development (Zynick et al. 1993), normal development and differentiation (Dearolf et al. 1988; Biggs et al. 1988, Wallet et al. 1990; Yamashiro et al. 1994), proliferation (Keim et al. 1992; Ohneda et al. 1994), and cell motility (Kantor et al. 1993). All *nm 23* genes encode the metabolic enzyme nucleoside diphosphate kinase (NDPK), whose principal known function is to provide nucleoside triphosphate precursors for DNA and RNA synthesis. It has been proposed

Department of Molecular Biology, Princeton University, Princeton, NJ 08646–1014, USA that NM23/NDPK proteins contain additional biochemical activities, for neither the developmental function of the *Drosophila* NM23/Awd protein (BIGGS et al. 1988; LASCU et al. 1992), or the metastasis suppression effect of human NM23 (SASTRE GARAU et al. 1992; STEEG et al. 1993) correlates directly with NDPK levels or its activity.

Recent evidence from this laboratory suggests that NM23/NDPK can function in gene transcription. The second human NM23 variant, NM23-H2/NDPK-B, was cloned as a DNA-binding protein with a transactivating function that is identical to the previously described *c-myc* transcription factor PuF (purine-binding factor; POSTEL et al. 1993). A mutational study of the catalytic residue of recombinant NM23-H2/NDPK-B/PuF indicated that the phosphotransferase activity of the protein is not necessary for DNA binding and transcription in vitro (POSTEL and FERRONE 1994). These findings suggest that, at least in the case of NM23-H2, some of the associated regulatory effects may be mediated by the transcriptional activity, particularly since one target of the Nm23-H2/PuF protein is the *c-myc* gene, itself a "master regulator" of cellular proliferation and differentiation (COLE 1986; SPENCER and GROUDINE 1991; MARCU et al. 1992; PRENDERGAST and ZIFF 1992).

The present chapter will provide a summary of our current knowledge of NM23-H2 as a transcriptional activator of the *c-myc* gene. In order to present a framework for understanding this new NM23 function, the relevant biological and biochemical properties of NM23, as well as the functions and regulation of the *c-myc* gene, are briefly reviewed. Throughout the text, effort is made to discuss and interpret controversial data and a final model is proposed to explain the transacting effect of NM23-H2 on the *c-myc* gene.

1.1 NM23 in Normal Development and Cellular Proliferation

The *nm23/ndpk* gene family is highly conserved among species (SHIMADA et al. 1993; MORERA et al. 1994). The two human genes *nm23-H1* (ROSENGARD et al. 1989) and *nm23-H2* (STAHL et al. 1991) are 88% homologous and are closely linked on chromosome 17q21 (KELSELL et al. 1993; BACKER et al. 1993). Both genes encode 152-amino acid peptides, but they differ in isoelectric points and mobilities in sodium dodecyl sulfate polyacrylamide electrophoresis (SDS-PAGE) gels. Both the murine genes *nm23 M1* and *M2* (URANO et al. 1992) and the rat *ndpk*-β and -α (KIMURA et al. 1990; SHIMADA et al. 1993) are 80% and 90% homologous, respectively, to human *nm23-H1* and *H2*. In fact, the isoforms are more closely related across species than to each other, indicating that they have diverged before the species (SHIMADA et al. 1993). The homology between human *nm23* and *Drosophila awd* gene (BIGGS et al. 1990) and *Dictyostelium ndpk* (LACOMBE et al. 1990) is also a remarkable 78%. Yeast (FUKUCHI et al. 1993) and bacterial (MUNOZ-DORADO et al. 1990) *ndpk* genes also display significant homology. This high degree of conservation across species has suggested that *nm23/awd/ndpk* genes perform functions that are crucial to the survival of the organism.

The involvement of NM23 in normal differentiation and development was first suggested by the homology between NM23 and Awd, a developmentally regulated *Drosophila* protein encoded by the *awd (abnormal wing disc)* gene (ROSENGARD et al. 1989). Additional evidence for a developmental function arose through the coincidence of NM23 protein accumulation with epithelial tissue development and differentiation, during the course of mouse organogenesis (LAKSO et al. 1992). The identification of a differentiation inhibiting factor (I factor) in cell lysates of mouse myeloid leukemia cells with a high degree of homology to NM23-H2 (OKABE-KADO 1992) also supports this notion. An NM23 role in proliferation is suggested by description of a proliferation-related increase in the activity of NM23-H1 in human lymphocytic and neuroblastoma tumors (KEIM et al. 1992). Moreover, the fact that a target of transactivation by NM23-H2 is the c-*myc* gene, whose influence on cellular proliferation and differentiation is well established (COLE 1986; SPENCER and GROUDINE 1991; MARCU et al. 1992), further implies that the biological activity of NM23 proteins must be related to regulation of cell growth and differentiation.

1.2 NM23 Is the Enzyme Nucleoside Diphosphate Kinase

Without exception, all NM23/Awd proteins display NDPK activity. NDPKs exchange γ-phosphates between nucleoside tri and diphosphates via a ping-pong mechanism (AGARWAL et al. 1978):

$$N_1TP + E \Leftrightarrow N_1DP + E \sim P$$
$$E \sim P + N_2DP \Leftrightarrow N_2TP + E$$

The phosphorylated intermediate in this reaction is a conserved histidine residue (AGARWAL et al. 1978) that corresponds to His-118 in human NDPK (GILLES et al. 1991; POSTEL and FERRONE 1994). The chain A polypeptide of human NDPK was identified as NM23-H1, and the chain B as NM23-H2 (GILLES et al. 1991). Endogenous NDPK isolated from *Drosophila* and human red blood cells is a hexamer (BIGGS et al. 1990; GILLES et al. 1991; CHIADMI et al. 1993), although neither the subunit composition nor the significance of the two distinct mammalian polypeptides is known yet. GILLES et al. (1991) have shown that human erythrocyte NDPK, when renatured together in vitro, can randomly associate to form isozymes of mixed composition (e.g., A6, A5B, A2B4, B6), leading LASCU (1996) to propose that an A to B subunit composition of NDPK isoenzymes may play a large role in regulating the biological activity of NDPK in vivo.

X-ray crystallography of the *Dictyostelium* NDPK (DUMAS et al. 1992; MORERA et al. 1994) and the *Drosophila* Awd protein (CHIADMI et al. 1993) have confirmed that the quaternary structure of eukaryotic NDPK is hexameric, consisting of six identical 17-kDa subunits that are arranged in a dihedral threefold symmetry. Recombinant *Dictyostelium* NDPK is also hexameric (LASCU et al. 1993), as is the recombinant NM23-H2/NDPK-B protein before it binds to DNA (E.H. POSTEL unpublished data).

The relationship of the enzymatic function to the metastatic suppressive behavior of NM23 has been the focus of many studies. At least in breast carcinoma cells, transfection studies indicate that NDPK activity is not the biochemical mechanism of suppression (STEEG et al. 1993). Despite the presence of a condition, dominant, and lethal mutation in the *Drosophila awd* gene known as the *killer of prune (k-pn)*, Awd protein levels and NDPK enzyme activity in the embryo are normal (BIGGS et al. 1988; LASCU et al. 1992), also suggestive of disparate functions for NM23/Awd. One might suppose that this function is the modulation of gene transcription, particularly since the transcriptional activity of NM23-H2 appears to be independent, at least in vitro, from the NDPK phosphotransferase activity (POSTEL and FERRONE 1994). This lack of correlation between enzymatic activity and regulatory properties could be related to the state of oligomerization of the protein subunits in the in vivo structure. For example, smaller-order oligomers, such as monomers, dimers, and trimers, could still function as enzymes; however, only the highest-order oligomers (hexamers) might have the capacity to bind DNA and regulate gene expression. Thus, what may ultimately regulate the biological activity of NM23 proteins in vivo is their subunit arrangements in a given circumstance. Additionally, associated protein factors could also modulate regulatory activity without a change in NDPK protein levels. Control of eukaryotic gene transcription is tightly regulated by such processes, and the availabilities of associated proteins would be determined by cell type, differentiation state, and the status of the cell cycle.

1.3 NM23-H2 in Metastasis and Oncogenic Development

The *nm23-1* gene was originally identified in rodent melanoma cells on the basis of its reduced expression in tumors with high metastatic potential (STEEG et al. 1988). The metastasis suppression effect of NM23 was subsequently described in other rodent and human malignancies; the strongest correlation between reduced expression and metastatic spread was observed in human breast carcinoma (BEVILAQUA et al. 1989; NAKAYAMA et al. 1992, 1993; HENESSY et al. 1991; STEEG et al. 1993). Reduced expression of *nm23*, however, is not always associated with metastasis, as many instances have been reported where no correlation between *nm23* expression and metastasis were seen (MYEROFF and MARKOWITZ 1993; FARLEY et al. 1993). In fact, elevated expression of NM23 has been reported in some cancers (HAILAT et al. 1991; KEIM et al. 1992; ROYDS et al. 1993). As an example of additional complexity, *nm23* overexpression was linked in one study with development of gastric carcinomas, while its decreased expression was correlated with metastasis spread (NAKAYAMA et al. 1993). Thus, *nm23* expression in metastasis appears to be related to the development and differentiation of cell types.

Because the expression of human *nm23-H1* has been the focus of most of the early studies, less is known about metastasis suppression by NM23-H2. In the few cases that have been reported, a correlation between reduced expression and increased metastasis has been observed (Fishman et al. 1994), but it was less pronounced than was the case with *nm23-H1* (Stahl et al. 1991; Yamaguchi et al. 1993). Overexpression of *nm23-H2* in oncogenic development has also been reported (Engel et al. 1993; Mandai et al. 1994). Again, these inconsistencies probably arose from the tissue and cell-type specificity of *nm23-H2* expression. From the studies in which differential effects were seen between *nm23-H1* and *nm23-H2* expression, it was concluded that the expression of the two variants is regulated independently (Stahl et al. 1991; Iizuka et al. 1995). Elucidation of these differences and the clinical and biological significance of the independence of *nm23-H1* and *H2* expression must rely in future more on the use of isoform specific probes, in quantitative RNase protection and reverse transcription polymerase chain reaction (PCR), to identify different RNA species (Fishman et al. 1994; Iizuka et al. 1995), combined with the use of monoclonal antibodies specific to each variant and to each function. Identification of proteins that interact with and modulate NM23 activity in the cell should provide additional insights into cell-specific modulation of NM23.

1.4 NM23-H2 Functions as a Transcription Factor

The role of NM23/Awd in tumor metastasis and development has thus far eluded a biochemical explanation. The suggestion of a leucine zipper-like motif (Stahl et al. 1991) and the identification of nuclear localization of the protein (Rosengard et al. 1989) have raised the possibility that NM23/Awd may contain a regulatory activity in addition to the NDPK function. A missense mutation discovered in a potential regulatory region of NM23-H2 in one case of aggressive childhood neuroblastoma (Leone et al. 1993) has raised the possibility that the tumorigenic activity of NM23-H2 may be a consequence of its regulatory function. Our laboratory has recently presented evidence showing that NM23-H2 is indeed a DNA-binding protein capable of stimulating transcription of the human c-*myc* gene in vitro (Postel et al. 1993) and that it can transactivate c-*myc* in a cell transfection assay (Berberich and Postel 1995). Experiments designed that address the role of potential regulatory regions relevant to the transcriptional activity of NM23-H2 are currently in progress in this laboratory.

An interesting new finding lending support to a transcriptional role for NM23 is that human hepatitis B virus X protein is similar to human NM23/NDPK in terms of its amino acid sequence, length, and NDPK activity (De-Medina and Shaul 1994). Protein X has long been recognized as a major cellular and viral transactivator (reviewed in Gilman et al. 1993), and it was shown recently that X can stimulate transcription of the adenovirus ML promoter in vitro albeit without a DNA-binding activity (De-Medina and Shaul 1994). Interestingly, the chromatographic fraction containing the PuF factor identified subsequently as NM23-H2 (Postel et al. 1993) was shown to contain an activity that bound to the ML promoter and directed correct initiation of ML RNA synthesis in a cell-free in vitro transcription system (Postel et al. 1989).

2 NM23-H2 Is the c-*myc* Transcription Factor PuF

2.1 Control of c-*myc* Expression in Human Cells

The c-*myc* gene (name derived from its involvement in the induction of avian myelocytomatosis) belongs to a small family of oncogenes that affect cell cycle progression, proliferation, and differentiation, and, as a transcription factor, the MYC protein is capable of either directly or indirectly activating or repressing other cellular genes (COLE 1986; SPENCER and GROUDINE 1991; MARCU et al. 1992). Regulation of c-*myc* expression occurs at the level of transcription initiation, elongation, and stability, by exceedingly complicated and as yet largely unresolved mechanisms. The c-*myc* gene has a three-exon structure: the first exon contains regulatory sequences, while the major polypeptide open reading frame resides in the second and third exons. The two major c-*myc* promoters, P1 and P2, are positioned 161 base pairs apart, and their relative expression depends on the cell type and physiological state of the cell (SPENCER and GROUDINE 1990; MARCU et al. 1992; POSTEL et al. 1993). Cell transfection assays established both positive and negative regulatory elements, but the data are controversial regarding boundaries, cell-type specificity, and chimera dependence. The region encompassing base-pairs-160 to–101 relative to P1 is absolutely necessary for expression in cell transfection assays (HAY et al. 1987; LIPP et al. 1987) and is required in a cell-free transcription assay (POSTEL et al. 1989). This element, termed NHE (nuclease-hypersensitive element), is the site of interaction with the PuF factor (POSTEL et al. 1989) that was recently cloned as n*m23-H2* (POSTEL et al. 1993; see below).

2.2 c-*myc* in Malignancy and Metastasis

As is the case with other oncogenes, c-*myc* can, if inappropriately expressed, contribute to neoplasia (COLE 1987; MARCU et al. 1992). The cause of such deregulation of normal regulatory mechanisms in tumor cells in unknown. Although there is a prevailing bias toward the involvement of the MYC protein in tumor progression, documentation is sporadic and pertains only to certain cell types and only when c-*myc* is expressed in collaboration with other oncogenes (MARCU et al. 1992). In one study, for example, no tumors were observed in rat embryo cells when c-*myc* alone was expressed; however, the cooperative action of c-Ha-ras produced phenotypic conversion of normal cells to transformed cells, with tumorigenic and metastatic potential (STORER et al. 1988).

A number of other studies have demonstrated a positive correlation between deregulation of c-*myc* (amplification) and the biological behavior of tumors, e.g., high proliferative capacity (KREIPE et al. 1993). In one retrospective study on breast cancer c-*myc* amplification and oncogene complementation were shown to play an intrinsic role in early tumor progression (ROUX DOSSETO et al. 1989). In the metastatic breast tumor cell line BSMZ (WATANABE et al. 1992), c-*myc* expression was increased two to five fold compared to the arguably nonmetastatic MCF-7 line, an increase that was commensurate with the five fold amplification of the gene in BSMZ. In another study, estrogen-associated proliferation of the MCF-7 breast cancer cell line was critically dependent on c-*myc* (DUBICK and SHIU 1992). Clearly, one mechanism by which high c-*myc* expression is associated with malignancy is amplification of the gene, but it is unclear whether *myc* is the primary target of selection or is merely a coincidental bystander in a general regulatory disturbance. It is also unclear whether other molecular defects that enable tumor cells to proliferate more rapidly are responsible for these effects or whether c-*myc* expression is simply necessary for cell proliferation, whether it be normal or cancerous.

In contrasting work, c-*myc* elevation was observed in only a minority of nonmetastatic and metastatic human breast cancer cell lines (KOZBOR and CROCE 1984). Moreover, from a clinical study on the role of mutations in human breast cancer, it was concluded that c-*myc* amplification was not significantly associated with a high risk of relapse with poor survival from breast cancer (BARRETT and ANDERSON 1993). These discrepancies among findings confirm the view that the complexity of metastasis and malignant development arises from differences at the cellular level, such as the identity of cell type and state of differentiation, in addition to host-related factors. For the establishment of c-*myc* regulation by NM23, it will be important to examine expression levels of both c-*myc* and *nm23* as a function of growth and differentiation in both normal and malignant cells.

2.3 Human c-*myc* Promoter Interacts with the PuF Transcription Factor

Mutational analysis has indicated that the P1 promoter element encompassing base pairs-160 to-101 is important to c-*myc* expression in both transfection assays (HAY et al. 1987; LIPP et al. 1987) and in in vitro transcription with partially purified PuF (POSTEL et al. 1989, 1993). The 160/–101 sequence, named NHE, is coincidental to the DNAsel hypersensitivity site III_1 (SIEBENLIST et al. 1984) and S1 nuclease-sensitive sites (BOLES and HOGAN 1987) in the region. Nuclease-hypersensitive sites frequently occur in front of eukaryotic promoters and in many instances function to upregulate gene expression (WELLS 1988; ADAMS and WORKMAN 1993). The 60-base pair NHE is a complex DNA segment that consists of an homopurine/homopyrimidine sequence replete with factor-binding motifs, inverted and direct repeats, and palindromes (POSTEL et al. 1989). Such pur/pyr sequences are capable of assuming non-B-DNA conformations (WELLS 1988) and secondary structures such as three-standed intramolecular H-DNA and can produce distortions in DNA (MIRKIN et al. 1987; KINNIBURG 1989; POSTEL et al. 1989; POSTEL 1992).

Intermolecular triple helix (triplex) formation with G-rich oligonucleotides has already been demonstrated with the NHE, wherein a colinear triplex formed between a G-rich site-specific oligonucleotide and the NHE repressed c-*myc* transcription in both a cell-free extract and in cultured HeLa cells (COONEY et al.

1988; Postel et al. 1991). One possible explanation for these data is that triplex formation inhibits the binding of one or more transcription factors whose natural affinity is for a duplex or a possibly non-B-DNA structure. Nonetheless, these findings confirmed the importance of the NHE sequence to c-*myc* gene activity and the proposition that the NHE is a critical determinant to c-*myc* transcriptional regulation. These data have also implied that the NHE may constitute a site for antisense control.

PuF was identified in HeLa cell nuclear extracts by an in vitro transcription assay and was subsequently purified partially by heparin-agarose affinity and diethylaminoethanol (DEAE)-sepharose ion-exchange chromatography. Electrophoretic mobility shift assays (EMSA) established sequence-specific DNA-binding to the −160 to −101 NHE region something which was also confirmed by methylation interference analysis. The latter identified the contact sequence as the CCCACCC/GGGTGGG motif that is repeated three times within the NHE (Postel et al. 1989).

In vitro transcription from the two c-*myc* promoters was greatly stimulated by the partially purified PuF protein when supplemented with a chromatographic fraction containing RNA polymerases and, most likely, additional factors (Postel et al. 1989, 1992). Both pol II (Postel et al. 1989) and pol III (E.H. Postel, unpublished data) could be utilized in this assay.

A comprehensive mutational analysis is in progress to evaluate the relative contributions of the three CCCACCC repeats and the various other motifs embedded in the complex NHE sequence and to establish the importance of individual nucleotides for PuF binding and transcription. Data obtained thus far suggest that the PuF function is less sensitive to point mutations than to deletions, implying further that the tertiary structure of this DNA region is important for recognition by the PuF protein.

The role of NM23-H2/PuF in c-*myc* activation was recently confirmed by Ji et al. (1995). They have observed that NM23-H2 from B-cell nuclear extracts in vitro binds to the PuF binding site (Postel et al. 1989) and more importantly, that the protein bound over the same site in vivo in the translocated c-*myc* allele in a Burkitt's lymphoma B cell line. The PuF/NM23-H2 footprint was absent in the normal allele in the same cell line, suggesting that NM23-H2 is involved in the deregulation of the c-*myc* gene in Burkitt's lymphoma, and thus, in the pathogenesis of this tumor. Ji et al. (1995) also confirmed that the PuF/NHE site of the c-*myc* promoter functions as a positive regulatory region in B cells as well.

Interaction of NM23-H2 with NHE may be central to its regulation of c-*myc* and of differentiation. One might suppose that, in the absence of NM23, c-*myc* expression is reduced, which alone could be sufficient to signal cells to differentiate. This is in fact supported by recent studies comparing in vivo protein occupancy of undifferentiated and differentiated HL60 cells (Arcinas and Boxer 1994). Dimethyl sulfate (DMS) protection over the NHE was observed in undifferentiated cells only, with the pattern similar to the in vitro PuF methylation interference footprint (Postel et al. 1989).

2.4 Interactions of PuF/NM23-H2 with Other Responsive Gene Promoters

Numerous protein factors have been shown to interact with regions of the c-*myc* gene, as established by in vitro DNA-protein binding methods and by the activity of exogenously transfected reporter genes (reviewed in MARCU et al. 1992). Proteins that have been shown to bind to the NHE (also called NSE or CT element) in vitro include nuclease-sensitive element protein-1 (NSEP-1; KOLLURI et al. 1992) and heterogeneous nuclear ribonucleoprotein K (hnRNP K; TAKIMOTO et al. 1993), both of which bind to single and double-stranded DNA. In addition to PuF (which also binds to single-stranded DNA; E.H. POSTEL, unpublished), hnRNP-K has also shown a functional correlation with promoter-binding activity (TAKIMOTO et al. 1993).

Although factor binding has not yet been demonstrated, additional potential binding sites encoded by the NHE include a TCCCCA AP2 consensus sequence (IMAGAWA et al. 1987), a –100 CACCCC β-globin element involved in erythroid differentiation (DEBOER et al. 1988), an epithelial growth factor (EGF) receptor-binding site (JOHNSON et al. 1988), transforming growth factor (TGF)-β_3 control region (LAFYATIS et al. 1991) and the plasminogen activator inhibitor PAI-1 element (DESCHEEMAEKER et al. 1992; AMAROSA 1993). The c-*myc* P2 and adenoviral IVa2 and ML promoters also contain potential PuF recognition sequences (POSTEL et al. 1989); in fact, transactivation by partially purified PuF of the adenoviral and of the human β-globin promoters have already been demonstrated in vitro (POSTEL et al. 1989).

A possible *nm23/PAI-1* connection is intriguing, since PAI-1 is elevated in breast tumors (SCHMITT et al. 1992). A model whereby NM23-H2 is responsible for the activation of *PAI-1* expression, by blocking the proteolytic pathway that leads to matrix degradation and metastasis, is worthy of consideration. *PAI-1* requires a PuF-like sequence element (TGGGTGGGGC) at –78 to –69 for promoter activity in cell transfection assays (DESCHEEMAEKER et al. 1992; AMAROSA 1993). Since urokinase (uPA) inhibitors such as PAI-1 may play a large role in determining whether a tumor cell has the ability to invade surrounding tissue or metastasize, it is possible that NM23-H2, PAI-1, and c-MYC (which can also regulate *PAI-1*; see PRENDERGAST et al. 1990) may all be involved in a common regulatory pathway.

2.5 Functional Cloning and Identification of PuF as NM23-H2/Nucleoside Diphosphate Kinase-β

PuF was cloned by screening a HeLa cell plasmid expression library with a 105-base pair long c-*myc* NHE promoter fragment containing three PuF binding sites (POSTEL et al. 1993). Sequence analysis of the cDNA has identified it as *nm23-H2/NDPK-β* (STAHL et al. 1991; GILLES et al. 1991). The purified, bacterially expressed recombinant PuF (rPuF) bound to the NHE fragment in EMSA and also stimulated in vitro transcription from the *c-myc* promoters, although less efficiently than did the HeLa cell endogenous PuF complex. An obvious explanation for this difference in affinities between rPuF and the human PuF (hPuF) endogenous protein complex is that the DNA binding specificity of PuF/NM23-H2 is increased by a second protein or proteins present in the HeLa cell endogenous hPuF complex.

Evidence that NM23-H2 is the PuF protein was provided by experiments performed with polyclonal antisera. First, antibody raised against rPuF inhibited both DNA binding and in vitro transcription. Second, the specificity of complex formation was verified by fragment competitions and antibody supershifting. While rPuF formed one complex in EMSA, the partially purified fraction containing (hPuF) formed three complexes. The different mobility shift bands generated by hPuF most likely represent cooperative binding by PuF to all three recognition sites. Finally, when hPuF was removed from the chromatographic fraction by anti-NM23-H2 antibody treatment, all DNA binding was abolished, indicating that hPuF was present in all of the bound complexes. When rPuF was added to hPuF-depleted fractions, the complex DNA-binding pattern was restored. Similar results were obtained in in vitro transcription assays: rPuF induced accurate *c-myc* transcription, but it was less active than hPuF and could be supplemented with hPuF-depleted proteins (alone inactive) for maximal transcriptional activity. These data have thus identified the PuF complex as NM23-H2, and, while it is clear that DNA binding and transcription are dependent on the NM23-H2 polypeptide, the data also suggest that at least two other proteins must interact with the promoter-PuF complex and either directly or indirectly stimulate transcription. Such a protein or proteins are present in the hPuF-depleted chromatographic fraction and may represent one or more activating, stabilizing, or modifying cellular factors (Postel et al. 1993). Data on the strong squelching effect of recombinant NM23-H2/PuF on RNA polymerase suggest that the unknown protein or proteins in that fraction may be a component of the RNA polymerase complex, possibly an associated factor or a subunit of the polymerase (Postel and Ferrone 1994). Both genetic and detailed biochemical studies will be necessary to analyze this interaction with RNA polymerase and additional regulatory factors. A search is in progress to identify such NM23-H2-associated factors and the protein domains responsible for their interaction.

2.6 Nucleoside Diphosphate Kinase Phosphotransferase Activity of NM23-H2/PuF Is Not Required for DNA-Binding and Transcription In Vitro

Regulation of transcription factor activity by phosphorylation is a common mechanism controlling DNA binding, transactivation, protein-protein contact, subcellular localization, and protein stability. As NDPKs act through phosphoprotein intermediates (Agarwal et al. 1978) and hence contain phosphate-incorporating activity, the question arises whether this activity is relevant to the DNA-binding and transcriptional functions. Additionally, is the phosphorylation of nucleoside 5'-diphosphates to corresponding triphosphates necessary for transactivation? The available data indicate that the DNA-binding and the in vitro transcriptional stimulatory activities of recombinant NM23-H2/NDPK-B are not dependent on these properties of NDPK.

Recombinant PuF/NM23-H2 is a functional NDPK being able to incorporate γ-phosphate from [^{32}P]γ-adenosine triphosphate (ATP) as well as transfer the phosphate from the His-118 catalytic residue to NDP substrates in coupled enzyme assays (POSTEL and FERRONE 1994). To confirm the role of His-118 in the NDPK autophosphorylation reaction, His-118 was mutated to phenylalanine, a nonphosphorylable residue. The H118F mutant protein was nonsphosphorylable, and thus mutation of the catalytic residue His-118 of PuF/NM23-H2/NDPK-B to phenylalanine resulted in a catalytically inactive NDPK.

Whether the NDPK enzymatic activity mediates or is relevant to the DNA-binding and transactivation function was examined with the H118F mutant protein in both EMSA and in vitro transcription assays. These data indicated that the H118F protein binds to DNA, as does the wild type, and that the in vitro transcriptional activity of rPuF was not altered by the phosphorylation status of His-118 or by the subsequent phosphate transfer. Whether the multimeric, complex DNA-binding behavior seen in reconstitutions with hPuF-depleted HeLa cell fraction (described above) is affected by this mutation is not yet known, nor has the possibility been excluded that the substrate-binding site of the catalytic domain is involved in DNA binding.

2.7 NM23-H2 Transactivates c-*myc* in Cell Transfection Assays via the NHE Element

Transactivation of the c-*myc* gene by NM23-H2 as discussed above was demonstrated in in vitro assays that involve proving that the purified protein can increase RNA polymerase transcription in a partially reconstituted system. Important to the relevance of NM23-H2 transactivation in vitro is the demonstration that NM23-H2 can also upregulate the c-*myc* gene and also accomplish other cellular functions in vivo. Cell transfection assays generally serve as model systems for in vivo transcription, although they are more complicated, because transfections depend on a cell producing the transcription factor from one plasmid and then using it to transcribe a second plasmid. Moreover, interpretation of transfection results can be complicated by the fact that untransfected cells already express c.-*myc* and *nm23* to some degree, and one can only augment expression of these genes above the unknown and intrinsically regulated levels.

It was recently demonstrated that NM23-H2/PuF can transactivate c-*myc* in transient co-transfection assays (BERBERICH and POSTEL 1995). In these experiments, overexpression of the *nm23/PuF* gene was accomplished by cloning the cDNA into an expression vector, where PuF is transcribed under the control of the cytomegalovirus promoter. When a *myc* CAT (chloramphenicol acetyl transferase) reporter plasmid was cotransfected into murine A31 Balb/c fibroblasts with the *nm23/PuF* expression vector, CAT activity was elevated three-to fourfold above the endogenously expressed c-*myc* levels. In contrast, no increase in CAT activity was observed by a *myc* CAT reporter plasmid in which the NHE element was deleted or where the *myc* promoter DNA was presented in the antisense orientation. Overall, these results suggest that expression of *nm23-H2/PuF* increase c-*myc* promoter activity in transient transfections and that this transactivation occurs via a functional interaction with the NHE promoter element (Berberich and Postel 1995).

Together with observations made in vitro, these transfection data strengthen the proposal that the NHE element upstream of the P1 promoter of the human c-*myc* gene is central to its function and to its response to upregulation by PuF/NM23-H2. To understand the rules that govern NM23 - promoter interactions in vivo, NHE promoter mutants have been constructed to study the sequence and structural requirements of PuF/NM23-H2 DNA binding in vitro. Work is also in progress testing mutated proteins with loss of activation function in vitro; such mutants, together with the NDPK-defective H118F protein, should promote our understanding of the structural and functional relationships between this c-*myc* promoter DNA and the NM23/PuF activator protein in vivo.

2.8 DNA Binding May Be a Function of NM23-H2 but not NM23-H1

It remains to be determined whether the NM23-H1 protein, whose involvement in the metastasis process is more closely correlated with inhibition, is also involved in gene transcription. Experiments performed in M. Veron's laboratory suggest that recombinant NM23-H1/NDPK-A expressed from the prokaryotic pPROK-1 vector binds weakly to c-*myc* promoter DNA (Hildebrandt et al. 1995). Comparisons of human erythrocyte NDPK-A and B preparations, and of recombinant NM23-H1/NDPK-A and NM23-H2/NDPK-B proteins expressed from the pET3C vector, confirmed that NM23-H1/NDPK-A binds poorly, if at all, to c-*myc* NHE promoter DNA fragments (E.H. Postel and C.A Ferrone, unpublished data). NDPK-A expressed as a glutathione-*S* transferase (GST) fusion protein apparently binds to the same c-*myc* sequence, albeit significantly less well than does NDPK-B (Chang et al. 1995). Stimulation of transcription by the A and B proteins was also observed by Chang et al. (1995). It is possible that NM23-H1 will recognize other DNA targets better, or that it may be able, like protein X (de-Medina et al. 1994), to transactivate a gene without a DNA-binding function.

3 How Does NM23-H2 Recognize DNA and Activate *c-myc* Transcription?

3.1 Structure and Function Relationships

Prior to the solution of the three-dimensional structure for *Dictyostelium* NDPK (DUMAS et al. 1992), it was proposed by STAHL et al. (1991) that NM23 may be a regulatory protein by virtue of the presence of a leucine zipper-like structure (LANDSCHULTZ et al. 1988) that could mediate DNA-binding and protein dimerization functions. While a heterodimerization-based transcriptional activation mechanism is consistent with the behavior of NM23-H2/PuF in DNA binding and transcription (POSTEL et al. 1993), several factors argue against a basic region-leucine zipper (bZIP) model. First, there is lack of homology with known leucine zipper sequences in an alignment program (LUPAS et al. 1991) that predicts a low probability score for NM23-H2 coiled coil interactions. Additionally, the crystal structures of the *Dictyostelium* and Awd NDPK argue against this model, because the proposed "basic" residues are located in the protein interface-oligomerization domain between adjacent monomers (DUMAS et al. 1992; CHIADMI et al. 1993; MORERA et al. 1994). These amino acids could therefore not be involved in DNA recognition directly, unless the DNA-binding structure is composed of subunits held together or stabilized by DNA. Such a model was in fact proposed by JANIN and coworkers for the Awd NDPK with Awd undergoing major conformational changes and possibly dissociating into dimers with DNA as the effector molecule (CHIADMI et al. 1993).

With the aim of fully understanding the DNA-binding specificity of NM23-H2/PuF, this laboratory is engaged in a systematic and comprehensive mutational analysis of NM23-H2 by constructing mutations that affect its DNA binding and mutations that affect domains of interaction with other proteins. Work in progress already suggests that the likely "business end" of the DNA binding structure of NM23-H2 is novel. In the final analysis, a full understanding of the residues that specify DNA-binding activity will require structural information that will be forthcoming from cocrystallization experiments.

3.2 A Model for Transcriptional Activation by PuF/NM23/Nucleoside Diphosphate Kinase

The *c-myc* DNA sequence to which PuF binds was established as nuclease hypersensitive in chromatin, and data suggested that DNA-binding to the NHE element is both structure and sequence dependent. In our current model, PuF first functions as a sensor for this unusual chromatin structure in undifferentiated cells. By unwinding or melting the proposed tertiary structure, which, as a transcription factor, NM23/PuF would be expected to do and for which there is evidence (HILDEBRANDT et al. 1995; E.H. POSTEL, unpublished), PuF can then bind to the DNA with the sequence specificity of a transcription factor (which has also been demonstrated; see POSTEL et al. 1989, 1993). Sequence-specific DNA binding could be initially accomplished through single, low-affinity DNA-binding protein interactions. These low-affinity complexes subsequently become stabilized through multiple interactions between multiple recognition helices of an oligomeric protein and the repeated PuF-binding motifs. In order for high affinity DNA binding to take place, PuF must also interact with other activator proteins referred to here as PAF (PuF associated factors; also supported by evidence; see POSTEL et al. 1993; POSTEL and FERRONE 1994). Through cooperation with PAF, PuF could loop or wrap the NHE into a large complex (seen in EMSA), which is then brought into contact with the transcription initiation machinery and RNA polymerase (data also support interaction with the transcription complex; see POSTEL and FERRONE 1994).

Why does one protein carry out two seemingly different functions? Do the NDPK enzymatic and transcriptional activities perform both distinct and common functions such that their biological activities overlap, or are these two biochemical activities mutually exclusive? Assuming that the melting of the complex NHE region requires energy, i.e., is an ATP-dependent function, NDPK could provide the ATP in a form bound to the substrate nucleotide-binding domain (CHIADMI et al. 1993). Additionally, the ATP-hydrolytic activity of NDPK might be used in transcription for dephosphorylating the C-terminal domain (CTD) of RNA polymerase, a step known to be required for RNA polymerase II to begin each new round of transcription (CHAMBERS and DAHMUS 1994). NDPKs could thus function in transcription as CTD phosphatases, enzymes about which very little is known at the moment. According to this model, the DNA-binding function is merely a vehicle promoting the phosphatase to enter the RNA polymerase–transcription complex.

3.3 The c-*myc* Connection

The NM23-c-*myc* connection in vivo is likely to be found in the regulatory pathway that governs cellular growth and differentiation. The c-*myc* gene is known as a primary response gene whose induction can be stimulated in the absence of DNA synthesis (MARCU et al. 1992). In response to the appropriate stimulus, NDPK can migrate from the cytoplasm into the nucleus as part of a large complex with other nucleic acid metabolic enzymes (PREM VEER REDDY and PARDEE 1980). In the present model, once inside the nucleus, NM23/NDPKs turn c-*myc* on and thereafter maintain the growing state through stabilization of the c-*myc* transcriptional complex. In the absence of NM23-H2/NDPK, *myc* is *off*, and consequently cells are induced to differentiate. Hence, the role of NM23 via c-*myc* is proposed to be the prevention of differentiation. Although identification of potential regulatory proteins of NM23 will clearly be important for the establishment of this pathway, numerous observations are consistent with this model:

1. As cited above, c-*myc* expression is antithetical to differentiation, and NM23 also behaves as a differentiation inhibitor. In addition, both MYC and NM23

levels have shown inverse correlation with the differentiation state, although their coordinate expression has not yet been demonstrated.

2. NM23-H2 transactivates c-*myc* and must therefore be upstream of *myc*.
3. Complementary studies have demonstrated that the c-*myc* NHE region is regulated in a differentiation-specific manner.

In the first of these, GROSSO and PITOT (1985) have shown that decreased S1 sensitivity in the NHE region after induced differentiation in HL60 cells is accompanied by a decrease in c-*myc* transcription. In other studies, downregulation of c-*myc* in DMSO-induced cells was shown to occur at the c-*myc* promoters (KOHLHUBER et al. 1993; SKERKA et al. 1993). Most importantly, and of particular relevance to NM23/PuF, is a recent description of a PuF-like protein occupancy pattern over the c-*myc* NHE region that appears to be regulated in a differentiation-specific manner (ARCINAS and BOXER 1994). These authors demonstrated a strong correlation between the presence of an in vivo footprint over the c-*myc* DNase hypersensitivity site III_1 (which is centered in the NHE) and the undifferentiated state of HL60 cells. The sequence of the protected DMSO footprint in undifferentiated cells in vivo was identical to the sequence reported by methylation interference for PuF in vitro (POSTEL et al. 1989). The in vivo protein occupancy pattern was observed in proliferating cells only, and the footprint disappeared after differentiation was induced. Thus, apparently, among the one or more proteins occupying the c-*myc* NHE region in vivo is the PuF/NM23 complex, strongly suggesting that PuF/NM23, together with other protein factors, drives c-*myc* expression in proliferating cells. Of unknown significance as yet is the suggestion from earlier in vivo data that initiation of RNA synthesis by NM23/PuF is favored from the P1 initiation site, as transcription from the P1 promoter was more responsive to NHE regulation than was the P2 promoter (HAY et al. 1987; POSTEL et al. 1991).

4 Conclusions and Prospects for the Future

Numerous observations over the years have suggested that NDPKs work both as housekeeping enzymes to maintain nucleoside triphosphate pools and as regulatory proteins in a variety of cellular processes that have no known biochemical basis yet. The recent identification of NM23-H2 as a transcription factor opens up the possibility that some of these biological effects may be mediated via the transcriptional activity of NM23. Additionally, the finding that NM23-H2 is a transcription factor for c-*myc* provides an added dimension, as it suggests that some of the developmental or metastatic functions attributed to NM23 may be mediated via c-*myc* control. Because both NM23 proteins may be inhibitors of breast tumor metastasis in humans and the relevance of c-*myc* expression in inducing or augmenting breast tumor spread is controversial, understanding c-*myc* regulation by NM23 is fundamentally important and may provide a useful tool in the development of new anticancer strategies.

The role of NM23 in metastasis is no less controversial. A systematic mutational analysis and identification of NM23/PuF-associated factors, in addition to providing useful information on the molecular mechanism of NM23-mediated transcriptional control, may ultimately lead to an understanding of the paradox that high NM23 expression in one cancer (e.g., neuroblastoma) is associated with metastasis, whereas in others (eg., breast cancer) NM23 appears to be a metastasis suppressor. It is likely that NM23 proteins form different hetero-complexes in different cell types, which in turn modulate their regulatory activity and phenotypic expression.

Acknowledgments. I wish to thank S. Berberich, J. Flint, I. Lascu, and WBP for their helpful comments on this manuscript, C. Chang, M. Hildebrandt, I. Lascu, and M. Veron for sharing with me unpublished information, I. Lascu for contributing the erythrocyte NDPK preparations, and M.L. Lacombe for providing the NDPK-A/NM23-H1 cDNA clone.

References

Adams CC, Workman JL (1993) Nucleosome displacement in transcription. Cell 72: 305–308

Agarwal RP, Robinson B, Parks RE (1978) Nucleoside diphosphokinase from erythrocytes. Method Enzymology 51: 376–386

Amorosa VK (1993) The role of the matrix metalloproteinases and the urokinase-type plasminogen activator in tumor invasion. Thesis, Princeton University

Arcinas M, Boxer LM (1994) Differential protein binding to the c-myc promoter during differentiation of hematopoietic cell lines. Oncogene 9: 2699–2706

Backer JK, Mendola CE, Kovesdi I, Fairhurst JL, O'Hara B, Eddy RL Jr, Shows TB, Mathew S, Murty VVVS, Chaganti RSK (1993) Chromosomal localization and nucleoside diphosphate kinase activity of human metastasis-suppresssor genes NM23-1 and NM23-2. Oncogene 8: 497–502

Barrett JC, Anderson M (1993) Molecular mechanisms of carcinogenesis in humans and rodents. Mol Carcinogen 7: 1–13

Berberich SJ, Postel EH (1995) PuF/NM23-H2 transactivates the human c-myc gene promoter-CAT via a functional nuclease hypersensitive element. Oncogene 10: 2343–2347

Bevilacqua G, Sobel ME, Liotta LA, Steeg PS (1989) Association of low nm23 RNA levels in human primary infiltrating ductal breast carcinomas with lymph node involvement and other histopathological indicators of high metastatic potential. Cancer Res 49: 5185–5190

Biggs J, Hersperger E, Steeg PS, Liotta LA, Shearn A (1990) A Drosophila gene that is homologous to a mammalian gene associated with tumor metastasis codes for a nucleoside diphosphate kinase. Cell 63: 933–940

Biggs J, Tripoulas N, Hersperger E, Dearolf C, Shearn A (1988) Analysis of the lethal interaction between the prune and killer of prune mutations of Drosophila. Genes Dev 2: 1333–1343

Boles C, Hogan M (1987) DNA structure equilibria in the human c-myc gene. Biochemistry 26: 367–376

Cambers RS, Dahmus ME (1994) Purification and characterization of a phosphatase from HeLa cells which dephosphorylates the C-terminal domain of RNA polymerase II. J Biol Chem 269: 26243–26248

Chang CL, Zhu XX, Thoraval DH, Ungar D, Rawwas J, Hora N, Strahler JR, Hanash SM, Radany E (1994) nm23-H1 mutation in neuroblastoma. Nature 370: 335–336

Chang CL, Hora N, Thoraval DH, He L (1996) Nucleoside diphosphate kinase A binds to c-MYC regulatory sequences and activates transcription through the P1 promoter in vitro (submitted)

Chiadmi M, Moréra S, Lascu I, Dumas C, Le Bras G, Véron M, Janin J (1993) Crystal structure of the Awd nucleotide diphosphate kinase from Drosophila. Structure 1: 283–293

Cole MD (1986) The myc oncongene: its role in transformation and differentiation. Annu Rev Genet 20: 361–384

Cooney M, Czernuszewicz G, Postel E, Flint SJ, Hogan ME (1988) Site-specific oligonucleotide binding represses transcription of the human c-myc gene in vitro. Science 241: 456–459

Dearolf CR, Tripoulas N, Biggs J, Shearn A (1988) Molecular consequences of awd^{b3}, a cell-autonomous lethal mutation of Drosophila induced by hybrid dysgenesis. Dev Biol 129: 169–178

deBoer E, Antoniou M, Mignotte V, Grosveld F (1988) The human β-globin promoter: nuclear protein factors and erythroid specific induction of transcription. EMBO J 7: 4203–4212

De-Medina T, Shaul Y (1994) Functional and structural similarity between the X protein of hepatitis B virus and nucleoside diphosphate kinases. FEBS Lett 351: 423–426

Descheemaeker KA, Wyns S, Nells L, Auwerx J, Ny T, Collen D (1992) Interaction of AP-1-, AP-2-, and Sp1-like proteins with two distinct sites in the upstream regulatory region of the plasminogen activator inhibitor-1 gene mediates the phorbol 12-myristate 13-acetate. J Cell Biol 267: 15086–15091

Dubick D, Shiu RPC (1992) Mechanism of estrogen activation of c-myc oncogene expression. Oncogene 7: 1587–1594

Dumas C, Lascu I, Moréra S, Glaser P, Fourme R, Wallet V, Lacombe M-L, Véron M, Janin J (1992) X-ray structure of nucleoside disphosphate kinase. EMBO J 11: 3203–3208

Engel M, Theisinger B, Seib T, Seitz G, Huwer H, Zang KD, Welter C, Dooley S (1993) High levels of nm23-H1 and nm23-H2 messenger RNA in human squamous-cell lung carcinoma are associated with poor differentiation and advanced tumor stages. Int J Cancer 55: 375–379

Farely DR, Eberhardt NL, Grant CS, Schaid DJ, van Heerden JA, Hay ID, Khosla S (1993) Expression of a potential metastasis suppressor gene (nm23) in thyroid neoplasms. World J Surg 17: 615–621

Fishman JR, Gumerlock PH, Meyers FJ, Devere White RW (1994) Quantitation of nm23 expression in human prostate tissues. J Urol 152: 202–207

Fukuchi T, Nikawa J, Kimura N, Watanabe K (1993) Isolation, overexpression and disruption of a Saccharomyces cerevisiae YNK gene encoding nucleoside diphosphate kinase. Gene 129: 141–146

Gilles A-M, Presecan E, Vonica A, Lascu I (1991) Nucleoside diphosphate kinase from human erythrocytes. J Biol Chem 266: 8784–8789

Gilman M (1993) Pleiotropy and henchman X. Nature 361: 687–688

Grosso LE, Pitot HC (1985) Chromatin structure of the c-myc gene in HL-60 cells during alterations of transcriptional activity. Cancer Res 45: 5035–5041

Hailat N, Keim DR, Melhem RF, Zhu X, Eckerskorn C, Brodeur GM, Reynolds CP, Seeger RC, Lottspeich F, Strahler JR, Hanash SM (1991) High levels of p19/nm23 protein in neuroblastoma are associated with advanced stage disease and with N-myc gene amplification J Clin Invest 88: 341–345

Hay N, Bishop JM, Levens D (1987) Regulatory elements that modulate expression of the human c-myc gene. Genes Dev 1: 659–671

Hennessy C, Henry JA, May FEB, Westley BR, Angus B, Lennard TWJ (1991) Expression of the antimetastatic gene nm23 in human breast cancer: an association with good prognosis. J Natl Cancer Inst 83: 281–285

Hildebrandt M, Lacombe M-L, Meshildrey S, Véron M (1995) Human NDP-kinase B specifically binds to single-stranded poly-pyrimidine sequences. Nucleic Acids Res 23: 3858–3864

Iizuka N, Oka M, Noma T, Nakazawa A, Hirose K, Suzuki T (1995) NM23-H1 and NM23-H2 messenger RNA abundance in human hepatocellular carcinoma. Cancer Res 55: 652–657

Imagawa M, Chiu R, Karin M (1987) Transcription factor AP-2 mediates induction by two different signal transduction pathways: protein kinase C and cAMP. Cell 51: 251–260

Ji L, Arcinas M, Boxer LM (1995) The transcriptions factor, NM 23-H2, binds to and activates the translocated c-MYC allele in Burkitt lymphoma. J Biol Chem 270: 13392–13398

Johnson AC, Jinno Y, Merlino GT (1988) Modulation of epidermal growth factor receptor proto-oncogene transcription by a promoter site sensitive to S1 nuclease. Mol Cell Biol 8: 4174–4184

Kantor JD, McCormick B, Steeg PS, Zetter BR (1993) Inhibition of cell motility after nm23 transfection of human and murine tumor cells. Cancer Res 53: 1971–1973

Keim D, Hailat N, Melhem R, Zhu XX, Lascu I, Veron M, Strahler J, Hanash SM (1992) Proliferation-related expression of p19/nm23 nucleoside diphosphate kinase. J Clin Invest 89: 919–924

Kelsell DP, Black DM, Solomon E, Spurr NK (1993) Localization of a second NM23 gene, NME2, to chromosome 17q21–q22. Genomics 17: 522–524
Kimura N, Shimada N, Nomura K, Watanabe K (1990) Isolation and characterization of a cDNA clone encoding rat nucleoside disphosphate kinase. J Biol Chem 265: 15744–15749
Kinniburgh AJ (1989) A cis-acting transcription element of the c-myc gene can assume an H-DNA conformation. Nucleic Acids Res 17: 7771–7778
Kohlhuber F, Strobl LJ, Eick D (1993) Early down-regulation of c-myc in dimethylsulfoxide-induced mouse erythroleukemia (MEL) cells is mediated at the P1/P2 promoters. Oncogene 8: 1099–1102
Kolluri R, Torrey TA, Kinniburg AJ (1992) A CT promoter element binding protein: definition of a double-strand and a novel single-strand DNA binding motif. Nucleic Acids Res 20: 111–116
Kozbor D, Croce CM (1984) Amplification of the c-myc oncogene in one of five human breast carcinoma cell lines. Cancer Res 44: 438–441
Kreipe H, Feist H, Fisher L, Felgner J, Heidorn K, Mettler L, Parwaresch R (1993) Amplification of c-myc but not of c-erb B-2 is associated with high proliferative capacity in breast cancer. Cancer Res 53: 1956–1961
Lacombe M-L, Wallet V, Troll H, Véron M (1990) Functional cloning of a nucleoside diphosphate kinase from Dictyostelium discoideum. J Biol Chem 265: 10012–10018
Lafyatis R, Denhez F, Williams T, Sporn M, Roberts A (1991) Sequence specific protein binding to and activation of the TGF-β3 promoter through a repeated TCCC motif. Nucleic Acids Res 19: 6419–6425
Lakso M, Steeg PS, Westphal H (1992) Embryonic expression of nm23 during mouse organogenesis. Cell Growth Differ 3: 873–879
Landschultz WH, Johnson PT, McKnight SL (1988) The leucine zipper: a hypothetical structure common to a new class of DNA binding proteins. Science 240: 1759–1764
Lascu I (1996) Mammalian nucleoside diphosphate kinases: A, B and Nm23 (submitted)
Lascu I, Chaffotte A, Limbourg-Bouchon B, Véron M (1992) A Pro/Ser substitution in nucleoside diphosphate kinase of Drosophila melanogaster (mutation killer of prune) affects stability but not catalytic efficiency of the enzyme. J Biol Chem 267: 12775–12781
Lascu I, Deville-Bonne D, Glaser P, Véron M (1993) Equilibrium dissociation and unfolding of nucleoside diphosphate kinase from Dictyostelium discoideum. J Biol Chem 268: 20268–20275
Leone A, Seeger RC, Hong CM, Hu YY, Arboleda MJ, Brodeur GM, Stram D, Slamon DJ, Steeg PS (1993) Evidence for nm23 RNA overexpression, DNA amplification and mutation in aggressive childhood neuroblastomas. Oncogene 8: 855–865
Lipp M, Schilling R, Wiest S, Laux G, Bornkamm GW (1987) Target sequences for cis-acting regulation within the dual promoter of the human c-myc gene. Mol Cell Biol 7: 1393–1400
Lupas A, Dyke MV, Stock J (1991) Predicting coiled coils from protein sequences. Science 252: 1162–1164
Mandai M, Konishi I, Koshiyama M, Mori T, Arao S, Tashiro H, Okamura H, Nomura H, Hiai H, Fukumoto M (1994) Expression of metastasis-related nm23-H1 and nm23-H2 genes in ovarian carcinomas: correlation with clinicopathology, EGFR, c-erbB-2, and c-erbB-3 genes, and sex steroid receptor expression. Cancer Res 54: 1825–1830
Marcu KB, Bossone SA, Patel AJ (1992) myc function and regulation. Annu Rev Biochem 61: 809–860
Mirkin SM, Lyamichev VI, Drushlyak KN, Dobrynin VN, Filippov SA, Frank-Kamenetskii MD (1987) DNA H form requires a homopurine-homopyrimidine mirror repeat. Nature 330: 495–497
Moréra S, LeBras G, Lascu I, Lacombe ML, Véron M, Janin J (1994) Refined X-ray structure of Dictyostelium discoideum nucleoside diphosphate kinase at 1.8 A resolution. J Mol Biol 243: 873–890
Muñoz-Dorado J, Inouye M, Inouye S (1990) Nucleoside diphosphate kinase from Myxococcus xanthus, I. Cloning and sequencing of the gene. J Biol Chem 265: 2702–2706
Myeroff LL, Markowitz SD (1993) Increased nm23-H1 and nm23-H2 messenger RNA expression and absence of mutations in colon carcinomas of low and high metastatic potential. J Natl Cancer Inst 85: 147–152
Nakayama T, Ohtsuru A, Nakao K, Shima M, Nakata K, Watanabe K, Ishii N, Kimura N, Nagataki S (1992) Expression in human hepatocellular carcinoma of nucleoside diphosphate kinase, a homologue of the nm23 gene product. J Natl Cancer Inst 84: 1349–1354
Nakayama H, Yasui W, Yokozaki H, Tahara E (1993) Reduced expression of nm23 is associated with metastasis of human gastric carcinomas. Jpn J Cancer Res 84: 184–190

Ohneda K, Fukuda M, Shimada N, Ishikawa N, Ichou T, Kaji K, Toyota T, Kimura N (1994) Increased expression of nucleoside diphosphate kinases/nm23 in human diploid fibroblasts transformed by SV40 large T antigen or^{60}Co irradiation. FEBS Lett 348: 273–277

Okabe-Kado J, Kasukabe T, Honma Y, Hayashi M, Henzel WJ, Hozumi M (1992) Identity of a differentiation inhibiting factor for mouse myeloid leukemia cells with NM23/nucleoside diphosphate kinase. Biochem and Biophys Res Comm 182: 987–994

Postel EH (1992) Modulation of c-myc transcription by triple helix formation. Ann N.Y Acad Sci 660: 57–63

Postel EH, Berberich SJ, Flint SJ, Ferrone CA (1993) Human c-myc transcription factor PuF identified as nm23-H2/nucleoside diphosphate kinase, a candidate suppressor of tumor metastasis. Science 261: 478–480

Postel EH, Flint SJ, Kessler DJ, Hogan ME (1991) Evidence that a triplex-forming oligodeoxyribonucleotide binds to the c-myc promoter in HeLa cells, thereby reducing c-myc mRNA levels. Proc Natl Acad Sci USA 88: 8227–8231

Postel EH, Ferrone CA (1994) Nucleoside diphosphate kinase enzyme acitivity of NM23-H2/PuF is not required for its DNA binding and in vitro transcriptional functions. J Biol Chem 269: 8627–8630

Postel EH, Mango SE, Flint SJ (1989) A nuclease-hypersensitive element of the human c-myc promoter interacts with a transcription initiation factor. Mol Cell Biol 9: 5123–5133

Prem veer Reddy G, Pardee AB (1980) Multienzyme complex for metabolic channeling in mammalian DNA replication. Proc Natl Acad Scid USA 77: 3312–3316

Prendergast GC, Ziff EB (1992) A new bind for Myc. TIG 8: 91–97

Prendergast GC, Diamond LE, Dahl D, Cole MD (1990) The c-myc-regulated gene mrl encodes plasminogen activator inhibitor 1. Mol Cell Biol 10: 1265–1269

Rosengard AM, Krutzsch HC, Shearn A, Biggs JR, Barker E, Margulies IMK, King CR, Liotta LA, Steeg PS (1989) Reduced Nm23/Awd protein in tumour metastasis and aberrant Drosophila development. Nature 342: 177–180

Roux-Desseto M, Martin PM (1989) A paradigm for oncogene complementation in human breast cancer. Res Virol 140: 571–591

Royds JA, Stephenson TJ, Rees RC, Shorthouse AJ, Silcocks PB (1993) Nm23 protein expression in ductal in situ and invasive human breast carcinoma. J Natl Cancer Inst 85: 727–731

Sastre-Garau X, Lacombe ML, Jouve M, Véron M, Magdelénat H (1992) Nucleoside diphosphate kinase/NM23 expression in breast cancer: lack of correlation with lymph-node metastasis. Int J Cancer 50: 533–538

Schmitt M, Jänicke F, Pache L, Harbeck N, Graeff H (1992) Both the serine protease urokinasse (uPA) and its inhibitor (PAI-1) are strong independent phognostic factors in node-negative breast cancer patients. Breast Cancer Res Treat 23: 177

Shimada N, Ishikawa N, Munakata Y, Toda T, Watanabe K, Kimura N (1993) A second form (β isoform) of nucleoside diphosphate kinase from rat. J Biol Chem 268: 2583–2589

Siebenlist U, Henninghausen L, Battey J, Leder P (1984) Chromatin structure and protein binding in the putative regulatory region of the c-myc gene in Burkitt lymphoma. Cell 37: 381–391

Skerka C, Zipfel PF, Siebenlist U (1993) Two regulatory domains are required for down-regulation of c-myc transcription in differentiating U937 cells. Oncogene 8: 2135–2143

Spencer CA, Groudine M (1991) Control of c-myc regulation in normal and neoplastic cells. Adv Cancer Res 56: 1–48

Stahl JA, Leone A, Rosengard AM, Proter L, King CR, Steeg PS (1991) Identification of a second human nm23 Gene, nm23–H2. Cancer Res 51: 445-449

Steeg PS, Bevilacqua G, Kopper L, Thorgerisson UR, Talmadge JE, Liotta LA, Sobel ME (1988) Evidence for a novel gene associated with low tumor metastatic potential. J Natl Cancer Inst 80: 200–205

Steeg PS, De La Rosa A, Flatow U, MacDonald NJ, Benedict M, Leone A (1993) Nm23 and breast cancer matastasis. Breast Cancer Res Treat 25: 175–187

Storer RD, Allen HL, Kraynak AR, Bradely MO (1988) Rapid induction of an experimental metastatic phenotype in first passage rat embryo cells by cotransfection of EJ c-Ha-ras and c-myc oncogenes. Oncogene 2: 141–147

Takimoto M, Tomonaga T, Matunis M, Avigan M, Krutzsch H, Dreyfuss G, Levens D (1993) Specific binding of heterogeneous ribonucleoprotein particle protein K to the human c-myc promoter, in vitro. J Biol Chem 268: 18249–18258

Urano T, Takamiya K, Furukawa K, Shiku H (1992) Molecular cloning and functional expression of the second mouse nm23/NDP kinase gene, nm23-M2. FEBS Lett 309: 358–362

Wallet V, Mutzel R, Troll H, Barzu O, Wurster B, Véron M, Lacombe M-L (1990) Dictyostelium nucleoside diphosphate kinase highly homologous to Nm23 and Awd proteins involved in mammalian tumor metastasis and drosophila development. J Natl Cancer Inst 82: 1199–1202
Watanabe M, Tanaka H, Kamada M, Okano H, Takahashi H, Uchida K, Iwamura A, Zeniya M, Ohno T (1992) Establishment of the human BSMZ breast cancer cell line, which overexpresses the erb-2 and c-myc genes. Cancer Res 52: 5178–5182
Wells RD (1988) Unusual DNA structures. J Biol Chem 263: 1095–1098
Yamaguchi A, Urano T, Goi T, Takeuchi K, Niimoto S, Nakagawara, G, Furukawa K, Shiku H (1993) Expression of human nm23-H1 and nm23-H2 proteins in hepatocellular carcinoma. Cancer 73: 2280–2284
Yamashiro S, Urano T, Shiku H, Furukawa K (1994) Alteration of nm23 gene expression during the induced differentiation of human leukemia cell lines. Oncogene 9: 2461–2468
Zinyk DL, McGonnigal BG, Dearolf CR (1993) Drosophila awd$^{k\text{-}pn}$, a homologue of the metastasis suppressor gene nm23, suppresses the Tum-1 haematopoietic oncogene. Nature Genet 4: 195–201

Role of Tiam1 in Rac-Mediated Signal Transduction Pathways

J.G. Collard, G.G.M. Habets, F. Michiels, J. Stam, R.A. van der Kammen, and F. van Leeuwen

1 Introduction 253
2 Identification of *Tiam1* by Proviral Tagging 254
2.1 Homology Domains in Tiam1 254
2.2 Chromosomal Localization and Expression Pattern of Mouse and Human *Tiam1* 255
2.3 Tiam1 Activates Rac In Vitro as Well as In Vivo 256
2.4 V12Rac1 But Not V14RhoA Induces Invasion in T Lymphoma Cells 256
2.5 Tiam1 and V12Rac1 Induce an Oncogenic Phenotype in NIH3T3 Cells 257
3 Role of Tiam1 in Signal Transduction Pathways 257
3.1 Rho Function 258
3.2 Rac Function 259
3.3 Connections Between the Ras and the Rho/Rac Signalling Pathways 260
4 Conclusions 261
References 262

1 Introduction

Metastasis is a multistep process that requires the complex interplay of a number of gene products. Many tumour cell properties have been correlated with metastatic capacity, but direct evidence for a role of specific genes in metastasis is scarce. Products of oncogenes, such as Ras, can confer metastatic capacity (see Collard et al. 1988), as can specific splice variants of CD44, of which the expression correlates with metastasis in model systems and certain human tumours (Günthert et al. 1991; Wielenga et al. 1993). In contrast, other gene products such as Nm23, which shows homology to nucleoside diphosphate kinases, can supress metastasis (Leone et al. 1991), and downmodulation of Nm23 in mammary carcinomas has been associated with poor prognosis (Bevilacqua et al. 1989). Proteins involved in invasion also influence metastasis, as has been shown for proteases and their inhibitors (Liotta et al. 1991) and for adhesion molecules that play a role in cell–cell and cell–matrix interactions, e.g. E-cadherin (Behrens et al. 1990; Vleminckx et al. 1991) and certain integrins

The Netherlands Cancer Institute, Division of Cell Biology, Plesmanlaan 121, 1066 CX Amsterdam, The Netherlands

(ROOSEIN et al. 1989; CHAN et al. 1991). In order to identify genes specifically involved in the acquisition of the invasive and metastatic phenotype of tumorigenic cells, we have used proviral tagging in combination with in vitro selection for invasive T lymphoma cells. These studies have led to the identification of the *Tiam1* gene, which encodes a protein that regulates the activation of Rac-mediated signalling pathways.

2 Identification of *Tiam1* by Proviral Tagging

In earlier studies we found that the capacity of BW5147 murine T lymphoma cells to infiltrate into monolayers of hepatocytes or fibroblasts correlates well with the formation of experimental metastases (ROOS et al. 1985; COLLARD et al. 1987a, 1987b, 1988,1989; HABETS et al. 1990, 1992, 1994). Invasiveness in T lymphoma cells can be induced by introduction of constitutive active V12Ras (CALLARD et al. 1987a) and by fusion with inherently invasive normal mouse or human T cells (ROOS et al. 1985; COLLARD et al. 1987b). Moreover, invasive variants can be selected in vitro from 5-azacytidine-treated BW5147 cells at low frequency, suggesting that endogenous invasion-inducing mouse genes are activated by DNA hypomethylation (HABETS et al. 1990). To identify such genes we have induced invasion by proviral insertional mutagenesis and selected low-frequency invasive variants in vitro on monolayers of fibroblasts (HABETS et al. 1994). In 40% (12 out of 30) of the generated invasive variants, proviral insertions were found within coding exons of the *Tiam 1* gene, resulting in both truncated 5'-end and 3'-end transcripts that give rise to N- and C- terminal *Tiam 1* protein fragments. In one invasive variant, an amplification of the *Tiam 1* locus was observed with concomitant increase in the amount of normal Tiam 1 protein, suggesting that invasion can be induced by high amounts of normal Tiam 1 or by protein truncation. Cell clones that are invasive in vitro produce experimental metastases in nude mice. Transfection of truncated *Tiam1* cDNA into non-invasive cells makes these cells invasive, establishing the invasion-inducing capacity of Tiam1 (HABETS et al. 1994).

2.1 Homology Domains in Tiam1

Tiam 1 encodes a protein of 1591 amino acid residues and contains different domains that show homology to other proteins (Fig. 1; HABETS et al. 1994). At the N terminus, a potential myristoylation site is present that could anchor the protein to membranes. In addition, two PEST domains are present, sequence motifs that in other proteins have been associated with protein instability. In the C-terminal half of the protein, Tiam 1 contains a Dbl homology (DH) domain. This DH domain is shared with a number of structurally different genes, including *Dbl, Bcr, Cdc24, Vav, Ost,* and *Ect-2* (see HABETS et al. 1994; ZHENG et al. 1994; HORII et al. 1994).

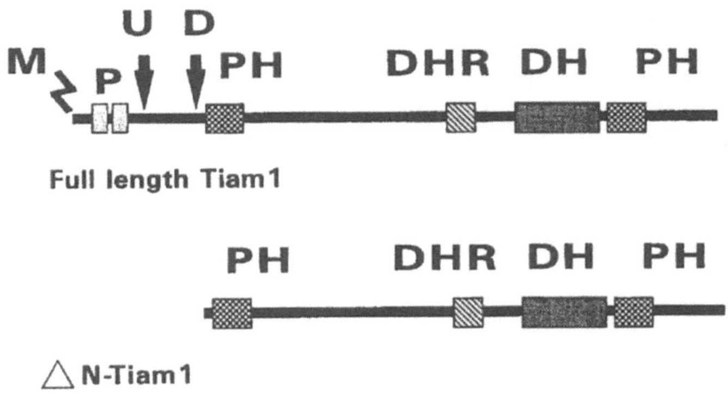

Fig. 1. Domains in full-length Tiam1. *M*, putative myristoylation site; *P*, PEST domain; *PH*, Pleckstrin homology domain; *DH*, Dbl homology domain; *DHR*, DHR domain or GLGF sequence motif. *Arrows* indicate the position of proviral insertions in Tiam1 as found in retrovirally induced invasive T lymphoma cells (HABETS et al. 1994). N-terminally truncated Tiam1 lacks the putative myristoylation site and the PEST domains

Most of these gene products modulate the activity of Rho-like GTPases and exhibit transforming properties after N-terminal truncation. Tiam1 also contains two Pleckstrin homologous (PH) domains (Fig. 1), a recently identified protein module which is present in many signalling molecules and is thought to play a role in either specific protein–protein interactions, comparable to SH2 and SH3 domains (GIBSON et al. 1994), or in interactions with phosphatidylinositol derivatives (HARLAN et al. 1994). One of the PH domains of Tiam1 is located C-terminally adjacent to the DH domain as found in other guanosine diphosphate (GDP)-dissociation stimulator (GDS) proteins, whereas the second PH domain is more N-terminally located. Between the N-terminal PH and the DH domain, Tiam1 harbours another protein motif, the so-called DHR domain (PONTING and PHILIPPS 1995). DHR domains are predominantly found in proteins that associate with cytoskeletal elements. Based on the sequence homologies, we hypothesized that Tiam1 might act as an activator of Rho-like small guanosine triphosphate (GTP)-binding proteins and could be involved in the regulation of the cytoskeletal organisation. It has been shown that small GTPases of the Rho subfamily transduce extracellular signals and exert effect on the cytoskeleton which determines the morphology, adhesion and motility of cells (HALL 1992).

2.2 Chromosomal Localization and Expression Pattern of Mouse and Human *Tiam1*

Tiam 1 maps to the distal end of murine chromosome 16. The human homologue of *Tiam 1* maps to 21q22, centromeric of the Aml1 gene (HABETS et al. 1995a). *Tiam 1* is highly conserved during evolution. Mouse and human *Tiam 1* are 95% identical at the amino acid level. The gene is highly expressed in brain and testis and at low levels in most other murine and human tissues (Habets et al.1994, 1995b). Tiam 1 is also expressed in virtually all tumour cell lines tested, including B and T cell lymphomas, melanomas, neuroblastomas and carcinomas of the breast, lung, ovary, bladder and pancreas (Habets et al. 1995b). Whether *Tiam 1* expression levels correlate with the invasive and metastatic capacity of human tumours is currently being investigated. The evolutionary conservation as well as the broad expression pattern of *Tiam 1* in normal tissues and tumorigenic cells suggests an important general function in cellular signalling processes.

2.3 Tiam1 Activates Rac In Vitro as Well as In Vivo

Using recombinant Tiam1-GST (glutathione S-transferase) and ^{3}H-GDP-loaded RhoA and Rac1 fusion proteins, we found that Tiam1 stimulates the exchange of ^{3}H-GDP for unlabelled GTP on Rac1. A similar, but less potent, effect was found on RhoA. Similarly, Tiam1 acted as a more potent activator for Rac1 than for RhoA when the exchange of preloaded GDP for GTPγ^{35}S was measured, indicating that Tiam1 can act as a GDS protein for Rho-like GTPases in vitro (Michiels et al. 1995).

In order to address whether Tiam1 can also activate Rho-like GTPases in vivo, full-length and 5'-truncated Tiam1 constructs (see Fig. 1) were transiently expressed in NIH3T3 cells, and the resulting transfectants were compared with V12Rac1- and V14RhoA-expressing cells. Cells transiently expressing full-length and N-terminally truncated Tiam1 exhibit a flat, pancake-shaped or sickle-shaped morphology with extensive membrane ruffling and many pinocytotic vesicles, similar to what was found for V12Rac1-transfected fibroblasts (Michiels et al. 1995). Expression of V14RhoA induces the formation of large bundles of stress fibers in NIH3T3 cells, which was not seen in Tiam1-transfected cells that contained even less stress fibres than untransfected control cells. The membrane ruffling in established Tiam1-transfected cells is inhibited by dominant-negative N17Rac1, indicating that Rac1 acts downstream of Tiam1 (Michiels et al. 1995). Treatment with C3 transferase, which inactivates RhoA, does not affect membrane ruffling, whereas it causes dissociation of most stress fibres in control and Tiam1-transfected cells, indicating that Tiam1-induced membrane ruffling is independent of RhoA activity. Like Rac1, the Tiam1 protein is present in the cytoplasm and co-localizes with F actin in membrane ruffles (Michiels et al. 1995). These studies strongly suggest that Tiam1 acts as an activator (GDS) of Rac1 and not RhoA in vivo in fibroblasts.

2.4 V12Rac1 But Not V14RhoA Induces Invasion in T Lymphoma Cells

To analyse whether activation of Rac by Tiam1 is sufficient to explain the invasive capacity of Tiam1-transfected T lymphoma cells, we have constructed retroviruses that contained cDNA encoding V12Rac1, V14RhoA or LacZ.

Retroviral transduction of V12Rac1 into non-invasive T lymphoma cells yields invasive cells, whereas no invasive cells are obtained after transduction of V14RhoA or LacZ (MICHIELS et al. 1995). The infiltrated cells strongly express the exogenous V12Rac1 protein, whereas non-invasive cells do not. C3 transferase inhibits invasion of the V12Rac-transduced cells, indicating that RhoA function is required but not sufficient for invasion. Thus, Tiam1-induced invasiveness of T lymphoma cells is mimicked by expression of constitutive active V12Rac1. This suggests that the invasion-inducing capacity of Tiam1 is caused by activation of Rac and implicates the Tiam1-Rac signalling pathway in the process of invasion and metastasis of tumour cells.

2.5 Tiam1 and V12Rac1 Induce an Oncogenic Phenotype in NIH3T3 Cells

Most Tiam-related genes such as *Dbl, Vav, Ost* and *Ect-2* have been identified as 5'-end truncated transforming genes by transfection of genomic DNA or cDNA libraries into NIH3T3 cells (EVA et al. 1988; KATZAV et al. 1989, 1991; MIKI et al. 1993; HORII et al. 1994). The transforming and oncogenic activity of these proteins is caused by truncation of the N-terminal region and is located in the DH domain (HART et al. 1994). In order to test whether N-terminally truncated Tiam1 also exhibits oncogenic potential, *Tiam1* cDNA constructs encoding full-length and N-terminally truncated Tiam1 (see Fig. 1) were transfected into NIH3T3 cells, and G418-resistant clones were isolated, expanded and injected subcutaneously into nude mice. Only the cells transfected with the truncated *Tiam1* construct produced tumours in virtually all injected mice, whereas cells transfected with the empty vector or full-length *Tiam1* did not (VAN LEEUWEN et al. 1995). Apparently, N-terminal truncation of Tiam1 activates its oncogenic potential, as has been found for the Tiam1-related genes *Dbl, Vav, Ost* and *Ect2*.

Since Tiam1 activates Rac in fibroblasts (MICHIELS et al. 1995), we also analysed V12Rac1 for oncogenic capacity in NIH3T3 cells. Indeed, similar to truncated Tiam1, V12Rac1-transfected cells produce tumours upon injection into nude mice, suggesting that the oncogenic potential of Tiam1 in NIH3T3 cells is also caused by activation of the Rac signalling pathway (VAN LEEUWEN et al. 1995).

3 Role of Tiam1 in Signal Transduction Pathways

Tiam1 thus appears to encode a GDS protein that activates the small GTP-binding Rho-like Rac protein in vivo. Small GTP-binding proteins, including Ras-, Rho- and Rab-like proteins, constitute a superfamily of proteins that participate in signalling pathways regulating many cellular functions including cell cycle progression, differentiation, cytoskeletal organization, protein transport and secretion (BOURNE et al. 1990; HALL 1990). All these proteins cycle between the active GTP-bound state and the inactive GDP-bound state (BOURNE et al. 1991), and their activity is influenced by three different classes of proteins. Activation is catalysed by GDS, such as Sos and Tiam1, whereas inactivation occurs by GTPase-activating proteins (GAP), such as Bcr, p120RasGAP and p190RhoGap, which stimulate the weak intrinsic GTPase activity of these proteins (BOGUSKI and MCCORMICK 1993). The guanine nucleotide dissociation inhibitors (GDI) are thought to prevent the activation and/or inactivation of small GTPases. Further regulation of the activity of the GTPases is achieved by changes in the intracellular localization of the implicated proteins, which may again be controlled by other regulatory proteins.

Tiam 1 thus activates Rho-like GTPases, and in particular Rac, in fibroblasts. In mammalian cells, nine Rho-like GTPases have been identified, including RhoA, -B,-C and -G, Rac1 and -2, TC10 and Cdc42Hs1 and -2 (CHARDIN et al. 1989; DOWNWARD 1992). Whether Tiam 1 is also capable of activating other members of the Rho-like GTPases is currently being investigated.

3.1 Rho Function

Similar to what has been found for Ras, overexpression or mutational activation of Rho A can result in the transformation of fibroblasts in vitro. These transformed cells show a weak oncogenic potential after injection into nude mice (AVRAHAM and WEINBERG 1989; PERONA et al. 1993), suggesting a connection between the Rho and the oncogenic Ras pathway. In fibroblasts, RhoA activation has been associated with the formation of actin filament bundles (stress fibres) and the assembly of focal contacts (membrane-associated structures connecting integrin adhesion molecules with the actin cytoskeleton) (RIDLEY and HALL 1992). In tissue culture cells, introduction of C3 adenosine diphosphate (ADP) ribosyltransferase, which ribosylates RhoA and thereby inactivates the protein (SEKINE et al. 1989), results in a complete loss of actin stress fibres (CHARDIN et al. 1989; PATERSON et al. 1990). Conversely, micro-injection of RhoA, or constitutively active Vl4RhoA, induces the assembly of focal contacts and stress fibres in serum-starved fibroblasts, an effect that is also seen upon exposure of the cells to the serum factor lysophosphatidic acid (LPA) (RIDLEY and HALL 1992). LPA is thought to interact with an as yet unknown G protein-coupled receptor and stimulates a set of different signalling cascades (VAN CORVEN et al. 1989), including the Ras-MAPK (mitogen-activated protein kinase) cascade by activation of Ras (VAN DER BEND et al. 1992).

The motility of fibroblasts is inhibited by inactivation of RhoA by C3 transferase (TAKAISHI et al. 1993). Constitutive active V14RhoA, however, does not induce motility, indicating that RhoA is necessary but not sufficient for cell motility (TAKAISHI et al. 1994). In vitro invasiveness of hepatoma (IMAMURA et al. 1993) and lymphoma cells (our unpublished results) is also inhibited by C3 transferase treatment and thus dependent on RhoA. We have, however, not been able to confer invasiveness onto T lymphoma cells by introduction of V14RhoA (MICHIELS et al. 1995). This suggests that RhoA function is required, but not sufficient for invasiveness of lymphoma cells, similar to the effects of RhoA on cell motility.

Recent findings in lymphoid cells indicate that lymphocyte function-associated antigen (LFA)-1 integrin-mediated aggregation is dependent on RhoA (TOMINAGA et al. 1993). This is of interest, since antibodies against LFA-1 interfere with the invasive capacity of T lymphoma cells and LFA-1-deficient mutants show impaired metastatic potential (ROOSSIEN et al. 1989). Rho activates a phosphatidylinositol 4-phosphate 5-kinase in mammalian cells, leading to the production of phosphatidylinositol 4,5-biphosphate (PtdIns(4,5)-P2; CHONG et al. 1994). Antibodies directed against certain integrins also trigger the production of PtdIns(4,5)-P2, and PtdIns(4,5)-P2 can stimulate actin-binding proteins such as α-actinin and profilin (LASSING and LINDBERG 1985; FUKAMI et al. 1992). These results implicate Rho in an integrin-signalling pathway and raise the possibility that PtdIns(4,5)-P2 mediates the effects of Rho on the cytoskeleton by regulating the function of actin-binding proteins. PtdIns(4,5)-P2 is also a substrate for P13 kinase and may therefore link the Rho to the Rac signalling pathway (see Sect. 3.2).

3.2 Rac Function

In contrast to LPA-induced stress fibre formation, most other growth factors such as platelet-derived growth factor (PDGF), epithelial growth factor (EGF), insulin and thrombin rapidly induce membrane ruffling in certain fibroblast cells, a process controlled by Rac1 (RIDLEY 1994, RIDLEY et al. 1992). Injection of constitutive active Rac1 (V12Rac1) into fibroblasts increases the amount of cortical F actin and rapidly induces the formation of membrane ruffles and pinocytotic vesicles, whereas dominant-negative N17Rac1 inhibits growth factor-induced membrane ruffling (RIDLEY et al. 1992). The rapid Rac-mediated membrane ruffling is in some cases followed by the appearence of stress fibres (RIDLEY et al. 1992), suggesting that Rho acts downstream of Rac. PI3 kinase, a dimer consisting of a p85 and a p110 subunit, is one of the major components involved in Ras- and growth factor-stimulated formation of membrane ruffles in fibroblasts. Stimulation of several distinct cell surface receptors leads to a rapid activation of PI3 kinase (PARKER and WATERFIELD 1992; FRY and WATERFIELD 1993). The enzyme stimulates synthesis of PtdIns (3,4,5)-P3 from PtdIns(4,5)-P2 in target cells, and PtdIns(3,4,5)-P3 is thought to act as a second messenger. The fungal metabolite wortmannin, which potently inhibits PI3 kinase activity in vitro by binding to p110 (ARCARO and WYMANN 1993; YANO et al. 1993), inhibits membrane ruffling induced by PDGF but not by V12Rac (NOBES et al. 1995; WENNSTROM et al. 1994), indicating that PI3 kinase acts upstream of Rac. Moreover, mutant p85, which lacks a region that is required for tight association with p110, also inhibits PDGF-induced membrane ruffling (WENNSTROM et al. 1994). All these data strongly suggest that activation of PI3 kinase, and the simultaneous generation of PtdIns(3,4,5)-P3, play a major role in PDGF- and other growth factor-induced membrane ruffling which depends on activation of Rac. Like V12Rac1, Tiam1 induces also membrane ruffles and pinocytotic vesicles in fibroblasts, and N17Rac1 inhibits Tiam1-induced membrane ruffling, indicating that Rac1 acts downstream of Tiam1 (Michiels et al. 1995). We are currently investigating whether Tiam1 is involved in growth factor-induced Rac-mediated formation of membrane ruffles and whether P13 kinase is involved in the regulation of Tiam1 activity.

3.3 Connections Between the Ras and the Rho/Rac Signalling Pathways

With respect to interconnections between the identified signalling pathways, p190 RhoGAP is of particular interest. The GTPase-activating protein p190 is tyrosine phosphorylated in growth factor-stimulated cells and forms a tight complex with p120RasGAP (Settleman et al. 1992a,b). This suggests that both proteins function in coupling the Ras and Rho signalling pathways. This notion is supported by the findings that a mutant N-terminal RasGAP (GAP-N), when expressed in cells, is constitutively complexed with p190 and leads to dissociation of stress fibres and focal contacts, a phenomenon also observed after down-regulation of Rho in cells (McGlade et al. 1993).

Evidence is accumulating that the Ras and Rac pathways are also interconnected. Recent studies indicate that Ras might activate Rac. Co-expression of N17Rac1 inhibits focus formation by oncogenic V12Ras, but not by RafCAAX, in NIH3T3 fibroblasts (Qiu et al. 1995). Furthermore, V12Rac1 strongly synergizes with RafCAAX in the focus formation assay, suggesting that oncogenic V12Ras drives both the Rac and MAPK pathways (Qiu et al. 1995). Our earlier findings that oncogenic V12Ras can induce invasion of T lymphoma cells (Collard et al. 1987a) might thus be explained by the activation of Rac by Ras. The fact that N-terminally truncated Tiam1 and V12Rac1 (Qiu et al. 1995; van Leeuwen et al. 1995) induce an oncogenic phenotype in NIH3T3 fibroblasts implies that activated Rac either influences the Ras/MAPK pathway or actuates a novel pathway separate from the Ras/MAPK.

Little is known about the downstream effectors of Rac. The GTPase-activating proteins Bcr and P190RhoGAP might serve as downstream effectors of Rac (Ridley et al. 1993). Another potential effector is the serine-threonine protein kinase $p65^{pak}$. The protein $p65^{pak}$ complexes specifically with activated GTP-bound Cdc42HS and Rac (Manser et al. 1994). The observation that $p65^{pak}$ binds to, and is activated by, Rac suggests that this kinase, or a homologue thereof, is a downstream sffector of Rac, comparable to Ras and Raf. Rip (Rac-interacting protein) was identified by the "yeast two-hybrid" interaction cloning strategy (L. van Aelst, Cold Spring Harbor Laboratory, NY, USA; personal communication). Rip interacts with Rac and induces membrane ruffling in fibroblasts. Moreover, dominant-negative Rip inhibits growth factor-induced membrane ruffling in fibroblasts (L. van Aelst, personal communications). Our preliminary results indicate that dominant-negative Rip also inhibits Tiam1-induced membrane ruffling. Recently, a novel myosin (Myr5) has been identified which harbours a domain that is conserved in Rho-specific GAP (RHEINHARD et al. 1995). Fusion proteins harbouring this GAP domain stimulate GTPase activity of RhoA and Cdc42HS and, to a lesser extent, of Rac1 in vitro. Myr5 may provide a direct link in the regulation of the actin cytoskeletal organization mediated by Rho and Rac signalling pathways.

4 Conclusions

Based on our current knowledge, we speculate that Tiam1 acts in vivo as a GDS protein towards Rac, similar to Sos towards Ras (Fig. 2). The involvement of Sos in pathways by which signals are transmitted from receptor tyrosine kinase (e.g. PDGF, EGF receptor) to Ras has been largely elucidated (EGAN et al. 1993; GALE et al. 1993; SIMON et al. 1993; BUDAY and DOWNWARD 1993). The activated receptor recruits the Grb2–Sos complex to its phosphorylated tyrosine via the SH2 domain of Grb2. The receptor-associated Sos provokes GDP–GTP exchange on the membrane-associated Ras and triggers a cascade of serine-threonine kinases via Raf (Fig. 2). Modified Raf, targeted to the membrane by a signal for farnesylation,

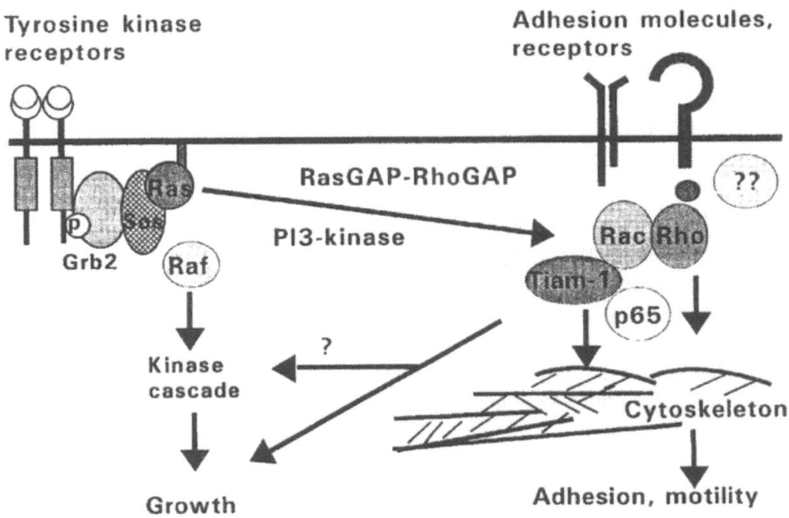

Fig. 2. Model of the role of Tiam1 in identified signal transduction pathways. Activation of tyrosine kinase receptors recruits the Grb2–Sos complex to its phosphorylated tyrosine. Sos, a GDS protein for Ras, provokes guanosine diphosphate–triphosphate (GDP–GTP) exchange on the membrane-associated Ras and triggers a cascade of serine-threonine kinases via Raf. Tiam1, a GDS for Rac, activates Rac upon receptor stimulation, including tyrosine kinase receptors, possibly adhesion molecules and unknown receptors, leading to a cytoskeletal reorganization with consecutive changes in morphology, adhesion and motility of cells. p120RasGAP and p190RhoGAP are thought to interconnect the Ras and Rho pathway, whereas Tiam1 and PI3 kinase might be involved in the interconnection of the Ras and Rac pathway. A novel pathway, most likely distinct to the classical Ras-MAPK (mitogen-activated protein kinase) pathway, makes a connection to the oncogenic pathway downstream of Rac

is constitutively activated and independent of Ras, suggesting that active Ras-GTP is a plasma membrane-targeting signal for Raf (LEEVERS et al. 1994). The membrane targeting of Sos and Raf is thus an essential event in the mitogenic signal cascade.

It is intriguing that the invasion-inducing Tiam1 gene, which was identified by a functional assay, encodes a GDS protein that activates Rac-mediated signalling pathways (Fig. 2). Rac and Rho proteins have been implicated in the regulation of the actin cytoskeleton in response to distinct extracellular signals from various types of receptors, including typosine kinase- and heterotrimeric G protein-coupled receptors and integrins. The activation of these pathways determines the adhesion, morphology and motility of cells, processes involved in invasion and metastasis. Like Sos, Tiam1 seems to translocate to the plasma membrane, which may leed to activation of membrane-associated or -translocated Rac, which in turn results in morphological transformation of fibroblasts accompanied by the formation of membrane ruffles. The findings that Rac acts downstream of Ras, and that activation of the Rac pathway is required for Ras transformation (QIU et al. 1995), indicates a direct connection between the Ras and the Rac pathway. We hypothesize that Tiam1 is involved in receptor- and Ras-induced signalling towards Rac. The activity of Tiam1 may be mediated by PI3 kinase and/or membrane translocation. Activation of PI3 kinase, as well as Ras, occurs upon stimulation of tyrosine kinase receptors and heterotrimeric G protein-coupled receptors (see COOK and MCCORMICK 1994). The proposed Ras signalling towards the Tiam1–Rac pathway could explain our earlier observations that V12Ras is also capable of inducing an invasive phenotype when highly expressed in T lymphoma cells (COLLARD et al. 1987a). The induction of an oncogenic phenotype in fibroblasts by mutant Tiam1 and V12Rac1 (VAN LEEUWEN et al. 1995) suggests that an oncogenic phenotype can be induced by a novel pathway downstream of Rac, which is distinct from the classical oncogenic Ras/MAPK pathway.

Taken together, our results implicate the Tiam1–Rac pathway in processes of tumor formation, invasion and metastasis. Aberrations in the Tiam1–Rac pathway may, therefore, have consequences for tumorigenesis and metastasis in general.

Acknowledgment. This work was supported by grants from The Dutch Cancer Society and from The Netherlands Organization of Scientific Research to J.G. Collard.

References

Arcaro A, Wymann MP (1993) Worthmannin is a potent phosphatidylinositol 3-kinase inhibitor: the role of phosphatidylinositol 3,4,5-triphosphate in neutrophil responses. Biochem J 296: 297–301

Avraham H, Weinberg RA (1989) Characterization and expression of the human rho 12 gene product. Mol Cell Biol 9: 2058–2066

Behrens J, Mareel MM, Van Roy FM, Birchmeier W (1990) Dissecting tumor cell invasion: epithelial cells acquire invasive properties after loss of Uvomorulin-mediated cell-cell adhesion. J Cell Biol 108: 2435–2447

Bevilacqua G, Sobel ME, Liotta LA, Steeg PS (1989) Association of low nm23 RNA levels in human primary infiltrating ductal breast carcinomas with lymph node involvement and other histopathological indicators of high metastatic potential. Cancer Res 49: 5185–5190

Boguski MS, McCormick F (1993) Proteins regulating Ras and its relatives. Nature 366: 643–654

Bourne HR, Sanders DA, McCormick F (1990) The GTPase superfamily: a conserved switch for diverse cell functions. Nature 348: 125–132

Bourne HR, Sanders DA, McCormick F (1991) The GTPase superfamily: conserved structure and molecular mechanism. Nature 349: 117–127

Buday L, Downward J (1993) Epidermal growth factor regulates p21(ras) through the formation of a complex of receptor, Grb2 adapter protein, and Sos nucleotide exchange factor. Cell 73: 611–620

Chan BMC, Matsuura N, Takada Y, Zetter BR, Hemler ME (1991) In vitro and in vivo consequences of VLA-2 expression on rhabdomyosarcoma cells. Science 251: 1600–1602

Chardin P, Boquet P, Madaule P, Popoff MR, Rubin EJ, Gill DM (1989) The mammalian G protein rho C is ADP-ribosylated by clostridium botulinum exoenzyme C3 and affects actin microfilaments in vero cells. EMBO J 8: 1087–1092

Chong LD, Traynorkaplan A, Bokoch GM, Schwartz MA (1994) The small GTP-binding protein Rho regulates a phosphatidylinositol 4-phosphate 5-kinase in mammalian cells. Cell 79: 507–513

Collard JG, Schijven JF, Roos E (1987a) Invasive and metastatic potential induced by ras-transfection into mouse BW5147 T lymphoma cells. Cancer Res 47: 754–759

Collard JG, Van de Poll M, Scheffer A, Roos E, Hopman AHM, Geurts van Kessel AHM, Van Dongen JJM (1987b) Location of genes involved in invasion and metastasis on human chromosome 7. Cancer Res 47: 6666–6670

Collard JG, Roos E, La Rivière G, Habets GGM (1988) Genetic analysis of invasion and metastasis. Cancer Surv 7: 692–710

Collard JG, Habets GGM, Van der Kammen R, Scholtes E (1989) Genetic basis of T lymphoma invasion. Invasion Metastasis 9: 379–390

Cook S, McCormick F (1994) Signal transduction–Ras blooms on sterile ground. Nature 369: 361–362

Downward J (1992) Signal transduction. Rac and Rho in tune. Nature 359: 273–274

Egan SE, Giddings BW, Brooks MW, Buday L, Sizeland AM (1993) Association of Sos Ras exchange protein with Grb2 is implicated in tyrosine kinase signal transduction and transformation. Nature 363: 45–51

Eva A, Vecchio G, Rao CD, Tronick SR, Aaronson SA (1988) The predicted DBL oncogene product defines a distinct class of transforming proteins. Proc Natl Acad Sci USA 85: 2061–2065

Fry MJ, Waterfield MD (1993) Structure and function of phosphatidylinositol 3-kinase: a potential second messenger system involved in growth control. Philos Trans R Soc Lond Biol 340: 337–344

Fukami K, Furuhashi K, Inagaki M, Endo T, Hatano S, Takenawa T (1992) Requirement of phosphatidylinositol 4,5-bisphosphate for alpha-actinin function. Nature 359: 150–152

Gale NW, Kaplan S, Lowenstein EJ, Schlessinger J, Barsagi D (1993) Grb2 Mediates the EGF-dependent activation of guanine nucleotide Exchange on Ras. Nature 363: 88–92

Gibson TJ, Hyvonen M, Birney E, Musacchio A, Saraste M (1994) PH domain–the first anniversary. Trends Biochem Sci 19: 349–353

Günthert U, Hofmann M, Rudy W, Reber S, Zöller M, Haussman I, Matzku S, Wenzel A, Ponta H, Herrlich P (1991) A new variant of glycoprotein CD44 confers metastatic potential to rat carcinoma cells. Cell 65: 13–24

Habets GGM, Van der Kammen R, Scholtes WHM, Collard JG (1990) Induction of invasive and metastatic potential in mouse T lymphoma cells (BW5147) by treatment with 5-azacytidine. Clin Exp Metastasis 8: 567–577

Habets GGM, Van der Kammen RA, Willemsen V, Balemans M, Wiegant J, Collard JG (1992) Sublocalization of an invasion-inducing locus and other genes on human chromosome 7. Cytogenet Cell Genet 60: 200–205

Habets GGM, Scholtes EHM, Zuydgeest D, Van der Kammen RA, Stam JC, Berns A, Collard JG (1994) Identification of an invasion-inducing gene, Tiam-1, that encodes a protein with homology to GDP-GTP exchangers for rho-like proteins. Cell 77: 537–549

Habets GGM, Van der Kammen RA, Jenkins NA, Gilbert DJ, Copeland NG, Hagemeijer A, Collard JG (1995a) The invasion-inducing TIAM-1 gene maps to human chromosome band 21q22 and mouse chromosome 16. Cytogenet Cell Genet 70: 48–51

Habets GGM, van der Kammen RA, Stam JC, Michiels F, Collard JG (1995b) Sequence of the human invasion-inducing TIAM1 gene, its conversion in evolution and its expression in tumor cell lines of different tissue origin. Oncogene 10: 1371–1376

Hall A (1990) The cellular functions of small GTP-binding proteins. Science 249: 635–640
Hall A (1992) Ras-related GTPases and the cytoskeleton. Mol Biol Cell 3: 475–479
Harlan JE, Hajduk PJ, Yoon HS, Fesik SW (1994) Pleckstrin homology domains bind to phosphatidylinositol-4,5-bisphosphate. Nature 371: 168–170
Hart MJ, Eva A, Zangrilli D, Aaronson SA, Evans T, Cerione RA, Zheng Y (1994) Cellular transformation and guanine nucleotide exchange activity are catalyzed by a common domain on the dbl oncogene product. J Biol Chem 269: 62–65
Horii Y, Beeler JF, Sakaguchi K, Tachibana M, Miki T (1994) A novel oncogene, ost, encodes a guanine nucleotide exchange factor that potentiality links Rho and Rac signaling pathways. EMBO J 13: 4776–4786
Imamura F, Horai T, Mukai M, Shinkai K, Sawada M, Akedo H (1993) Induction of in vitro tumor cell invasion of cellular monolayers by lysophosphatidic acid or phospholipase D. Biochem Biophys Res Commun 193: 497–503
Katzav S, Martin-Zanca D, Barbacid M (1989) Vav, a novel human oncogene derived from a locus ubiquitously expressed in hematopoietic cells. EMBO J 8: 2283–2290
Katzav S, Cleveland JL, Heslop HE, Pulido D (1991) Loss of the amino-terminal helix-loop-helix domain of the vav proto-oncogene activates its transforming potential. Mol Cell Biol 11: 1912–1920
Lassing I, Lindberg U (1985) Specific interaction between phosphatidylinositol 4,5 bisphosphate and profilactin. Nature 314: 472–474
Leevers SJ, Paterson HF, Marshall CJ (1994) Requirement for Ras in Raf activation is overcome by targeting Raf to the plasma membrane. Nature 369: 411–420
Leone A, Flatow U, King CR, Sandeen MA, Margulies IM, Liotta LA, Steeg PS (1991) Reduced tumor incidence, metastatic potential, and cytokine responsiveness of nm23-transfected melanoma cells. Cell 65: 25–35
Liotta LA, Steeg P, Stetler Stevenson WG (1991) Cancer metastasis and angiogenesis: an imbalance of positive and negative regulation. Cell 64: 327–336
Manser E, Leung T, Salihuddin H, Zhao ZS, Lim L (1994) A brain serine threonine protein kinase activated by Cdc42 and Rac1. Nature 367: 40–46
McGlade J, Brunkhorst B, Anderson D, Mbamalu G, Settleman J, Dedhar S, Rozakis-Adcock M, Chen LB, Pawson T (1993) The N-terminal region of GAP regulates cytoskeletal structure and cell adhesion. EMBO J 12: 3073–3081
Michiels F, Habets GGM, Stam JC, Van der Kammen RA, Collard JG (1995) A role for Rac in Tiam1-induced membrane ruffling and invasion. Nature 375: 338–340
Miki T, Smith CL, Long JE, Eva A, Fleming TP (1993) Oncogene ect2 is related to regulators of small GTP-binding proteins. Nature 362: 462–465 (erratum published in Nature 364: 737)
Nobes C, Hawkins P, Stephens L, Hall A (1995) Activation of the small GTP-binding proteins Rho and Rac by growth factor receptors. J Cell Sci 108: 225–233
Parker PJ, Waterfield MD (1992) Phosphatidylinositol 3-kinase: a novel effector. Cell Growth Differ 3: 747–752
Paterson HF, Self AJ, Garrett MD, Just I, Aktories K, Hall A (1990) Microinjection of recombinant p21rho induces rapid changes in cell morphology. J Cell Biol 111: 1001–1007
Perona R, Esteve P, Jimenez B, Ballestero RP, Cajal SR (1993) Tumorigenic activity of rho genes from aplysia-californica. Oncogene 8: 1285–1292
Ponting CP, Philips C (1995) DHR domains in synthrophins, neuronal NO synthases and other intracellular proteins. Trends Biochem Sci 20: 102–103
Qiu R, Kirn D, McCormick F, Symons M (1995) An essential role for Rac in Ras transformation. Nature 374: 457–459
Reinhard J, Scheel AA, Diekmann D, Hall A, Ruppert C, Bahler M (1995) A novel type of myosin implicated in signalling by Rho family GTPases. EMBO J 14: 101–108
Ridley AJ (1994) Membrane ruffling and signal transduction. Bioessays 16: 321–327
Ridley AJ, Hall A (1992) The small GTP-binding protein rho regulates the assembly of focal adhesions and actin stress fibres in response to growth factors. Cell 70: 389–399
Ridley AJ, Paterson HF, Johnston CL, Diekmann D, Hall A (1992) The small GTP-binding protein rac regulates growth factor-induced membrane ruffling. Cell 70: 401–410
Ridley A, Self AJ, Kasmi F, Paterson HF, Hall A, Marshall CJ, Ellis C (1993) Rho family GTPase activating proteins p190, Bcr and RhoGAP show distinct specificities in vitro and in vivo. EMBO J 12: 5151–5160
Roos E, La Rivière G, Collard JG, Stukart MJ, de Baetselier P (1985) Invasiveness of T cell hybridomas in vitro and their metastatic potential in vivo. Cancer Res 45: 6238–6243

Roossien FF, de Rijk D, Bikker A, Roos E (1989) Involvement of LFA-1 in lymphoma invasion and metastasis demonstrated with LFA-1-deficient mutants. J Cell Biol 108: 1979–1985
Sekine A, Fujiwara M, Narumiya S (1989) Asparagine residue in the rho gene product is the modification site for botulinum ADP-ribosyltransferase. J Biol Chem 264: 8602–8605
Settleman J, Albright CF, Foster LC, Weinberg RA (1992a) Association between GTPase activators for Rho and Ras families. Nature 359: 153–154
Settleman J, Narashimhan V, Foster LC, Weinberg RA (1992b) Molecular cloning of cdnas encoding the gap-associated protein p190: implications for a signaling pathway from ras to the nucleus. Cell 69: 539–549
Simon MA, Dodson GS, Rubin GM (1993) An SH3-SH2-SH3 protein is required for p21 (Ras1) activation and binds to sevenless and Sos proteins in vitro. Cell 73: 169–177
Takaishi K, Kikuchi A, Kuroda S, Kotani K, Sasaki T, Takai Y (1993) Involvement of rho p21 and its inhibitory GDP/GTP exchange protein (rho GDI) in cell motility. Mol Cell Biol 13: 72–79
Takaishi K, Sasaski T, Kato M, Yamochi W, Kuroda S, Nakamura T, Takeichi M, Takai Y (1994) involvement of Rho P21 small GTP-binding protein and its regulator in the HGF-induced cell motility. Oncogene 9: 273–279
Tominaga T, Sugie K, Hirata M, Morii N, Fukata J, Uchida A, Narumiya S (1993) Inhibition of PMA-induced, lfa-1-dependent lymphocyte aggregation by ADP ribosylation of the small molecular weight GTP binding protein, rho. J Cell Biol 120: 1529–1537
van Corven EJ, Groenink A, Jalink K, Eichholtz T, Moolenaar WH (1989) Lysophosphatidate-induced cell proliferation: identification and dissection of signaling pathways mediated by G proteins. Cell 59: 45–54
van der Bend RL, Brunner J, Jalink K, van Corven EJ, Moolenaar WH, van Blitters WJ (1992) Identification of a putative membrane receptor for the bioactive phospholipid, lysophosphatidic acid. EMBO J 11: 2495–2501
van Leeuwen FN, van der Kammen RA, Habets GGM, Collard JG (1995) Oncogenic activity of Tiam1 and Rac1 in NIH3T3 cells. Oncogene 11: 2215–2221
Vleminckx K, Vakaet L Jr, Mareel M, Fiers W, Van Roy F (1991) Genetic manipulation of E-cadherin expression by epithelial tumor cells reveals an invasion suppressor role. Cell 66: 107–119
Wennstrom S, Hawkins P, Cooke F, Hara K, Yonezawa K, Kasuga M, Jackson T, Claesson-Welsh L, Stephens L (1994) Activation of phosphoinositide 3-kinase is required for PDGF-stimulated membrane ruffling. Curr Biol 4: 385–393
Weilenga VJM, Heider KH, Offerhaus GJA, Adolf GR, Van Den berg FM, Ponta H, Herrlich P, Pals ST (1993) Expression of CD44 variant proteins in human colorectal cancer is related to tumor progression. Cancer Res 53: 4754–4756
Yano H, Nakanishi S, Kimura K, Hanai N, Saitoh Y, Fukui Y, Nonomura Y, Matsuda Y (1993) Inhibition of histamine secretion by wortmannin through the blockade of phosphatidylinositol 3-kinase in RBL-2H3 cells. J Biol Chem 268: 25846–25856
Zheng Y, Cerione R, Bender A (1994) Control of the yeast bud-site assembly GTPase Cdc42 – catalysis of guanine nucleotide exchange by Cdc24 and stimulation of GTPase activity by Bem3. J Biol Chem 269: 2369–2372

Influence of *c-myc* on the Progression of Human Breast Cancer

P.H. Watson[1,2], R. Singh[1], and A.K. Hole[1]

1 Introduction 267

2 Tumor Progression in Breast Cancer 268

3 The *c-myc* Gene 269
3.1 Function and Physiology 269
3.2 Regulation and Molecular Pathology 270

4 The *c-myc* Gene in Breast Cancer 272
4.1 Importance of *c-myc* in Breast Cell Lines 272
4.2 Structural Alteration of *c-myc* in Breast Tumors 272
4.3 Functional Alteration of *c-myc* in Breast Tumors 274

5 The *c-myc* Gene and Invasion in Early Breast Cancer 275
5.1 Downstream Genes 275
5.2 Cell Adhesion 276
5.3 Influence of nm23 276

6 Conclusion 277

References 277

1 Introduction

The clinical management of breast cancer patients is guided by measures of tumor growth and metastatic potential (McGuire 1987). These clinical–pathological measures involve a number of parameters including the tumor size, histological type and grade, the presence of steroid receptors, and perhaps most importantly the status of the axillary lymph nodes (Simpson and Page 1992; Fisher 1992). The scope of this assessment is, however, significantly limited in patients with node-negative tumors and those with preinvasive lesions.

Improved prediction of intrinsic biological behavior is needed for these subgroups. This improvement may in part come from better methods for the application and the integration of known prognostic factors, but the identification

[1]Department of Pathology, Faculty of Medicine, University of Manitoba, 770 Bannatyne Avenue, Winnipeg, Manitoba, R3E OW3, Canada
[2]Department of Physiology, Faculty of Medicine, University of Manitoba, 770 Bannatyne Avenue, Winnipeg, Manitoba, R3E OW3, Canada

of new biological markers is also needed to improve out ability to predict metastatic behavior of tumors in these patients (McGuire and Clark 1992). This in turn is dependent on a better understanding of the molecular determinants of tumor progression and, in particular, the molecular mechanisms behind growth, invasion, and metastasis in the breast cancer system.

In this review we discuss the concept of tumor progression as it relates to breast cancer and then consider the role that the *c-myc* oncogene might play in influencing the many facets of this phenomenon.

2 Tumor Progression in Breast Cancer

The term tumor progression encompasses a range of incremental biological changes that typically occur in tumors over time (Foulds 1958). These changes affect overall tumor growth, response to signals encouraging cell proliferation, programmed cell death, cell differentiation, and respect for position within the organ and tissue environment of the host. Together, these become manifest in the clinically important attributes of unrestricted growth, invasion, and metastasis. One view is that these attributes are the products of distinctive cellular phenotypes and that the clinical evolution of tumors is conditional on the dynamic and transient expression of these phenotypes (Mareel et al. 1990). An alternative view is that the range and extent of cellular properties required for the metastatic process is incompatible with acquisition through a series of independent genetic events (Liotta 1988). It is proposed instead that the genomic instability inherent in the tumor cell leads to disruption of only a relatively small number of critical regulatory genes common to several cellular processes (Egan et al. 1991).

This view is supported by the demonstration that metastasis may be controlled in experimental systems by altering the expression of oncogenes which are also known to have critical regulatory roles in growth. In cell models, these include the *ras* genes and a number of serine/threonine and tyrosine kinase genes (Bradley et al. 1986; Egan et al. 1987). Nuclear oncogenes such as *c-myc* and p53 are often ineffective in fibroblast systems, and as a result their possible roles in metastatic progression have not been as actively pursued. However, N-*myc* has been implicated in the regulation of metastasis in neuroblastoma cell lines (Bernards et al. 1986), and the influence of several nuclear oncogenes on invasiveness, in cooperation with v-ras^{H}, has recently been shown in experiments with a series of immortalized normal breast cell lines. H-*ras* oncogenes can also stimulate invasion in some malignant breast cell lines in vitro, but this effect does not persist in in vivo models. (Ochieng et al. 1991; Gelmann et al. 1992).

Study of human breast cancers has identified a number of specific genetic and molecular lesions that may play a role in more than one aspect of the process of tumor progression (Dickson and Lippman 1992). For example, the emergence of hormone-independent growth is one clinically important component of progression in breast cancer. The molecular changes that underlie resistance to endocrine therapies may involve specific alterations in several components of the mechanism of estrogen action and related growth pathways (WATSON and BARAL 1991). These include altered expression of the estrogen receptor and changes in the regulation of estrogen-responsive target genes such as c-*myc* and specific growth factors (MURPHY 1990; MCGUIRE et al. 1991; DUBIK and SHIU 1992) in addition to lesions in other growth-related genes such as p53, *erb-B*, and *erb-B2/neu* and growth factors, which are also prevalent (DICKSON and LIPPMANN 1987; CALLAHAN 1992). Invasiveness may be influenced by some of the same genes that affect the hormone-responsive phenotype in addition to alteration in the expression of specific "invasion" genes (CLARKE et al. 1993). The latter genes include cell adhesion molecules such as the cadherins, CD44, the integrins and the laminin receptor, proteases such as cathepsin D, metalloproteinases, plasminogen activators and their inhibitors, motility factors such as hepatocyte growth factor, and cytoskeletal molecules such as vimentin (DICKSON and LIPPMAN 1992; CASTRONOVO et al. 1989; TANDON et al. 1990; ROCHEFORT 1992; DUFFY 1992). Most of these factors will also necessarily influence metastasis together with many additional factors, including, angiogenesis and the loss of the expression of metastasis suppressor genes (KERBEL 1989). Alterations of most of those oncogenes that are commonly associated with the metastatic phenotype in laboratory models, such as *ras*, are rare in human breast cancers (BOS 1989).

In addition to these more defined lesions, the occurrence of loss of heterozygosity (LOH) at several loci has been demonstrated in breast cancer (ALI et al. 1989; DEVILEE and CORNELISSE 1990; BIECHE et al. 1992), and some of these lesions have been shown to have prognostic significance. Yet the biological function of the majority of these genes and deleted loci is only poorly understood, and this limits their application and the determination of their relevance to prognostic assessment.

3 The c-*myc* Gene

3.1 Function and Physiology

The c-*myc* gene has been strongly implicated in both the cause and progression of breast cancer. The structural design of this archetypal nuclear oncogene has long suggested that it encodes a *trans*-acting nuclear protein (LUSCHER and EISENMAN 1990). This has now been affirmed by the determination of its functional association with another nuclear protein, *max* (BLACKWOOD and EISENMAN 1991) that permits c-*myc* to transactivate target genes in vitro by specific binding to DNA target elements. These are characterized by the core sequence CAnnTG, which is common to many helix–loop–helix DNA-binding proteins. The *myc* and *max* genes show a propensity to bind to the sequences CACGTG and CACCTG in particular, and these short core sequences are present in the promoter regions of a number of potential target genes. However, the contribution of the adjacent nucleotides which flank the core sequence and which are also likely to determine specificity has yet to be established (Dang 1991; Halazonetis and Kandil 1991). The c-*myc* gene can also repress the transcription of target genes, and it is believed that this effect can be mediated in some instances by a mechanism that is dependent on initiator (Inr) elements within the basal promotors of target genes (Philipp et al. 1994; Li et al. 1994).

Both c-*myc* and the closely related N-*myc* have been implicated in the regulation of only a handful of genes, summarized in Table 1 (Jansen Durr et al. 1993; Philipp et al. 1994; Daksis et al. 1994; Reisman et al, 1993; Penn et al. 1990b; Facchini et al. 1995; Li et al. 1994; Eilers et al. 1991; Gaubatz et al. 1994; Mai and Jalava 1994; Pena et al. 1993; Bello Fernandez et al. 1993; Benvenisty et al. 1992; Suen and Hung 1991; Akeson and Bernards 1990; Versteeg et al. 1988; Peltenburg and Schrier 1994; Peltenburg et al. 1993; Inghirami et al. 1990; Shtivelman and Bishop 1991; Gross et al. 1994; Yang et al. 1991; Dickinson Laing et al. 1995; Prendergast et al. 1990; Prendergast and Cole 1989; Wasfy et al. 1995), and direct evidence for a role in transcriptional regulation exists for only very few of these. It is therefore hardly surprising that the biological function of the c-*myc* gene is still poorly understood after more than a decade of intense study, but it has been implicated in the control of a wide range of normal cellular processes including cell growth, differentiation, and apoptosis in various cell systems (Studzinski et al. 1986; Heikkila et al. 1987; Evan et al. 1992).

3.2 Regulation and Molecular Pathology

The c-*myc* gene is regulated at multiple levels, and alteration of expression might therefore occur, at least in theory, by a number of mechanisms. For example, both positive, negative, and autoregulation occur at the level of transcription initiation and elongation (Spencer and Groudine 1990; Penn et al. 1990a). Once transcribed, the persistence of the mRNA transcript (and protein) is tightly controlled by the short half-life of the message (in the order of 20 min), and this instability is partly conferred by an AU-rich sequence in the 3' portion of the molecule (Hann and Eisenman 1984; Jones and Cole 1987). Translation of two proteins is controlled from adjacent translational start sites, and the functional attributes of these proteins may be influenced by phosphorylation, which is conferred by several kinase enzymes (Luscher et al. 1989; Luscher and Eisenman 1992).

Alteration of the c-*myc* gene has already been associated with a range of human neoplasms, including lymphomas (Karp and Broder 1991) and carcinoma of the lung, colon, and breast (Little et al., 1983; Erisman et al. 1985; Escot et al. 1986). In hematopoietic neoplasms, altered regulation has classically been attributed to a change in the control of transcription brought about through

Table 1. Putative c-*myc* downstream genes

Gene	Expression	Proposed Mechanism	References
cyclin D1	Repression	Inr element in promoter	JANSEN-DURR et al. 1993 PHILIPP et al. 1994
cyclin D1	Induction	Three CACGTG elements upstream of the transcription start site	DAKSIS et al. 1994
cyclin A	Induction	Unknown	JANSEN-DURR et al. 1993
cyclin E	Induction	Unknown	JANSEN-DURR et al. 1993
p53	Induction	Consensus CACGTG sequence in exon 1	REISMAN et al. 1993
c-*myc*	Repression	Autoregulation	PENN et al. 1990 FACCHINI et al. 1995
C/EBPa	Repression	Inr element	LI et al. 1994
α-Prothymosin	Induction	CACGTG in intron 1	EILERS et al. 1991 GAUBATZ et al. 1994
Dihydrofolate reductase		Two CACCTG consensus sites separated by eight nucleotides	MAI and JALAVA1994
Ornithine decarboxylase	Induction	CACCTG in intron 1	PĚNA et al. 1993 BELLO-FERNANDEZ et al. 1993
ECA39	Induction	CACGTG in exon 1	BENVENISTY et al. 1992
neu (c-*erbB2*)	Repression	Transcriptional	SUEN et al. 1991
N-CAM	Repression	Inr element	AKESON and BERNARD 1990
MHC class I	Repression	Inr element	VEERSTEEG et al. 1988 LI et al. 1994 PELTENBURG and SCHRIER 1994
LFA-1	Repression	Transcriptional; inr element	INGHIRAMI et al. 1990 LI et al. 1994
CD44	Repression	Inr element	SHTIVELMAN and BISHOP 1991 PHILLIP et al. 1994; GROSS et al. 1994
Pro-α_1 (I) collagen Pro-α_2 (I) collagen Pro-α_3 (IV) collagen	Repression	Unknown	YANG et al. 1991
Albumin	Repression	Inr element	LI et al. 1994
Gelatinase B	Induction	Five potential binding sites, two in the promoter region;	DICKINSON LAING et al. 1995
PAI-1		Post-transcriptional	PRENDERGAST and COLE 1989 PRENDERGAST et al. 1990
MIG-1	Induction	Anonymous cDNA, unknown	WASFY et al. 1995
MIG-11	Induction	Anonymous cDNA, unknown	WASFY et al. 1995

N-CAM, neural cell adhesion molecule; MHC, major histocompatibility complex; Inr intiator; PAI, plasminogen activator inhibitor; LFA, lymphocyte function-associated antigen.

translocation and disruption of the regulatory elements or alteration in the stability of the mRNA message (AGHIB et al. 1990). More recently, it has been shown that point mutations with the potential to disrupt transcriptional control or protein phosphorylation sites can also occur (SCARPA et al. 1991; ZAJAC KAYE et al. 1988; AGHIB and BISHOP 1991). However, in solid tumors most studies have focused on alteration through amplification of the gene, which appears to be particularly prevalent in lung tumors and, as described below, breast cancer.

4 The c-*myc* Gene in Breast Cancer

Abnormal expression of c-*myc* transgenes in the mouse can induce mammary tumors (LEDER et al. 1986; SCHOENENBERGER et al. 1988), and alteration in the pattern of expression has also been found in pre-neoplastic lesions in the human breast (HEHIR et al. 1993; ESCOT et al. 1993). However, practical issues have focused most attention on the status of the c-*myc* gene in tumors and cell lines.

4.1 Importance of c-*myc* in Breast Cell Lines

Manipulation of the expression of this gene in cell lines has confirmed that c-*myc* is critical for the growth of breast cancer (WATSON et al. 1991). In breast cancer cells that are estrogen receptor positive and hormone dependent, c-*myc* is directly regulated by estrogen (DUBIK and SHIU 1988, 1992), in contrast, the regulation of c-*myc* appears to be altered in estrogen receptor-negative, hormone-independent cells, where expression of c-*myc* may be high and consitutive (DUBIK et al. 1987). The mechanisms that underlie this alteration in regulation are not always clear, but it is known that both c-*myc* amplification and altered mRNA stability can exist in breast tumor cell lines (DUBIK and SHIU 1988; WATSON et al. 1993). Thus both in vitro and in vitro studies indicate that alteration of c-*myc* may be an important factor not only in initiation, but also in some aspects of breast tumor progression. For example, this gene may play a role in the transition from hormone-dependent to hormone-independent growth.

4.2 Structural Alteration of c-*myc* in Breast Tumors

A summary of almost 30 studies carried out since 1986 that have studied 50 or more cases and reported c-*myc* amplification in breast cancer is shown in Table 2 (ESCOT et al. 1986; GUERIN et al. 1988; ADNANE et al. 1989; GARCIA et al. 1989; MACHOTKA et al. 1989; SESHADRI et al. 1989; TAVASSOLI et al. 1989; TSUDA et al. 1989; ZHOU et al. 1989; BROUILLET et al. 1990; MEYERS et al. 1990; TANG et al. 1990; PATERSON et al. 1991; BERNS et al. 1992a, 1992b; BORG et al. 1992; ROUX DOSSETO et al. 1992; GAFFEY et al. 1993; HENRY et al. 1993; KNYAZEV et al. 1993; KREIPE et al. 1993; OTTESTAD et al. 1993; SCORILAS et al. 1993; WATSON et al. 1993; BIECHE et al. 1994; BOLUFER et al. 1994; HARADA et al. 1994; BERNS et al. 1995). Overall, amplification has been found in 15% of over 5000 tumors encompassed by these studies, but the incidence ranges from 1% to 33%, with an even distribution that is unrelated to study size. This appears to be a reflection of the different genetic composition between study groups, but it may also be related to technical variation, such as selection of cases, tissue source and tumor cell content, the use of different control genes, and the definition of amplification. Interestingly, in two thirds of these studies the presence of c-*erbB2/neu* amplification was also

Table 2. Frequency of c-*myc* amplification in studies of breast cancer

Cases (–)	Amplication of myc (%)	Correlations and comments	Reference
121	32	Age (>50 years)	ESCOT et al. 1986
116	6	High RNA levels in 45% correlated with node positive	GUERIN et al. 1988
219	21	PR negative, high grade	ADNANE et al. 1989
125	18	Inflammatory carcinoma	GARCIA et al. 1989
90	15	Node positive	MACHOTKA et al. 1989
73	23		SESHADRI et al. 1989
63	21	High grade, aneuploidy	TAVASSOLI et al. 1989
176	4	Poor prognosis	TSUDA et al. 1989
171	6		ZHOU et al. 1989
140	23	High Cathepsin D	BROUILLET et al. 1990
100	1		MEYERS et tal. 1990
106	10	ER positive, lymphocytic infiltration	TANG et al. 1990
230	3		PATTERSON et al. 1991
1052	17	PR negative	BERNS et al. 1992b
282	20	Size, node positive, outcome for node-negative and ER-positive groups	BERNS et al. 1992a
311	8	High S phase, ErbB2 amplification, recurrence and survival	BORG et al. 1992
65	25	Relapse-free survival	ROUX DOSSETO et al. 1992
50	12	High grade (poor tubule formation)	GAFFEY et al. 1993
108	22	High grade	HENRY et al. 1993
60	3		KNYAZEV et al. 1993
60	33	High proliferative capacity (KI-67 leveL)	KREIPE et al. 1993
100	1		OTTESTAD et al. 1993
62	17	High Cathepsin D	SCORILAS et al. 1993
154	7		WATSON et al. 1993
122	23	LOH 1p32-pter	BIECHE et al. 1994
145	13	ER negative	BOLUFER et al. 1994
109	28	High grade and node positive with multiple alterations	HARADA et al. 1994
662	11	Relapse-free survival after a median 90-month follow-up period	BERNS et al. 1995

LOH, loss of heterozygosity; PR, progesterone receptor; ER, estrogen receptor.

assessed, and in this subset c-*myc* and *neu* amplification averaged 14% and 22%, respectively (with similar standard deviations of 8% and 6%). Heterogeneity within the tumor may be an additional factor in determining gene amplification, particulary for c-*myc*, as most groups have found only a modest increase in copy number (two- to five fold) in most cases. This is supported by direct analysis of gene amplification within individual cells in tumors by fluorescent in situ hybridization (SHAPIRO et al. 1993; KALLIONIEMI et al. 1992). The occurence of c-*myc* amplification may also be related to the frequency of alteration of a gene located on the short arm of chromosome 1. The clinical significance of c-*myc* amplification has been hard to determine because of the relatively small number of cases enrolled in most of these studies. The indication is that the presence of amplification correlates with a number of poor prognostic indicators, including high levels of cathepsin D, but possibly the most consistent and convincing association is with high pathological tumor grade (ADNANE et al. 1989; TAVASSOLI et al. 1989;

Gaffey et al. 1993; Henry et al. 1993; Harada et al. 1994). Several large and mostly retrospective studies have also concluded that alteration of c-*myc* is associated with short relapse-free survival and overall survival (Borg et al. 1992; Berns et al. 1992a, 1995).

All of the above studies have concentrated on the entire tumor, however, our own experience, based on microdissection and a polymerase chain reaction (PCR)-based assay, is that c-*myc* amplification can also occur at an early stage of tumor progression in the in situ component of invasive carcinoma (Watson et al. 1993). We have also shown that, while amplification occurs in both the in situ and invasive components of tumors, it does not always persist in the corresponding nodal metastases (Watson et al. 1993). Similar observations concerning c-*myc* and other amplified loci have been made by others (Tsuda et al. 1991; Isola et al. 1994).

4.3 Functional Alteration of c-*myc* in Breast Tumors

By contrast to these amplification studies, there have been only a few reports concerning expression of c-*myc* in vivo, and these have been mostly limited to only small numbers of breast tumors (Spandidos et al. 1989; Mariani Costantini et al. 1988; Walker et al. 1989). As described above, it is quite possible that c-*myc* expression might become altered by a number of mechanisms apart from amplification (Soini et al. 1994). These might include changes in chromatin structure in the promotor region (Miller et al. 1993) or gene structure through translocation. This has been described in some cases, but appears to be uncommon (Varley et al. 1987b; Escot et al. 1986). We have also screened for point mutations within the promotor and 5' region of c-*myc* by single-strand conformation polymorphism (SSCP) assay in a small number of tumors, but no mutations were detected (P.H. Watson, unpublished observation). However, the existence of altered regulation at other levels such as changes in the regulation of transcription, mRNA stability, or titration of the activity of the protein through levels of *max* and other binding partners has not been well investigated. This is partly due to the fact that many expression studies are difficult to interpret given the marked heterogeneity of c-*myc* expression that has often been documented (Varley et al. 1987a; Mariani Costantini et al. 1988; Soini et al. 1994), both in tumor cells and in relation to infiltrating lymphocytes, and the probable influence of tissue quality on the level of the unstable c-*myc* message. Nevertheless, it appears that tamoxifen therapy (Le Roy et al. 1991) and proliferative status (Escot et al. 1993) can influence the levels of c-*myc* mRNA and protein, and in one small study an interesting correlation between increased c-*myc* expression at the invasive edge and nodal metastasis was noted (Pavelic et al. 1992). The involvement of c-*myc* in apoptosis makes it possible that expression of this gene will also be one of many factors that affect response to endocrine or chemotherapies that rely on the induction of apoptosis, but this has not been explored as yet.

5 The c-*myc* Gene and Invasion in Early Breast Cancer

The pattern of alteration that we have observed is consistent with a role for c-*myc* in both early growth and in the development of the invasive cellular phenotype that is central to the transition from in situ to invasive carcinoma. Invasiveness is an essential, but not sufficient component of the more complex process of metastasis. Thus the in vivo association between alteration of c-*myc* with high-grade tumors (HENRY et al. 1993), early recurrence, and poor prognosis (ROUX DOSSETO et al. 1992; VARLEY et al. 1987a), but less consistently with nodal status or metastasis (ESCOT et al. 1986; BORG et al. 1992; WATSON et al. 1993), is in keeping with a specific role in invasion. Although this speculation is not supported by studies of invasion and metastasis in experimental systems, many of these studies are based on fibroblast rather than epithelial cell models (EGAN et al. 1991) and also do not distinguish between invasion and metastasis. Consideration of the c-*myc* gene as a factor that might influence invasion of breast cancer cells is consistent with the recent knowledge of genes that may be regulated by c-*myc* and the presence of alteration of c-*myc* expression in estrogen receptor-negative, estrogen-independent cells (DUBIK et al. 1987). This correlates with a number of other phenotypic traits in these cells that are associated with tumor progression, in addition to that of altered hormone response implicit in the classification. For example, estrogen receptor-negative cells show loss of differentiation, loss of homotypic cell–cell adhesion, and increased invasiveness in comparison with estrogen receptor-positive cells (THOMPSON et al. 1992; CLARKE et al. 1993).

5.1 Downstream Genes

The c-*myc* gene might influence invasion through direct or indirect mechanisms. As described above, c-*myc* is a primary target gene in the mechanism of estrogen action, but its expression can also be influenced by progestins and retinoic acid (SHEIKH et al. 1993; MUSGROVE et al. 1993). At the same time, all three steroid hormones have been shown to modulate invasiveness in estrogen receptor-positive breast cancer cells. (THOMPSON et al. 1989, 1991; SHI et al. 1993). While these hormonal effects might well be the result of direct effects of each steroid on genes such as collagenase and the laminin receptor (VAN DEN BRULE et al. 1992), it is also possible that the effect of all three hormones on invasion are partly mediated through c-*myc* and its downstream genes. For example, at least two candidates among the short list of putative *myc* downstream genes, gelatinase B (DICKINSON LAING et al. 1995) and plasminogen activator inhibitor (PAI-1) (PRENDERGAST and COLE 1989), are known to influence proteolysis in experimental systems and have been considered in the context of invasion of breast cancer (FOEKENS et al. 1994; LIOTTA and STETLER STEVENSON 1991). Other candidates for *myc* downstream genes include adhesion molecules such as integrins (INGHIRAMI et al. 1990) and CD44 (SHTIVELMAN and BISHOP 1991), and a general role of c-*myc* in the control of cell adhesion can be inferred from several indirect or correlative findings. For example, changes in *c-myc* expression have been seen to occur in parallel with changes in fibroblast cell adhesion (Dhawan and Farmer 1990), and in B lymphoblastoid cells *c-myc* can influence homotypic adhesion through regulation of the integrin LFA-1 (Inghirami et al. 1990). In estrogen receptor-positive breast cancer cells, estrogen stimulates both *c-myc* expression and adhesion to substratum (Millon et al. 1989).

5.2 Cell Adhesion

It is therefore interesting to speculate that *c-myc* may modulate some aspect of adhesion, perhaps through the direct regulation of a cadherin-like cell adhesion molecule or a CD44-like molecule in breast cancer cells. This speculation is based on the following observations. In human breast cancer cell lines, E-cadherin (E-CAM) expression is generally high in estrogen receptor-positive cells (Frixen et al. (1991), but decreased or absent in estrogen receptor-negative cells, where *c-myc* expression tends to be high and constitutively expressed. Similarly, a different profile of CD44 expression exists in human breast cell lines and tumors distinguished on the basis of estrogen receptor status (Watson et al. 1994). In neuroblastoma, amplification of the N-*myc* gene is associated with loss of CD44H expression in vivo, but not necessarily in cultured cell lines (Gross et al. 1994), and it has been suggested that N-*myc* can transcriptionally downregulate both CD44 and the expression of N-CAM through Inr sequences in this system (Philipp et al. 1994; Shtivelman and Bishop 1991). Examination of the E-CAM promotor has shown that it contains several potential *c-myc* target sequences (Behrens et al. 1991) and an initiator element at the transcription start site that resembles those of the N-CAM and CD44 genes. Interestingly, one of these target sequences lies within a region that has been shown to mediate the differential expression of this gene in estrogen receptor-positive and estrogen receptor-negative breast cancer cells (Behrens et al. 1991). Finally, it has been shown that increased *c-myc* expression can downregulate E-CAM in immortalized normal breast cells, but, paradoxically, *c-myc* overexpression also negatively influences invasion and chemotaxis in this particular system (Thompson et al. 1994). Nevertheless, the biological effects of *c-myc* are known to be strongly influenced by cell type and differentiation in hematopoietic systems (Klein and Klein 1986), and this might help to defend our speculation in the face of this paradox.

5.3 Influence of nm23

Perhaps more intriguing is the link between *c-myc* and the nm23 metastasis suppressor gene. The nm23 gene was initially believed to function as a nucleoside diphosphate (NDP) kinase, but an additional activity as a transactivator of *c-myc* has been demonstrated recently, at least for the nm23-H2 protein (Postel et al. 1993). It is clear that overexpression of nm23-H1 in breast cell lines can alter both tumorigenic and metastatic ability independently of NDP kinase activity (LEONE et al. 1993), reduce motility in response to growth factors, and induce differentiation as marked by acinar formation, induction of the secretion of basement membrane and mucin, and growth arrest (HOWLETT et al. 1994). In clinical studies, loss of nm23 expression has been associated with poor prognosis and invasive potential in in situ tumors (ROYDS et al. 1993; STEEG et al. 1993), but contradictory results have been reported for invasive tumors (HENNESSY et al. 1991; SAWAN et al. 1994). However, for the moment nm23 remains a good candidate for a metastasis suppressor gene in breast cancer, and the possibility that c-*myc* might mediate some of the biological functions of the nm23 proteins deserves investigation.

6 Conclusion

The c-*myc* gene is critical in growth, differentiation, and apoptosis, and it is perhaps not surprising that alteration of the gene is a common occurrence in human neoplasia. Elucidation of the downstream genes and the cellular circuits that are influenced by the *myc-max* hub will be important to extend our understanding of the role that c-*myc* plays in these aspects of progression in solid tumors such as breast cancer. However, given the known involvement of this gene in such a wide range of fundamental cellular processes, it is likely that with increasing knowledge its influence on progression in breast cancer will eventually encompass many of the units originally described by Foulds.

Acknowledgments. This work was supported by grants from the National Cancer Institute of Canada and the Medical Research Council of Canada. Dr. Peter Watson is a recipient of a Clinician-Scientist Award from the Medical Research Council.

References

Adnane J, Gaudray P, Simon MP, Simony Lafontaine J, Jeanteur P, Theillet C (1989) Protooncogene amplification and human breast tumor phenotype. Oncogene 4: 1389–1395

Aghib DF, Bishop JM (1991) A 3' truncation of myc caused by chromosomal translocation in a human Tcell leukemia is tumorigenic when tested in established rat fibroblasts. Oncogene 6: 2371–2375

Aghib DF, Bishop JM, Ottolenghi S, Guerrasio A, Serra A, Saglio G (1990) A 3' truncation of MYC caused by chromosomal translocation in a human Tcell leukemia increases mRNA stability. Oncogene 5: 707–711

Akeson R, Bernards R (1990) N-myc down regulates neural cell adhesion molecule expression in rat neuroblastoma. Mol Cell Biol 10: 2012–2016

Ali IU, Lidereau R, Callahan R (1989) Presence of two members of c-erbA receptor gene family (c-erbA beta and c-erbA2) in smallest region of somatic homozygosity on chromosome 3p21–p25 in human breast carcinoma. J Natl Cancer Inst 81: 1815–1820

Behrens J, Lowrick O, Klein Hitpass L, Birchmeier W (1991) The E-cadherin promoter: functional analysis of a G. C-rich region and an epithelial cell-specific palindromic regulatory element. Proc Natl Acad Sci USA 88: 11495–11499
Bello Fernandez C, Packham G, Cleveland JL (1993) The ornithine decarboxylase gene is a transcriptional target of c-Myc. Proc Natl Acad Sci USA 90: 7804–7808
Benvensity N, Leder A, Kuo A, Leder P (1992) An embryonically expressed gene is a target for c-Myc regulation via the c-Myc binding sequence. Genes Dev 6: 2513–2523
Bernards R, Dessain SK, Weinberg RA (1986) N-myc amplification causes down-modulation of MHC class I antigen expression in neuroblastoma. Cell 47: 667–674
Berns EM, Klijn JG, van Putten WL, van Staveren IL, Portengen H, Foekens JA (1992a) c-myc amplification is a better prognostic factor than HER2/neu amplification in primary breast cancer. Cancer Res 52: 1107–1113
Berns EM, Klijn JG, van Staveren IL, Portengen H, Noordegraaf E, Foekens JA (1992b) Prevalence of amplification of the oncogenes c-myc, HER2/neu, and int-2 in one thousand human breast tumours: correlation with steroid receptors. Eur J Cancer 28: 697–700
Berns EM, Klijn JG, Look MP, van Staveren IL, Portengen H, Foekens JA (1995) separate oncogene amplification for prediction of recurrent breast cancer: an analysis on 662 patients. Proc Am Assoc Cancer Res 36: 558 (abstr)
Bieche I, Champeme MH, Matifas F, Hacene K, Callahan R, Lidereau R (1992) Loss of heterozygosity on chromosome 7q and aggressive primary breast cancer. Lancet 339: 139–143
Bieche I, Champeme MH, Lidereau R (1994) A tumor suppressor gene on chromosome 1p32-pter controls the amplification of MYC family genes in breast cancer. Cancer Res 54: 4274–4276
Blackwood EM, Eisenman RN (1991) Max: a helix-loop-helix zipper protein that forms a sequence-specific DNA-binding complex with Myc. Science 251: 1211–1217
Bolufer P, Molina R, Ruiz A, Hernandez M, Vazquez C, Lluch A (1994) Estradiol receptors in combination with neu or myc oncogene amplifications might define new subtypes of breast cancer. Clin Chim Acta 229: 107–122
Borg A, Baldetorp B, Ferno M, Olsson H, Sigurdsson H (1992) c-myc amplification is an independent prognostic factor in postmenopausal breast cancer. Int J Cancer 51: 687–691
Bos JL (1989) ras oncogenes in human cancer: a review. Cancer Res 49: 4682–4689
Bradley MO, Kraynak AR, Storer RD, Gibbs JB (1986) Experimental metastasis in nude mice of NIH 3T3 cells containing various ras genes. Proc Natl Acad Sci USA 83: 5277–5281
Brouillet JP, Theillet C, Maudelonde T, Defrenne A, Simony Lafontaine J, Sertour J, Pujol H, Jeanteur P, Rochefort H (1990) Cathepsin D assay in primary breast cancer and lymph nodes: relationship with c-myc, c-erb-B-2 and int-2 oncogene amplification and node invasiveness. Eur J Cancer 26: 437–441
Callahan R (1992) p53 mutations, another breast cancer prognostic factor. J Natl Cancer Inst 84: 826–827
Castronovo V, Taraboletti G, Liotta LA, Sobel ME (1989) Modulation of laminin receptor expression by estrogen and progestins in human breast cancer cell lines. J Natl Cancer Inst 81: 781–788
Clarke R, Thompson EW, Leonessa F, Lippman J, McGarvey M, Frandsen TL, Brunner N (1993) Hormone resistance, invasiveness, and metastatic potential in breast cancer. Breast Cancer Res Treat 24: 227–239
Daksis JI, Lu RY, Facchini LM, Marhin WW, Penn LJ (1994) Myc induces cyclin D1 expression in the absence of de novo protein synthesis and links mitogen-stimulated signal transduction to the cell cycle. Oncogene 9: 3635–3645
Dang CV (1991) c-myc oncoprotein function. Biochim Biophys Acta 1072: 103–113
Devilee P, Cornelisse CJ (1990) Genetics of human breast cancer. Cancer Surv 9: 605–630
Dhawan J, Farmer SR (1990) Regulation of alpha 1 (I)-collagen gene expression in response to cell adhesion in Swiss 3T3 fibroblasts. J Biol Chem 265: 9015–9021
Dickinson Laing TMA, Gibson AW, Johnson RN, Edwards DR (1995) Role of c-myc in regulation of gelatinase B expression. Proc Am Assoc Cancer Res 36: 97 (abstr)
Dickson RB, Lippman ME (1987) Estrogenic regulation of growth and polypeptide growth factor secretion in human breast carcinoma. Endocri Inol Rev 8: 29–43
Dickson RB, Lippman ME (1992) Molecular determinants of growth, angiogenesis, and metastases in breast cancer. Semin Oncol 19: 286–298
Dubik D, Dembinski TC, Shiu RP (1987) Stimulation of c-myc oncogene expression associated with estrogen-induced proliferation of human breast cancer cells. Cancer Res 47: 6517–6521
Dubik D, Shiu RP (1988) Transcriptional regulation of c-myc oncogene expression by estrogen in hormone-responsive human breast cancer cells. J Biol Chem 263: 12705–12708

Dubik D, Shiu RP (1992) Mechanism of estrogen activation of c-myc oncogene expression. Oncogene 7: 1587–1594

Duffy MJ (1992) The role of proteolytic enzymes in cancer invasion and metastasis. Clin Exp Metastasis 10: 145–155

Egan SE, Wright JA, Jarolim L, Yanagihara k, Bassin RH, Greenberg AH (1987) Transformation by oncogenes encoding protein kinases induces the metastatic phenotype. Science 238: 202–205

Egan SE, Wright JA, Greenberg AH (1991) Molecular determinants of metastatic transformation. Environ Health Perspect 93: 91–95

Eilers M, Schirm S, Bishop JM (1991) The MYC protein activates transcription of the alpha-prothymosin gene. EMBO J 10: 133–141

Erisman MD, Rothberg PG, Diehl RE, Morse CC, Spandorfer JM, Astrin SM. (1985) Deregulation of c-myc gene expression in human colon carcinoma is not accompanied by amplification or rearrangement of the gene. Mol Cell Biol 5: 1969–1976

Escot C, Theillet C, Lidereau R, Spyratos F, Champeme MH, Gest J, Callahan R (1986) Genetic alteration of the c-myc protooncogene (MYC) in human primary breast carcinomas. Proc Natl Acad Sci USA 83: 4834–4838

Escot C, Simony Lafontaine J, Maudelonde T, Puech C, Pujol H, Rochefort H (1993) Potential value of increased MYC but not ERBB2 RNA levels as a marker of high-risk mastopathies [published erratum appears in Oncogene (7): 2023]. Oncogene 8: 969–974

Evan GI, Wyllie AH, Gilbert CS, Littlewood TD, Land H, Brooks M, Waters CM, Penn LZ, Hancock DC (1992) Induction of apoptosis in fibroblasts by c-myc protein. Cell 69: 119–128

Facchini LF, Chen S, Penn LJZ (1995) Myc repression of gene transcription: one pathway or many? Proc Am Assoc Cancer Res 36: 532 (abstr)

Fisher B (1992) The evolution of paradigms for the management of breast cancer: a personal perspective. Cancer Res 52: 2371–2383

Foekens JA, Schmitt M, Van Putten WL, Peters HA, Kramer MD, Janicke F, Klijn JG (1994) Plasminogen activator inhibitor-1 and prognosis in primary breast cancer. J Clin Oncol 12: 1648–1658

Foulds LJ (1958) The natural history of cancer. J Chronic Disease 8: 2–37

Frixen UH, Behrens J, Sachs M, Eberle G, Voss B, Warda A, Lochner D, Birchmeier W (1991) E-cadherin-mediated cell-cell adhesion prevents invasiveness of human carcinoma cells. J Cell Biol 113: 173–185

Gaffey MJ, Frierson HF Jr, Williams ME (1993) Chromosome 11q13, c-erbB-2, and c-myc amplification in invasive breast carcinoma: clinicopathologic correlations. Mod Pathol 6: 654–659

Garcia I, Dietrich PY, Aapro M, Vauthier G, Vadas L, Engel E (1989) Genetic alterations of c-myc, c-erbB-2, and c-Ha-ras protooncogenes and clinical associations in human breast carcinomas. Cancer Res 49: 6675–6679

Gaubatz S, Meichle A, Eilers M (1994) An E-box element localized in the first intron mediates regulation of the prothymosin alpha gene by c-myc. Mol Cell Biol 14: 3853–3862

Gelmann EP, Thompson EW, Sommers CL (1992) Invasive and metastatic properties of MCF-7 cells and rasH-transfected MCF-7 cell lines. Int J Cancer 50: 665–669

Gross N, Beretta C, Peruisseau G, Jackson D, Simmons D, Beck D (1994) CD44H expression by human neuroblastoma cells: relation to MYCN amplification and lineage differentiation. Cancer Res 54: 4238–4242

Guerin M, Barrois M, Terrier MJ, Spielmann M, Riou G (1988) Overexpression of either c-myc or c-erbB-2/neu proto-oncogenes in human breast carcinomas: correlation with poor prognosis. Oncogene Res 3: 21–31

Halazonetis TD, Kandil AN (1991) Determination of the c-MYC DNA-binding site. Proc Natl Acad Sci USA 88: 6162–6166

Hann SR, Eisenman RN (1984) Proteins encoded by the human c-myc oncogene: differential expression in neoplastic cells. Mol Cell Biol 4: 2486–2497

Harada Y, Katagiri T, Ito I, Akiyama F, Sakamoto G, Kasumi F, Nakamura Y, Emi M (1994) Genetic studies of 457 breast cancers Clinicopathologic parameters compared with genetic alterations. Cancer 74: 2281–2286

Hehir DJ, McGreal G, Kirwan WO, Kealy W, Brady MP (1993) C-myc oncogene expression: a marker for females at risk of breast carcinoma. J Surg Oncol 54: 207–209

Heikkila R, Schwab G, Wickstrom E, Loke SL, Pluznik DH, Watt R, Neckers LM (1987) A C-myc antisense oligodeoxynucleotide inhibits entry into S phase but not progress from GO to G1. Nature 328: 445–449

Hennessy C, Henry JA, May FE, Westley BR, Angus B, Lennard TW (1991) Expression of the antimetastatic gene nm23 in human breast cancer: an association with good prognosis. J Natl Cancer Inst 83: 281–285

Henry JA, Hennessy C, Levett DL, Lennard TW, Westley BR, May FE (1993) int-2 amplification in breast cancer: association with decreased survival and realationship to amplification of c-erbB-2 and c-myc. Int J Cancer 53: 774–780
Howlett AR, Petersen OW, Steeg PS, Bissell MJ (1994) A novel function for the nm23-H1 gene: overexpression in human breast carcinoma cells leads to the formation of basement membrane and growth arrest. J Natl Cancer Inst 86: 1838–1844
Inghirami G, Grignani F, Sternas L, Lombardi L, Knowles DM, Dalla Favera R (1990) Down-regulation of LFA-1 adhesion receptors by C-myc oncogene in human B lymphoblastoid cells. Science 250: 682–686
Isola J, DeVries S, Chu L, Ghazvini S, Waldman F (1994) Analysis of changes in DNA sequence copy number by comparative genomic hybridization in archival paraffin-embedded tumor samples. Am J Pathol 145: 1301–1308
Jansen Durr P, Meichle A, Steiner P, Pagano M, Finke K, Botz J, Wessbecher J, Draetta G, Eilers M (1993) Differential modulation of cyclin gene expression by MYC. Proc Natl Acad Sci USA 90: 3685–3689
Jones TR, Cole MD (1987) Rapid cytoplasmic turnover of c-myc mRNA: requirement of the 3'untranslated sequences. Mol Cell Biol 7: 4513–4521
Kallioniemi OP, Kallioniemi A, Kurisu W, Thor A, Chen LC, Smith HS, Waldman FM, Pinkel D, Gray JW (1992) ERBB2 amplification in breast cancer analyzed by fluorescence in situ hybridization. Proc Natl Acad Sci USA 89: 5321–5325
Karp JE, Broder S (1991) Acquired immunodeficiency syndrome and non-Hodgkins' lymphomas. Cancer Res 51: 4743–4756
Kerbel RS (1989) Towards an understanding of the molecular basis of the metastatic phenotype. Invasion Metastasis 9: 329–337
Klein G, Klein E (1986) Conditioned tumorigenicity of activated oncogenes. Cancer Res 46: 3211–3224
Knyazev PG, Imyanitov EN, Chernitsa OI, Nikiforova IF (1993) Loss of heterozygosity at chromosome 17p is associated with HER-2 amplification and lack of nodal involvement in breast cancer. Int J Cancer 53: 11–16
Kreipe H, Feist H, Fischer L, Felgner J, Heidorn K, Mettler L, Parwaresch R (1993) Amplification of c-myc but not of c-erbB-2 is associated with high proliferative capacity in breast cancer. Cancer Res 53: 1956–1961
LeRoy X, Escot C, Brouillet JP, Theillet C, Maudelonde T, Simony Lafontaine J, Pujol H, Rochefort H (1991) Decrease of c-erbB-2 and c-myc RNA levels in tamoxifen-treated breast cancer. Oncogene 6: 431–437
Leder A, Pattengale PK, Kuo A, Stewart TA, Leder P (1986) Consequences of widespread deregulation of the c-myc gene in transgenic mice: multiple neoplasms and normal development. Cell 45: 485–495
Leone A, Flatow U, VanHoutte K, Steeg PS (1993) Transfection of human nm23-H1 into the human MDA-MB-435 breast carcinoma cell line: effects on tumor metastatic potential, colonization and enzymatic activity. Oncogene 8: 2325–2333
Li LH, Nerlov C, Prendergast G, MacGregor D, Ziff EB (1994) c-Myc represses transcription in vivo by a novel mechanism dependent on the initiator element and Myc box II. EMBO J 13: 4070–4079
Liotta LA (1988) Growth autonomy: the only requirement for metastasis? J Natl Cancer Inst 80: 300
Liotta LA, Stetler Stevenson WG (1991) Tumor invasion and metastasis: an imbalance of positive and negative regulation. Cancer Res 51: 5054s–5059s
Little CD, Nau MM, Carney DN, Gazdar AF, Minna JD (1983) Amplification and expression of the c-myc oncogene in human lung cancer cell lines. Nature 306: 194–196
Luscher B, Eisenman RN (1990) New light on Myc and Myb I. Myc Genes Dev 4: 2025–2035
Luscher B, Eisenman RN (1992) Mitosis-specific phosphorylation of the nuclear oncoproteins Myc and Myb. J Cell Biol 118: 775–784
Luscher B, Kuenzel EA, Krebs EG, Eisenman RN (1989) Myc oncoproteins are phosphorylated by casein kinase II. EMBO J 8: 1111–1119
Machotka SV, Garrett CT, Schwartz AM, Callahan R (1989) Amplification of the proto-oncogenes int-2, c-erb B-2 and c-myc in human breast cancer. Clin Chim Acta 184: 207–217
Mai S, Jalava A (1994) c-Myc binds 5'flanking sequence motifs of the dihydrofolate reductase gene in cellular extracts: role in proliferation. Nucleic Acids Res 22: 2264–2273
Mareel MM, Van Roy FM, De Baetselier P (1990) The invasive phenotypes. Cancer Metastasis Rev 9: 45–62
Mariani Costantini R, Escot C, Theillet C, Gentile A, Merlo G, Lidereau R, Callahan R (1988) In situ c-myc expression and genomic status of the c-myc locus in infiltrating ductal carcinomas of the breast. Cancer Res 48: 199–205

McGuire WL (1987) Prognostic factors for recurrence and survival in human breast cancer. Breast Cancer Res Treat 10: 5–9
McGuire WL, Clark GM (1992) Prognostic factors and treatment decisions in axillary-node-negative breast cancer. N Engl J Med 326: 1756–1761
McGuire WL, Chamness GC, Fuqua SA (1991) Estrogen receptor variants in clinical breast cancer. Mol Endocrinol 5: 1571–1577
Meyers SL, O'Brien MT, Smith T, Dudley JP (1990) Analysis of the int-1, int-2, c-myc, and neuoncogenes in human breast carcinomas. Cancer Res 50: 5911–5918
Miller TL, Huzel NJ, Davie JR, Murphy LC (1993) C-myc gene chromatin of estrogen receptor positive and negative breast cancer cells. Mol Cell Endocrinol 91: 83–89
Millon R, Nicora F, Muller D, Eber M, Klein Soyer C, Abecassis J (1989) Modulation of human breast cancer cell adhesion by estrogens and antiestrogens. Clin Exp Metastasis 7: 405–415
Murphy LC (1990) Estrogen receptor variants in human breast cancer. Mol Cell Endocrinol 74: C83– 6
Musgrove EA, Hamilton JA, Lee CS, Sweeney KJ, Watts CK, Sutherland RL (1993) Growth factor, steroid, and steroid antagonist regulation of cyclin gene expression associated with changes in T-47D human breast cancer cell cycle progression. Mol Cell Biol 13: 3577–3587
Ochieng J, Basolo F, Albini A, Melchiori A, Watanabe H, Elliott J, Raz A, Parodi S, Russo J (1991) Increased invasive, chemotactic and locomotive abilities of c-Ha-ras-transformed human breast epithelial cells. Invasion Metastasis 11: 38–47
Ottestad L, Andersen TI, Nesland JM, Skrede M, Tveit KM, Nustad K, Borresen AL (1993) Amplification of c-erbB-2, int-2 and c-myc genes in node-negative breast carcinomas. Relationship to prognosis. Acta Oncol 32: 289–294
Paterson MC, Dietrich KD, Danyluk J, Paterson AH, Lees AW, Jamil N, Hanson J, Jenkins H, Krause BE, McBlain WA et al. (1991) Correlation between c-erbB-2 amplification and risk of recurrent disease in node-negative breast cancer. Cancer Res 51: 556–567
Pavelic ZP, Pavelic K, Carter CP, Pavelic L (1992) Heterogeneity of c-myc expression in histologically similar infiltrating ductal carcinomas of the breast. J Cancer Res Clin Oncol 118: 16–22
Peltenburg LT, Schrier PI (1994) Transcriptional suppression of HLA-B expression by c-Myc is mediated through the core promoter elements. Immunogenetics 40: 54–61
Peltenburg LT, Dee R, Schrier PI (1993) Downregulation of HLA class I expression by c-myc in human melanoma is independent of enhancer A. Nucleic Acids Res 21: 1179–1185
Pena A, Reddy CD, Wu S, Hickok NJ, Reddy EP, Yumet G, Soprano DR, Soprano KJ (1993) Regulation of human ornithine decarboxylase expression by the c-Myc. Max protein complex. J Biol Chem 268: 27277–27285
Penn LJ, Brooks MW, Laufer EM, Land H (1990a) Negative autoregulation of c-myc transcription. EMBO J 9: 1113–1121
Penn LJ, Brooks MW, Laufer EM, Littlewood TD, Morgenstern JP, Evan GI, Lee WM, Land H (1990b) Domains of human c-myc protein required for autosuppression and cooperation with ras oncogenes are overlapping. Mol Cell Biol 10: 4961–4966
Philip A, schneider A, Vasrik I, Finke K, Xiong Y, Beach D, Alitalo K, Eilers M (1994) Repression of cyclin D1: a novel function of MYC. Mol Cell Biol 14: 4032–4043
Postel EH, Berberich SJ, Flint SJ, Ferrone CA (1993) Human c-myc transcription factor PuF identified as nm23-H2 nucleoside diphosphate kinase, a candidate suppressor of tumor metastasis. Science 261: 478–480
Prendergast GC, Cole MD (1989) Posttranscriptional regulation of cellular gene expression by the c-myc oncogene. Mol Cell Biol 9: 124–134
Prendergast GC, Diamond LE, Dahl D, Cole MD (1990) The c-myc-regulated gene mr1 encodes plasminogen activatorr inhibitor 1. Mol Cell Biol 10: 1265–1269
Reisman D, Elkind NB, Roy B, Beamon J, Rotter V (1993) c-Myc trans-activates the p53 promoter through a required downstream CACGTG motif. Cell Growth Differ 4: 57–65
Rochefort H (1992) Cathepsin D in breast cancer: a tissue marker associated with metastasis. Eur J Cancer 28A: 1780–1783
Roux Dosseto M, Romain S, Dussault N, Desideri C, Piana L, Bonnier P, Tubiana N, Martin PM (1992) c-myc gene amplification in selected node-negative breast cancer patients correlates with high rate of early relapse. Eur J Cancer 28A: 1600–1604
Royds JA, Stephenson TJ, Rees RC, Shorthouse AJ, Silcocks PB (1993) Nm23 protein expression in ductal in situ and invasive human breast carcinoma. J Natl Cancer Inst 85: 727–731
Sawan A, Lascu I, Veron M, Anderson JJ, Wright C, Horne CH, Angus B (1994) NDP-K/nm23 expression in human breast cancer in relation to relapse, survival, and opther prognostic factors: an immunohistochemical study. J Pathol 172: 27–34

Scarpa A, Borgato L, Chilosi M, Capelli P, Menestrina F, Bonetti F, Zamboni G, Pizzolo G, Hirohashi S, Fiore Donati L (1991) Evidence of c-myc gene abnormalities in mediastinal large B-cell lymphoma of young adult age. Blood 78: 780–788

Schoenenberger CA, Andres AC, Groner B, van der Valk M, LeMeur M, Gerlinger P (1988) Targeted c-myc gene expression in mammary glands of transgenic mice induces mammary tumors with constitutive milk protein gene transcription. EMBO J 7: 169–175

Scorilas A, Yotis J, Gouriotis D, Keramopoulos A, Ampela K, Trangas T, Talieri M (1993) Cathepsin-D and c-erb-B2 have an additive prognostic value for breast cancer patients. Anticancer Res 13: 1895–1900

Seshadri R, Mathews C, Dobrovic A, Horsfall DJ (1989) The significance of oncogene amplification in primary breast cancer. Int J Cancer 43: 270–272

Shapiro DN, Valentine MB, Rowe ST, Sinclair AE, Sublett JE, Roberts WM, Look AT (1993) Detection of N-myc gene amplification by fluorescence in situ hybridization. Diagnostic utility for neuroblastoma. Am J Pathol 142: 1339–1346

Sheikh MS, Shao ZM, Chen JC, Ordonez JV, Fontana JA (1993) Retinoid modulation of c-myc and max gene expression in human breast carcinoma. Anticancer Res 13: 1387–1392

Shi YE, Torri J, Yieh L, Sobel ME, Yamada Y, Lippman ME, Dickson RB, Thompson EW (1993) Expression of 67kDa laminin receptor in human breast cancer cells: regulation by progestins. Clin Exp Metastasis 11: 251–261

Shtivelman E, Bishop JM (1991) Expression of CD44 is represented in neuroblastoma cells. Mol Cell Biol 11: 5446–5452

Simpson JF, Page DL (1992) Prognostic value of histopathology in the breast. Semin Oncol 19: 254–262

Soini Y, Mannermaa A, Winqvist R, Kamel D, Poikonen K, Kiviniemi H, Paakko P (1994) Application of fine-needle aspiration to the demonstration of ERBB2 and MYC expression by in situ hybridization in breast carcinoma. J Histochem Cytochem 42: 795–803

Spandidos DA, Yiagnisis M, Papadimitriou K, Field JK (1989) ras, c-myc and c-erbB-2 oncoproteins in human breast cancer. Anticancer Res 9: 1385–1393

Spencer CA, Groudine M (1990) Transcription elongation and eukaryotic gene regulation. Oncogene 5: 777–785

Steeg PS, De la Rosa A, Flatow U, MacDonald NJ, Benedict M, Leone A (1993) Nm23 and breast cancer metastasis. Breast Cancer Res Treat 25: 175–187

Studzinski GP, Brelvi ZS, Feldman SC, Watt RA (1986) Participation of c-myc protein in DNA synthesis of human cells. Science 234: 467–470

Suen TC, Hung MC (1991) c-myc reverses neu-induced transformed morphology by transcriptional repression. Mol Cell Biol 11: 354–362

Tandon AK, Clark GM, Chamness GC, Chirgwin JM, McGuire WL (1990) Cathepsin D and prognosis in breast cancer. N Engl J Med 322: 297–302

Tang RP, Kacinski B, Validire P, Beuvon F, Sastre X, Benoit P, dela Rochefordiere A, Mosseri V, Pouillart P, Scholl S (1990) Oncogene amplification correlates with dense lymphocyte infiltration in human breast cancers: a role for hematopoietic growth factor release by tumor cells? J Cell Biochem 44: 189–198

Tavassoli M, Quirke P, Farzaneh F, Lock NJ, Mayne LV, Kirkham N (1989) c-erbB-2/c-erbA co-amplification indicative of lymph node metastasis and c-myc amplification of high tumor grade, in human breast carcinoma. Br J Cancer 60: 505–510

Thompson EW, Katz D, Shima TB, Wakeling AE, Lippman ME, Dickson RB (1989) ICI 164, 384, a pure antagonist of estrogen-stimulated MCF-7 cell proliferation and invasiveness. Cancer Res 49: 6929–6934

Thompson EW, Lippman ME, Dickson RB (1991) Regulation of basement membrane invasiveness in human breast cancer model systems. Mol Cell Endocrinol 82: C203–8

Thompson EW, Paik S, Brunner N, Sommers CL, Zugmaier G, Clarke R, Shima TB, Torri J, Donahue S, Lippman ME et al. (1992) Association of increased basement membrane invasiveness with absence of estrogen receptor and expression of vimentin in human breast cancer cell lines. J Cell Physiol 150: 534–544

Thompson EW, Torri J, Sabol M, Sommers CL, Byers S, Valverius EM, Martin GR, Lippman ME, Stampfer MR, Dickson RB (1994) Oncogene-induced basement membrane invasiveness in human mammary epithelial cells. Clin Exp Metastasis 12: 181–194

Tsuda H, Hirohashi S, Shimosato Y, Hirota T, Tsugane S, Yamamoto H, Miyajima N, Toyoshima K, Yamamoto T, Yokota J et al. (1989) Correlation between long-term survival in breast cancer patients and amplification of two putative oncogene-coamplification units: hst-1/int-2 and c-erbB-2/ear-1. Cancer Res 49: 3104–3108

Tsuda H, Hirohashi S, Hirota T, Shimosato Y (1991) Alterations in copy number of c-erbB-2 and c-myc proto-oncogenes in advanced stage of human breast cancer. Acta Pathol Jpn 41: 19–23

van den Brule FA, Engel J, Stetler-Stevenson WG, Liu FT, Sobel ME, Castronovo V (1992) Genes involved in tumor invasion and metastasis are differently modulated by estradiol and progestin in human breast-cancer cells. Int J Cancer 52: 653–657

Varley JM, Swallow JE, Brammar WJ, Whittaker JL, Walker RA (1987a) Alterations to either c-erbB-2(neu) or c-myc proto-oncogenes in breast carcinomas correlate with poor short-term prognosis. Oncogene 1: 423–430

Varley JM, Wainwright AM, Brammar WJ (1987b) An unusual alteration in c-myc in tissue from a primary breast carcinoma. Oncogene 1: 431–438

Versteeg R, Noordermeer IA, Kruse Wolters M, Ruiter DJ, Schrier PI (1988) c-myc down-regulates class I HLA expression in human melanomas. EMBO J 7: 1023–1029

Walker RA, Senior PV, Jones JL, Critchley DR, Varley JM (1989) An immunohistochemical and in situ hybridization study of c-myc and c-erbB-2 expression in primary human breast carcinomas. J Pathol 158: 97–105

Wasfy G, Marhin W, Lu R, Daksis J, Penn LJZ (1995) Cloning and identification of myc regulated genes. Proc Am Assoc Cancer Res 36: 521 (abstr)

Watson PH, Baral E (1991) Endocrine therapy of breast cancer: research trailing the clinical edge. Can J Oncol 1: 42–44

Watson PH, Pon RT, Shiu RP (1991) Inhibition of c-myc expression by phosphorothioate antisense oligonucleotide identifies a critical role for c-myc in the growth of human breast cancer. Cancer Res 51: 3996–4000

Watson PH, Safneck JR, Le K, Dubik D, Shiu RP (1993) Relationship of c-myc amplification to progression or breast cancer from in situ to invasive tumor and lymph node metastasis. J Natl Cancer Inst 85: 902–907

Watson PH, Snell L, Millns AK (1994) CD44 expression and estrogen receptor status in human breast cancer. American Association for Cancer Research Conference on Molecules Genetics of Tumor Progression, February 1994, Big Sky, Montana, Abstr no A-30

Yang BS, Geddes TJ, Pogulis RJ, de Crombrugghe B, Freytag SO (1991) Transcriptional suppression of cellular gene expression by c-myc. Mol Cell Biol 11: 2291–2295

Zajac Kaye M, Gelmann EP, Levens D (1988) A point mutation in the c-myc locus of a Burkitt lymphoma abolishes binding of a nuclear protein. Science 240: 1776–1780

Zhou DJ, Ahuja H, Cline MJ (1989) Proto-oncogene abnormalities in human breast cancer: c-ERBB-2 amplification does not correlate with recurrence of disease. Oncogene 4: 105–108

Subject Index

acidic fibroblast growth factor 122
adenomatous polyposis coli tumor suppressor gene 121
adherens junctions 118
adhesion 13–23, 118, 255, 276
– molecules 13–23, 275
– – immunglobulin SPF 14, 21
– – – ICAM-1 18, 21, 23
– – – PECAM-1 21, 23
– – integrin 21
– – – antibody 22
– – – $\alpha_4\beta_7$ 21
– – – $\alpha_5\beta_1$ 22
– – – $\alpha_4\beta_1$/VLA-4 21, 22
– – intercellular 125
– – selectins 14, 18
– – – E-selectin 18, 19
– – – VCAM-1 18, 20–22
angiogenesis 31, 32
– assays
– – in vitro 37–39
– – in vivo 33
– associated diseases 58
– hemangioma 32
– phases 35
– physiological 32, 33
– regulators 34
– retinopathy 32
– therapy 59
– tumor 32
anit-TNF monoclonal antibody 17
armadillo 121
astroglial cells 108
autocrine loops 119
autocrine motolity factor 139
– intracellular signaling 141
– motility 155
– receptor 143 ff.
– – cellular locomotion 154
– – regulation 157–161
– – surface expression 148, 149
– response 141
awd 216, 221

B16 (murine) melanoma 16, 22, 217
basement membrane 119
basic fibroblast growth factor (bFGF) 5, 35
blood vessels 31, 32
bone marrow 16
brain 15
– invasion 102
– metastasis 90
breast carcinoma (cancer) 20, 124, 127, 267
– lobular 126
Burkitt lymphoma 240

C-CAM 70, 71
c-Met tyrosine kinase 122
c-myc 238, 239, 267–277
c-neu/HER2 126
c-*ret* 126
C3 transferase 256
Ca^{2+}-binding proteins 184
cadherin 69, 70, 71, 118
cAMP 225
capillary
– blood vessels 32
– morphogenesis 31, 33
carbohydrate 18, 19, 21
– sialyl *Lewis* 19
carcinogenesis 128
carcinoma(s)
– breast 126
– colon 16, 19
– colorectal 126
– hepatocellular 127
– human 125
– mammary, rat 217
– ovarian 126
– pancreatic 19
– prostate 126
– renal 16
– – cell 126
catenin 121
CD44 71
cell
– adhesion 118
– proliferation 223

cell-cell contacts 125
Chinese hamster ovar 17, 20
chromosome 17q21 216
colon 16, 18–20
– carcinoma 16, 18, 19
– HT-29
colorectal carcinomas 126
compaction 121
cyclin
– A 198–201, 203, 208
– B 198, 199, 203, 208
– D1 198–203
– E 198–200, 203
cytokine 13–23, 35
– biphasic effect 51
– contextual activity 58
– fibroblast growth factor 35
– interactions 36
– interferon-α 53
– interleukin-1 15–23
– – receptor antagonist 15, 16
– leukemia inhibitory factor 55
– oncostatin M 55
– synergism 43–46
– transforming growth factor-β 35
– tumor necrosis factor 15–23
– – α 35
– – antibodies 17, 18
– vascular endothelial growth factor 35
cytoskeleton 255

Dbl homology (DH) domain 254
desmocollin 118
desmoglein 118
desmosomes 118
DHR domain 255
differential colony hybridization 216
differentiation 221, 233, 235, 247
diffuse-type gastric carcinomas 125
downregulation of expression 128

ECM *see* extracellular matrix
embryogenesis 121, 222
endocrine therapies 268
endothelial cell 23, 32
– activation status 34
endothelium 13
– hepatic sinusoidal 15
epidermal growth factor 122
epithelia
– differentiation 119
– growth 119
– morphogenesis 119
– motolity 119
epithelial (cells) 119
– differentiation 125
– – loss of 128
– mesenchymal transitions 120
– morphogenesis 124
ErbB2 receptor 127

estrogen action 268
extracellular matrix (ECM) 119, 215
– components 220, 223

fibroblast
– growth factor 35
– – angiogenesis 35
– – immunoreactivity 42
– – isoforms 41
– – receptors 40, 41
– – synergism 43–46
– – rat embryo 217
– specific protein 128
fibrosarcoma 17, 22
focal contacts 258

gastric carcinoma, diffuse-type 125
growth
– acidic fibroblast 122
– factor 101, 259
– – basic fibroblast 5, 35
– – epidermal 122
– – hepatocyte 122, 123
– – hormone-independent 268
– – transforming 35
– – – β *see* transforming growth factor
– – vascular endothelial 35
– – – isoforms 39, 41, 46, 53
GTPase
– activating proteins (GAP) 258
– Rho-like 255
guanine nucleotide inhibitors (GDI) 258
guanosine diphosphate (GDP)-dissociation stimulator (GDS)- proteins 255

head and neck cancers 126
hemangioma 32
– angiogenesis 32
– interferon-α 54
– therapy 54
hepatic sinusoidal endothelium 15
hepatocellular carcinomas 127
hepatocyte growth factor 122
– SF/HGF 122
hereditary hemorrhagic telangectasia 52
12(S)-HETE 73, 80, 81
*Hha*I 182
homologous recombination 121
hormone-independent growth 268
*Hpa*II 183
HRCC *see* human renal cell carcinoma
HT-29 19
human carcinomas (*see also* carcinomas) 125
human renal cell carcinoma (HRCC) 4

ICAM-1 18, 21, 23
IFN *see* interferon
IL-1 15–17
– receptor antagonist 15
IL-1ra 16

IL-6 55
– angiogenesis 56
IL-8 5
immunoglobulin 14
– superfamily 21
– supergene family 69
immunoreactivity 42
in vitro assays 48
integrin 14, 21, 69
– $\alpha IIb\beta_3$ 72
– $\alpha_4\beta_1$ 21
– $\alpha_4\beta_7$ 21
– $\alpha_5\beta_1$ 22, 70, 71
– $\alpha v\beta_3$ 77
– receptor 119
intercellular adhesion 125
interferon
– α 6, 53
– β 6
– angiogenesis 54
– hemangioma 54
– isoforms 53
– receptors 53
invasion 138, 253, 268
invasive fronts 125
invasiveness 120, 124, 268

junctional deficiencies 128
junctions
– adherens 118
– splice 126
– tight 118

K-1735 TK melanoma 220
Kaplan-Meier analysis 126
Kaposi sarcoma 127
KGF 119
kringle domains 122

lectine 18
leukemia 127
– inhibitory factor 55
– – angiogenesis 56
leukocyte 18, 20
– adhesion 18, 20
– rolling 20
Lewis^a 19
Lewis^x 19
Li-Fraumeni syndrome 127
lipopolysaccharide (LPS) 15, 16
liver 16, 19
lobular breast carcinomas 126
LPS *see* lipopolysaccaride
lung 16, 19
– colonies 16
lymph node metastasis 126
lymphoma 16
– ESb 16, 17

mammary
– carcinoma, rat 217
– gland 124
melanoma(s) 5, 16, 19–23, 127
– B16 16, 22, 217
– K-1735 TK 220
– cell lines, murine 216
melantropic peptides 108
membrane ruffling 259
mesenchymal cells 119
metastasis 13–23, 137, 138, 253, 268
– bone marrow 16
– brain 15, 90
– homing 13
– lung 15, 16, 19, 22
– liver 15, 16, 19
– lymph node 126
– pathogenesis 1
– patterns 3
– specific gene *mts*-1 128
metastatic
– capacity 253
– process 216
morphogenesis, capillary 33
morphology 255
motility 255
mta-1 216
Mts-1-myosin complex 188
mts-1 gene 171–190
– expression 180
– – RNA 174
– mRNA 179
– transcription in mouse 173
mts1 protein 184–190
murine melanoma cell lines 216
mutants, transdominant 120
mutation 128
– targeted 121, 123
myc promoter 227

NDPK 216, 226
– activity 224, 225
neoplastic cells 139
neuregulin 119
neurotrophin 91 ff.
– signal transduction 96
NIH3T3 cells 256
nm23 215, 218, 219, 233–248
nm23-H1 216
Nm23-H1 221
nm23-H2 216
Nm23-H2 221
nuclease hypersensitive sites 238, 239, 243–245

oncogenes 253
– dominant 117
– recessive 117
oncostatin M 55
– angiogenesis 56

osteosarcoma 20, 21, 127
ovarian carcinoma 16, 20, 126

P13 kinase 259
p65[pak] 260
pancreatic carcinoma 19
paracrine signaling
– molecules 119
– system 123
patients' survival 126
PECAM-1 21, 23, 72
phosphoserine 226
– Nm23 227
PKC 76
plakoglobin 122
plasminogen 122
platelets 14, 16
Pleckstrin homologous (PH) domains 255
pp125[FAK] 76
prognosis 161, 162
prostate carcinomas 126
proviral insertional mutagenesis. 254
PuF 227

rabdosarcoma 21
Rac 254
– interacting protein (Rip) 260
radiation 205, 208
– UV 205
rat
– embryo fibroblast 217
– mammary carcinoma 217
Rb *see* retinoblastoma
recombination, homologous 121
renal carcinoma 16, 126
retinoblastoma (Rb) 200, 202, 204
retinopathy 32
RGD 23
Rho-like GTPases 255
RhoA 256, 258
Rip *see* Rac-interacting protein
Ros 127

S100 superfamily 184
sarcoma 16, 17, 22, 23
– CFS1 16
– M5076
scatter factor/hepatocyte growth factor 123
selectins 14, 18, 19
serine proteases 122
SF/HGF 122
signal transduction pathways 253–262
signaling process 220
splice junctions 126
src 128
stem cell technology 121
stress fibers 258
survival 126
– prediction, prognostic factor 127
synergism 43, 45, 46

T lymphoma cells 254
targeted mutation 121, 123
TGF *see* transforming growth factor
Tiam1 gene 254
tight junctions 118
TNF *see* tumor necrosis factor
transdominant mutants 120
transfection 183
– studies 217
transforming genes 257
transforming growth factor (TGF-β) 35, 51, 52, 223
– angiogenesis
– – in vitro 48
– – in vivo 47
– biphasic effect 51
– hereditary hemorrhagic telangectasia 52
– isoforms 46
– receptors 46, 52
transmembrane tyrosine kinase receptors 39, 40
triple helix 239
Trk 127
tumor angiogenesis 2
tumor cell
– endothelial cell interactions 77
– survival 99
tumor cell-matrix interactions 75
tumor necrosis factor (TNF) 15, 16, 35
– α 15, 35
– angiogenesis 39
– anti-TNF monoclonal antibody 17
– isoforms 39
– receptors 39
– synergism 43–46
tumor(s) 257
– progression 124, 268
– suppressor gene 126
tyrosine kinase receptors 128
tyrosine phosphorylated proteins 157
tyrosine phosphorylation 122

UV radiation 205

vascular endothelial growth factor (VEGF) 35
– isoforms 39, 41, 46, 53
vasculogenesis 31, 32
VCAM-1 18, 21, 22
– anti-VCAM-1 monoclonal antibodies 20, 22
VEGF *see* vascular endothelial growth factor
VLA-4 21, 22
– monoclonal antibodies to 22

WDNM1 216
WDNM2 216
Wnt 122
– signaling pathway 122

X-ray crystallography 235, 245

Current Topics in Microbiology and Immunology

Volumes published since 1989 (and still available)

Vol. 173: **Pfeffer, Klaus; Heeg, Klaus; Wagner, Hermann; Riethmüller, Gert (Eds.):** Function and Specificity of γ/δ TCells. 1991. 41 figs. XII, 296 pp. ISBN 3-540-53781-3

Vol. 174: **Fleischer, Bernhard; Sjögren, Hans Olov (Eds.):** Superantigens. 1991. 13 figs. IX, 137 pp. ISBN 3-540-54205-1

Vol. 175: **Aktories, Klaus (Ed.):** ADP-Ribosylating Toxins. 1992. 23 figs. IX, 148 pp. ISBN 3-540-54598-0

Vol. 176: **Holland, John J. (Ed.):** Genetic Diversity of RNA Viruses. 1992. 34 figs. IX, 226 pp. ISBN 3-540-54652-9

Vol. 177: **Müller-Sieburg, Christa; Torok-Storb, Beverly; Visser, Jan; Storb, Rainer (Eds.):** Hematopoietic Stem Cells. 1992. 18 figs. XIII, 143 pp. ISBN 3-540-54531-X

Vol. 178: **Parker, Charles J. (Ed.):** Membrane Defenses Against Attack by Complement and Perforins. 1992. 26 figs. VIII, 188 pp. ISBN 3-540-54653-7

Vol. 179: **Rouse, Barry T. (Ed.):** Herpes Simplex Virus. 1992. 9 figs. X, 180 pp. ISBN 3-540-55066-6

Vol. 180: **Sansonetti, P. J. (Ed.):** Pathogenesis of Shigellosis. 1992. 15 figs. X, 143 pp. ISBN 3-540-55058-5

Vol. 181: **Russell, Stephen W.; Gordon, Siamon (Eds.):** Macrophage Biology and Activation. 1992. 42 figs. IX, 299 pp. ISBN 3-540-55293-6

Vol. 182: **Potter, Michael; Melchers, Fritz (Eds.):** Mechanisms in B-Cell Neoplasia. 1992. 188 figs. XX, 499 pp. ISBN 3-540-55658-3

Vol. 183: **Dimmock, Nigel J.:** Neutralization of Animal Viruses. 1993. 10 figs. VII, 149 pp. ISBN 3-540-56030-0

Vol. 184: **Dunon, Dominique; Mackay, Charles R.; Imhof, Beat A. (Eds.):** Adhesion in Leukocyte Homing and Differentiation. 1993. 37 figs. IX, 260 pp. ISBN 3-540-56756-9

Vol. 185: **Ramig, Robert F. (Ed.):** Rotaviruses. 1994. 37 figs. X, 380 pp. ISBN 3-540-56761-5

Vol. 186: **zur Hausen, Harald (Ed.):** Human Pathogenic Papillomaviruses. 1994. 37 figs. XIII, 274 pp. ISBN 3-540-57193-0

Vol. 187: **Rupprecht, Charles E.; Dietzschold, Bernhard; Koprowski, Hilary (Eds.):** Lyssaviruses. 1994. 50 figs. IX, 352 pp. ISBN 3-540-57194-9

Vol. 188: **Letvin, Norman L.; Desrosiers, Ronald C. (Eds.):** Simian Immunodeficiency Virus. 1994. 37 figs. X, 240 pp. ISBN 3-540-57274-0

Vol. 189: **Oldstone, Michael B. A. (Ed.):** Cytotoxic T-Lymphocytes in Human Viral and Malaria Infections. 1994. 37 figs. IX, 210 pp. ISBN 3-540-57259-7

Vol. 190: **Koprowski, Hilary; Lipkin, W. Ian (Eds.):** Borna Disease. 1995. 33 figs. IX, 134 pp. ISBN 3-540-57388-7

Vol. 191: **ter Meulen, Volker; Billeter, Martin A. (Eds.):** Measles Virus. 1995. 23 figs. IX, 196 pp. ISBN 3-540-57389-5

Vol. 192: **Dangl, Jeffrey L. (Ed.):** Bacterial Pathogenesis of Plants and Animals. 1994. 41 figs. IX, 343 pp. ISBN 3-540-57391-7

Vol. 193: **Chen, Irvin S. Y.; Koprowski, Hilary; Srinivasan, Alagarsamy; Vogt, Peter K. (Eds.):** Transacting Functions of Human Retroviruses. 1995. 49 figs. IX, 240 pp. ISBN 3-540-57901-X

Vol. 194: **Potter, Michael; Melchers, Fritz (Eds.):** Mechanisms in B-cell Neoplasia. 1995. 152 figs. XXV, 458 pp. ISBN 3-540-58447-1

Vol. 195: **Montecucco, Cesare (Ed.):** Clostridial Neurotoxins. 1995. 28 figs. XI., 278 pp. ISBN 3-540-58452-8

Vol. 196: **Koprowski, Hilary; Maeda, Hiroshi (Eds.):** The Role of Nitric Oxide in Physiology and Pathophysiology. 1995. 21 figs. IX, 90 pp. ISBN 3-540-58214-2

Vol. 197: **Meyer, Peter (Ed.):** Gene Silencing in Higher Plants and Related Phenomena in Other Eukaryotes. 1995. 17 figs. IX, 232 pp. ISBN 3-540-58236-3

Vol. 198: **Griffiths, Gillian M.; Tschopp, Jürg (Eds.):** Pathways for Cytolysis. 1995. 45 figs. IX, 224 pp. ISBN 3-540-58725-X

Vol. 199/I: **Doerfler, Walter; Böhm, Petra (Eds.):** The Molecular Repertoire of Adenoviruses I. 1995. 51 figs. XIII, 280 pp. ISBN 3-540-58828-0

Vol. 199/II: **Doerfler, Walter; Böhm, Petra (Eds.):** The Molecular Repertoire of Adenoviruses II. 1995. 36 figs. XIII, 278 pp. ISBN 3-540-58829-9

Vol. 199/III: **Doerfler, Walter; Böhm, Petra (Eds.):** The Molecular Repertoire of Adenoviruses III. 1995. 51 figs. XIII, 310 pp. ISBN 3-540-58987-2

Vol. 200: **Kroemer, Guido; Martinez-A., Carlos (Eds.):** Apoptosis in Immunology. 1995. 14 figs. XI, 242 pp. ISBN 3-540-58756-X

Vol. 201: **Kosco-Vilbois, Marie H. (Ed.):** An Antigen Depository of the Immune System: Follicular Dendritic Cells. 1995. 39 figs. IX, 209 pp. ISBN 3-540-59013-7

Vol. 202: **Oldstone, Michael B. A.; Vitković, Ljubiša (Eds.):** HIV and Dementia. 1995. 40 figs. XIII, 279 pp. ISBN 3-540-59117-6

Vol. 203: **Sarnow, Peter (Ed.):** Cap-Independent Translation. 1995. 31 figs. XI, 183 pp. ISBN 3-540-59121-4

Vol. 204: **Saedler, Heinz; Gierl, Alfons (Eds.):** Transposable Elements. 1995. 42 figs. IX, 234 pp. ISBN 3-540-59342-X

Vol. 205: **Littman, Dan. R. (Ed.):** The CD4 Molecule. 1995. 29 figs. XIII, 182 pp. ISBN 3-540-59344-6

Vol. 206: **Chisari, Francis V.; Oldstone, Michael B. A. (Eds.):** Transgenic Models of Human Viral and Immunological Disease. 1995. 53 figs. XI, 345 pp. ISBN 3-540-59341-1

Vol. 207: **Prusiner, Stanley B. (Ed.):** Prions Prions Prions. 1995. 42 figs. VII, 163 pp. ISBN 3-540-59343-8

Vol. 208: **Farnham, Peggy J. (Ed.):** Transcriptional Control of Cell Growth. 1995. 17 figs. IX, 141 pp. ISBN 3-540-60113-9

Vol. 209: **Miller, Virginia L. (Ed.):** Bacterial Invasiveness. 1996. 16 figs. IX, 115 pp. ISBN 3-540-60065-5

Vol. 210: **Potter, M.; Rose, N. R. (Eds.):** Immunology of Silicones. 1996. 136 figs. XX, 430 pp. ISBN 3-540-60272-0

Vol. 211: **Wolff, L.; Perkins, A. S. (Eds.):** Molecular Aspects of Myeloid Stem Cell Development. 1996. 98 figs. XIV, 298 pp. ISBN 3-540-60414-6

Vol. 212: **Vainio, O.; Imhof, B. A. (Eds.):** Immunology and Developmental Biology of the Chicken. 1996. 43 figs. IX, 281 pp. ISBN 3-540-60585-1

Vol. 213/I: **Günthert, U.; Birchmeier, W. (Eds.):** Attempts to Understand Metastasis Formation I. 1996. 35 figs. XV, 293 pp. ISBN 3-540-60680-7

Vol. 213/III: **Günthert, U.; Schlag, P. M.; Birchmeier, W. (Eds.):** Attempts to Understand Metastasis Formation III. 1996. 14 figs. XV, 262 pp. ISBN 3-540-60682-3

Springer-Verlag and the Environment

We at Springer-Verlag firmly believe that an international science publisher has a special obligation to the environment, and our corporate policies consistently reflect this conviction.

We also expect our business partners – paper mills, printers, packaging manufacturers, etc. – to commit themselves to using environmentally friendly materials and production processes.

The paper in this book is made from low- or no-chlorine pulp and is acid free, in conformance with international standards for paper permanency.

MIX
Papier aus verantwortungsvollen Quellen
Paper from responsible sources
FSC® C105338

If you have any concerns about our products,
you can contact us on
ProductSafety@springernature.com

In case Publisher is established outside the EU,
the EU authorized representative is:
Springer Nature Customer Service Center GmbH
Europaplatz 3, 69115 Heidelberg, Germany

Printed by Libri Plureos GmbH
in Hamburg, Germany